Physics of Thin Films
Advances in Research and Development

VOLUME 9

CONTRIBUTORS TO THIS VOLUME

P. Chaudhari

J. J. Cuomo

R. J. Gambino

E. A. Giess

H. Raether

J. L. Vossen

Physics of Thin Films

Advances in Research and Development

Edited by

GEORG HASS

Night Vision Laboratory
U.S. Army Electronics Command
Fort Belvoir, Virginia

MAURICE H. FRANCOMBE

Research and Development Center
Westinghouse Electric Corporation
Pittsburgh, Pennsylvania

RICHARD W. HOFFMAN

Department of Physics
Case Western Reserve University
Cleveland, Ohio

VOLUME 9

1977

ACADEMIC PRESS   NEW YORK   SAN FRANCISCO   LONDON
A Subsidiary of Harcourt Brace Jovanovich, Publishers

Copyright © 1977, by Academic Press, Inc.
ALL RIGHTS RESERVED.
NO PART OF THIS PUBLICATION MAY BE REPRODUCED OR
TRANSMITTED IN ANY FORM OR BY ANY MEANS, ELECTRONIC
OR MECHANICAL, INCLUDING PHOTOCOPY, RECORDING, OR ANY
INFORMATION STORAGE AND RETRIEVAL SYSTEM, WITHOUT
PERMISSION IN WRITING FROM THE PUBLISHER.

ACADEMIC PRESS, INC.
111 Fifth Avenue, New York, New York 10003

United Kingdom Edition published by
ACADEMIC PRESS, INC. (LONDON) LTD.
24/28 Oval Road, London NW1

LIBRARY OF CONGRESS CATALOG CARD NUMBER: 63–16561

ISBN 0–12–533009–X

PRINTED IN THE UNITED STATES OF AMERICA

Contents

CONTRIBUTORS TO VOLUME 9	vii
PREFACE	ix
CONTENTS OF PREVIOUS VOLUMES	xi
ARTICLES PLANNED FOR FUTURE VOLUMES	xiii

Transparent Conducting Films
J. L. Vossen

I.	Introduction	1
II.	Physics of Conductivity and Transparency	4
III.	Processes for Film Deposition	30
IV.	Application of Materials and Processes	60
V.	Conclusion	64
	References	64

Metal–Dielectric Interference Filters

I.	Introduction	74
II.	Basic Theory	75
III.	General Considerations in Bandpass Filter Design	93
IV.	Single-Cavity Filter Design	102
V.	One-M Filter Design	112
VI.	Multiple Cavity and Other Designs	124
VII.	Reflection Filters	128
VIII.	The Production of Filters	132
IX.	Future Developments	142
	References	143

Surface Plasma Oscillations and Their Applications
H. Raether

I.	General Considerations on Surface Plasmons	145
II.	Excitation of Radiative Surface Plasmons	171
III.	Excitation of Nonradiative Surface Plasmons	199
	References	255

Magnetic Bubble Films

P. Chaudhari, J. J. Cuomo, R. J. Gambino, and E. A. Giess

I.	Introduction	263
II.	General Magnetics	266
III.	Materials for Magnetic Bubbles	272
IV.	Growth of Magnetic Bubble Materials	276
V.	Anisotropy in Magnetic Bubble Films	284
VI.	Defects in Films	288
VII.	Summary and Conclusions	293
	References	294

AUTHOR INDEX 299
SUBJECT INDEX 310

Contributors to Volume 9

Numbers in parentheses indicate the pages on which the authors' contributions begin.

P. CHAUDHARI (263), IBM Thomas J. Watson Research Center, Yorktown Heights, New York

J. J. CUOMO (263), IBM Thomas J. Watson Research Center, Yorktown Heights, New York

R. J. GAMBINO (263), IBM Thomas J. Watson Research Center, Yorktown Heights, New York

E. A. GIESS (263), IBM Thomas J. Watson Research Center, Yorktown Heights, New York

H. RAETHER (145), Institute of Applied Physics, University of Hamburg, Hamburg, Germany

J. L. VOSSEN (1), RCA Corporation, David Sarnoff Research Center, Princeton, New Jersey

Preface

The present, ninth volume of *Physics of Thin Films* contains four review articles dealing primarily with optical aspects of thin films.

The first review, by J. L. Vossen, describes the status and technical application of "Transparent Conducting Films." Metal films, such as Au, Ag, and Pt, and oxide films based upon, for example, SnO_2, In_2O_3, and CdO are considered. The preparative methods employed, the physics of conductivity and transparency, dependence of electrical and optical properties on conditions of deposition, and areas of application are treated.

The second article is titled "Metal–Dielectric Interference Filters," and has been compiled from the work of many physicists actively engaged in research in this field. The interesting and useful characteristics of multilayer filters in which one or more of the layers is absorbing (e.g., metallic) are discussed. Detailed design criteria are presented which demonstrate that such filters can display unusually low offband transmittance, while retaining high transmittance within a chosen narrow passband. Optimization of filter performance depends sensitively upon the ability to deposit metal layers which are structurally continuous at very low thickness.

The third review, "Surface Plasma Oscillations and Their Applications" by H. Raether, presents the first comprehensive discussion of this important field of surface physics. Raether discusses in detail the physical basis of surface plasmons on conductive media, their excitation by light and electrons, and the experimental work thus far available, mainly for thin film samples.

In the fourth and final article of this volume, P. Chaudhari and co-workers review the recent status of "Magnetic Bubble Films." Single-crystal garnet layers and more recently developed amorphous alloy films are discussed. This is an important memory technology which offers to supplant existing discs and tapes and holds the prospect of achieving bit densities in excess of $10^9/cm^2$.

G. Hass
M. H. Francombe
R. W. Hoffman

Contents of Previous Volumes

Volume 1

Ultra-High Vacuum Evaporators and Residual Gas Analysis
Hollis L. Caswell

Theory and Calculations of Optical Thin Films
Peter H. Berning

Preparation and Measurement of Reflecting Coatings for the Vacuum Ultraviolet
Robert P. Madden

Structure of Thin Films
Rudolf E. Thun

Low Temperature Films
William B. Ittner, III

Magnetic Films of Nickel-Iron
Emerson W. Pugh

AUTHOR INDEX · SUBJECT INDEX

Volume 2

Structural Disorder Phenomena in Thin Metal Films
C. A. Neugebauer

Interaction of Electron Beams with Thin Films
C. J. Calbick

The Insulated-Gate Thin-Film Transistor
Paul K. Weimer

Measurement of Optical Constants of Thin Films
O. S. Heavens

Antireflection Coatings for Optical and Infrared Optical Materials
J. Thomas Cox and Georg Hass

Solar Absorptance and Thermal Emittance of Evaporated Coatings
Louis F. Drummeter, Jr. and Georg Hass

Thin Film Components and Circuits
N. Schwartz and R. W. Berry

AUTHOR INDEX · SUBJECT INDEX

Volume 3

Film-Thickness and Deposition-Rate Monitoring Devices and Techniques for Producing Films of Uniform Thickness
Klaus H. Behrndt

The Deposition of Thin Films by Cathode Sputtering
Leon I. Maissel

Gas-Phase Deposition of Insulating Films
L. V. Gregor

Methods of Activating and Recrystallizing Thin Films of II–VI Compounds
A. Vecht

The Mechanical Properties of Thin Condensed Films
R. W. Hoffman

Lead Salt Detectors
D. E. Bode

AUTHOR INDEX · SUBJECT INDEX

Volume 4

Precision Measurements in Thin Film Optics
H. E. Bennett and Jean M. Bennett

Nucleation Processes in Thin Film Formation
J. P. Hirth and K. L. Moazed

Evaporated Single-Crystal Films
J. W. Matthews

The Growth and Structure of Electrodeposits
Kenneth R. Lawless

Thin Glass Films
W. A. Pliskin, D. R. Kerr, and J. A. Perri

Hot-Electron Transport and Electron Tunneling in Thin Film Structures
C. R. Crowell and S. M. Sze

AUTHOR INDEX · SUBJECT INDEX

Volume 5

Interference Photocathodes
D. Kossel, K. Deutscher, and K. Hirschberg

Design of Multilayer Interference Filters
Alfred Thelen

Oxide Layers Deposited from Organic Solutions
H. Schroeder

The Preparation and Properties of Semiconductor Films
M. H. Francombe and J. E. Johnson

The Preparation of Films by Chemical Vapor Deposition
W. M. Feist, S. R. Steele, and D. W. Ready

AUTHOR INDEX · SUBJECT INDEX

Volume 6

Anodic Oxide Films
C. J. Dell'Oca, D. L. Pulfrey, and L. Young

Size-Dependent Electrical Conduction in Thin Metal Films and Wires
D. C. Larson

Optical Properties of Metallic Films
F. Abelès

Interactions in Multilayer Magnetic Films
Arthur Yelon

Diffusion in Metallic Films
C. Weaver

AUTHOR INDEX · SUBJECT INDEX

Volume 7

Electron Diffraction Analysis of the Local Atomic Order in Amorphous Films
D. B. Dove

The Preparation and Use of Unbacked Metal Films as Filters in the Extreme Ultraviolet
W. R. Hunter

Properties and Applications of III–V Compound Films Deposited by Liquid Phase Epitaxy
H. Kressel and H. Nelson

Electromigration in Thin Films
F. M. d'Heurle and R. Rosenberg

Built-Up Molecular Films and Their Applications
V. K. Srivastava

AUTHOR INDEX · SUBJECT INDEX

Volume 8

Dielectric Film Materials for Optical Applications
Elmar Ritter

Inhomogeneous and Coevaporated Homogeneous Films for Optical Applications
R. Jacobsson

Discontinuous and Cermet Films
Z. H. Meiksin

Electrical Conduction in Disordered Nonmetallic Films
A. K. Jonscher and R. M. Hill

Topologically Structured Thin Films in Semiconductor Device Operation
H. C. Nathanson and J. Guldberg

SUBJECT INDEX

Articles Planned for Future Volumes

Evaporated Films for Space Applications
 G. Hass and W. R. Hunter

Thin Film IV–VI Photodiodes
 H. Holloway

Thin Films for Integrated Optics
 D. B. Ostrowsky

Correction of Optical Elements by Evaporated Films
 J. R. Kurdock and R. R. Austin

Laser Coatings, Effect of Surface Roughness
 H. E. Bennett and J. M. Bennett

Superconducting Thin Films
 M. Ashkin, J. A. Gavaler, M. A. Janocko, and J. H. Parker

Irradiation-Induced Buildup of Positive Space Charge in Silicon Dioxide Films: A Critical Review
 A. Holmes-Siedle

Surfaces and Surface Coatings for Photothermal Solar Energy Conversion
 B. O. Seraphin and R. E. Hahn

Scattering by All-Dielectric Multilayer Bandpass Filters and Mirrors for Lasers
 J. M. Eastman

Thin Silicon Films for Solar Cells
 T. L. Chu

Recent Advances in Molecular Beam Epitaxy
 A. Y. Cho

Ferroelectric Films
 M. H. Francombe

Dielectric Films for Passivation and Protection of Semiconductor Devices
 J. R. Szedon

High Resolution Image Recording and Image Storage Using Halide Films
 M. R. Tubbs

Physics of Thin Films
Advances in Research and Development

Volume 9

Transparent Conducting Films

J. L. VOSSEN

RCA Corporation
David Sarnoff Research Center
Princeton, New Jersey

I. Introduction	1
II. Physics of Conductivity and Transparency	4
A. Electrical Conduction in Thin Metal Films	4
B. Electrical Conduction in Semiconducting Oxide Films	9
C. Transparency	13
D. Bulk Electrical and Optical Properties of Semiconducting Oxides	22
III. Processes for Film Deposition	30
A. Chemical Vapor Deposition	31
B. Evaporation and Sputtering of Thin Metal Films	42
C. Reactive Evaporation	47
D. Reactive Sputtering	49
E. Sputtering of Oxide Targets	53
F. Miscellaneous Processes	59
IV. Application of Materials and Processes	60
V. Conclusion	64
References	64

I. Introduction

The first reported observation that a thin solid film was at once semitransparent to visible light and electrically conducting appears to have been made by Bädeker in 1907, who prepared CdO films by thermal oxidation of sputtered Cd films (1). The subject remained a scientific curiosity until the 1940s when the need for such films was generated by the aircraft industry which required transparent electrical heaters for windshield deicing. Because of the increasing interest in interactions of light with electricity and electronically active materials, the materials and techniques for producing semitransparent, electrically conducting films are of great interest at this time.

Published material in this field up to about 1955 has been reviewed by Holland (2). A substantial amount of new work has appeared since then. In this review, we shall discuss the physics of conductivity and transparency in the various materials used, the materials themselves, and the many processes that are used for film deposition. Most of the detailed information on thin film properties is organized according to the film deposition process used, rather than the film material, because ostensibly the same material deposited by two different techniques ordinarily yields widely differing sets of physical properties.

This is not surprising because of the complex physics and chemistry of the materials used for this purpose. Broadly speaking, there are two classes of materials: semiconducting oxides and very thin metals. The properties of the oxides depend critically upon their oxidation state (stoichiometry) and on the nature and quantity of impurities trapped in the film. Most of the metals used are very thin (≤ 50 Å), and so their properties depend on film nucleation and initial coalescence phenomena.

As a direct consequence of electromagnetic theory, perfect electrical conductivity and complete transparency in any material represent a contradiction because the incoming electromagnetic wave is dissipated by heating the charge carriers. Besides this, there are numerous other sources of light loss which depend upon the film material, its structure, and surface morphology. The practical consequence of this is that there is, for any given film material, a trade-off between electrical conductivity and visible light transmission. An example of this is shown in Fig. 1.

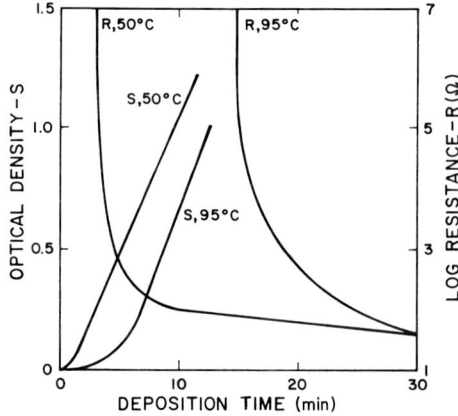

FIG. 1. The dependence of optical density and electrical resistivity on the time of condensation of thin Bi films at the indicated temperatures. [After Palatnik and Komnik (3).]

Many of the conduction mechanisms in the two classes of materials are quite different and so they will be treated separately. The metals that have been used as transparent conductors include Au, Pt, Rh, Ag, Cu, Fe, Ni, and many of these same metals deposited on top of various materials which affect the nucleation and growth of the metal films. The semiconducting oxides that have been used as transparent conductors include SnO_2, In_2O_3, CdO, Cd_2SnO_4, and these same materials doped at cation sites with Sb, In, Sn, Cd, Ti, Te, P, or W and/or at anion sites with F or Cl. Not all of these materials are in general use, having proven undesirable in one way or another. Most of the discussion of detailed film properties will concentrate on materials that are commonly used.

The required quality of the films increases with increasing sophistication of the application. Besides the use of these films as transparent heaters, there is increasing usage for a variety of display devices [e.g., liquid crystal displays (4)] and imaging tubes (e.g., vidicons). The increased pace of work on transparent conductors in recent years has been caused by the needs of these electronic devices. In addition to transparent electrodes for heaters and displays, the same materials are often put to other uses (e.g., transparent electrodes for electrochemical studies; infrared reflectors; antistatic coatings; thin film resistors; low temperature secondary thermometers; and coatings on glass bottles to provide scratch resistance, impact strength, and stability against alkali contamination).

The properties of the films that are considered important for sophisticated applications are electrical resistivity; optical transmission versus wavelength; environmental stability; life stability; the chemical nature, structure, and morphology of the film surface; and the chemical resistance or etchability of the films.

It is only in the fairly recent past that really detailed investigations of these films and the relationships between their properties and the method of deposition have been conducted. This probably accounts for the large number of contradictory results reported prior to the mid-1960s (especially in the patent literature). However, even with careful investigations, contradictory (or, at least, different) results are often obtained with very similar processes.

The processes that have been used to deposit transparent conducting films are hydrolysis of chlorides; pyrolysis; evaporation and sputtering of metals; reactive evaporation and sputtering; sputtering of oxide targets; screen printing; doctor-blading; ultraviolet reduction of Au salts; vapor transport; and glow discharge decomposition of organometallics or metal halides.

II. Physics of Conductivity and Transparency

As will become evident, the mechanisms of electrical conduction and optical transmission are very much interdependent. Further, the film deposition process employed can alter these mechanisms radically. Therefore, in the following discussion, there are regrettable but necessary references to later sections.

A. Electrical Conduction in Thin Metal Films

The electrical properties of thin metal films have been reviewed recently by Maissel (5). Only those features of this field pertinent to transparent conducting films will be outlined here.

1. *Discontinuous Thin Metal Films.* In the early stages of heterogeneous nucleation of thin metal films, it is well known that an island structure exists. While there is still some controversy over theoretical models for thin film nucleation (6, 7), there is a large body of experimental work, by direct observation of film nucleation in transmission electron microscopes (8–14), which has shown how nuclei form, how they grow into islands, and how these islands finally coalesce into a continuous film. In the early stages of nucleation, small nuclei are formed by several adsorbed atoms from the impinging vapor stream. These nuclei are believed to form at point defects in the substrate and both the defects and the nuclei are probably charged (14). Since, in nearly every case, atoms arrive with energies greater than kT (where T is the substrate temperature), it is found that some of the atoms reevaporate, some are directly reflected from the surface, and some lose their energy by moving about on the substrate surface until a small cluster, or island, is formed at a site occupied by a nucleus. An example of such an island film is shown in Fig. 2.

There are five effects that determine the island size and interisland spacing: thermal effects at the substrate, the kinetic energy of the arriving vapor, the arriving vapor flux density, the angle of incidence of arriving vapor, and electrostatic effects at the substrate. Increasing the substrate temperature increases the mobility of arriving vapor, resulting in larger islands and greater interisland separation. Increasing the kinetic energy of the arriving vapor has the same effect, since more energy must be dissipated by lateral motion of the arriving vapor. Increasing the arriving vapor flux density imparts more momentum at the surface and has the same effect. Increasing the angle of incidence of the vapor stream from normal to glancing incidence imparts a higher velocity component parallel to the substrate surface, also

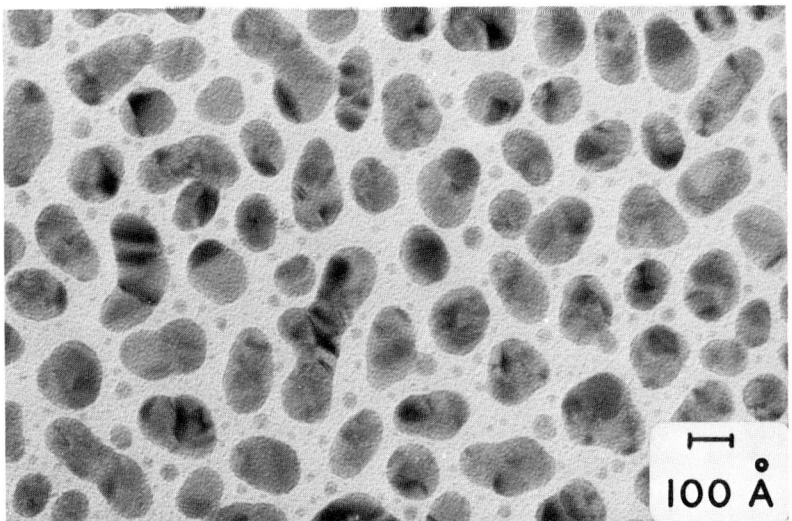

Fig. 2. Transmission electron micrograph of a 40-Å-thick Au film sputtered onto collodion at a rate of 35 Å/min with the substrate at room temperature.

resulting in greater lateral mobility and increased agglomeration. The electrostatic effects at the substrate are less well established, but a recent study of Pt, Au, Ag, and Ni nucleation on 25-Å-thick SiO_2 films on n- and p-type Si has shown that the metals carrying a positive charge (Pt and Ni) resulted in smaller, more closely spaced islands on n-type substrates, while the reverse was true on p-type substrates (14). The opposite appears to be true of Au and Ag, which apparently carry negative charge. Less agglomeration has been observed when a dc electric field is applied in the plane of the substrate (15–20) or when the substrate is charged by electron or ion bombardment (21).

The implications of an island structure to transparent conductors are threefold. First, the resistivity of such films is very high. Second, if the islands become quite large, they act to scatter incident light, rather than transmit it (22). Third, all other things being equal, a thicker film must be deposited to obtain sufficient electrical conductivity, but this results in more light absorption (Fig. 1).

Ideally, one would like the thinnest possible continuous film. Thus, agglomeration is undesirable. From the brief description above, it is clear that reduced agglomeration and early film coalescence are enhanced by low substrate temperature, low deposition rate, low vapor kinetic energy, vapor incidence normal to the substrate, and charging of the substrate and incident vapor. The example shown in Fig. 2 represents an intermediate situation (i.e., a moderate degree of agglomeration).

Largely because of the role of electric charge in nucleation processes and differences in vapor kinetic energy, differences have been noted in evaporation from resistance heated filaments, electron beam evaporation, and sputtering (12, 23). Differing charge densities of the vapor and substrate are expected with these processes. Electroless plating results in very large islands which do not coalesce until considerable thicknesses have been grown (24). Islands must grow to the order of several hundreds of angstroms before coalescence occurs, in comparison to the tens of angstroms that have been reported for evaporation and sputtering (25).

The above discussion is included primarily because it points the way to producing very thin continuous metal films. Island films are, in general, not used as transparent conductors, mainly because their resistivities are too high. All theories of conduction in discontinuous films require an activation term. Currently, there are four theories and they differ only in the source of activation. These transport theories are thermionic emission, activated-charge-carrier creation and tunneling, substrate-assisted tunneling, and tunneling between allowed states. All have been reviewed recently (5), and only the results are outlined below.

To have the thermionic emission theory conform to experiments, the well-known Richardson equation must be modified by a barrier lowering term. There are three possible reasons for barrier lowering: small particles have lower activation energies than bulk materials, the shape of the particles somehow reduces the work function, or there is an overlap of image-force potentials of closely spaced islands that lowers the activation energy by an amount $\gamma q^2/d$, where q is the island charge, d the island separation, and γ is a function of island size and separation (26).

This model usually requires the assumption of very low barrier heights. Rather than assuming that electrons jump from island to island directly, if one assumes that they are injected into the substrate, the barrier lowering term is simply the electron affinity of the substrate (27).

For large interisland separations, it is assumed that there are large numbers of trapping centers with energies between the valence and conduction bands (28). The electrons then tunnel from a trap of lower energy to one of higher energy (27), which is more probable than a single interisland jump.

In the model of Neugebauer and Webb (29) and Herman and Rhodin (30), the island structure is envisaged as an array of small metal islands, fairly closely spaced, some of which are charged and some of which are neutral. The activation energy is that required to transfer an electron from one neutral island to another, leaving the first island charged. Activation is thus assumed to be thermal and electrostatic, and the model predicts that the conductivity is independent of applied field and exponentially dependent upon temperature. The electrostatic part is modified by the dielectric constant of the medium through which tunneling occurs, air (29) or the substrate (30).

Tunneling between allowed states can occur if the islands and interisland spacings are very small because of quantized energy levels in small particles and because energy widths, between which electrons tunnel, can overlap (31).

Figure 3 gives a graphic illustration of the applicability of the various theories (5).

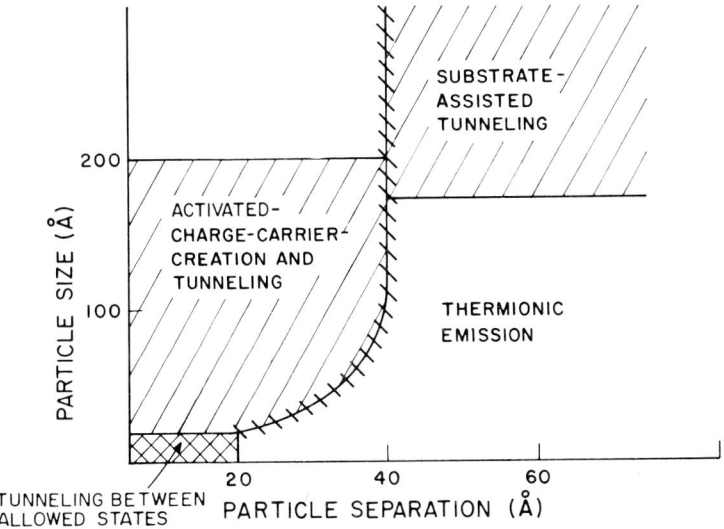

FIG. 3. Range of particle size and spacing over which various theories of conduction in discontinuous films are applicable. [After Maissel (5).]

2. *Continuous Metal Films.* Once metal films grow to the point where they are continuous (or semicontinuous), all of the normal bulk-metal conductivity phenomena apply, but these are modified somewhat by the so-called size effect [scattering of carriers by the surfaces of the film (32–34)] and the fact that impurity concentrations, far in excess of allowable equilibrium limits, can occur in thin films (35). Homogeneously dissolved point defects act as carrier scattering sites even when the valence of the impurity is the same as that of the host material (36). In general, the effect is greater in disordered solid solutions than in the cases where intermetallic compounds are formed. In either case, the effect is far more pronounced if the valence

of the impurity differs substantially from that of the host. It is all too common to find insulating or semiconducting phases (usually as oxide) dispersed through metal films, even in films prepared under relatively clean conditions. Metallic impurities can often increase the resistivity of a film to values as high as 5–10 times the bulk value. Increases in film over bulk resistivity of several orders of magnitude can occur when insulating impurities are present. The extreme example of this is the production of cermet films—purposeful mixtures of metals and insulators. In this case, when the volume fraction of the insulating phase becomes large enough ($\sim 10-25\%$) the insulator disrupts the continuity of the conducting phase and an abrupt increase in resistivity occurs, which can be as high as 8–10 orders of magnitude (37–39).

In thin films, contributions to the resistivity by grain boundaries, dislocations, etc. are very minor by comparison to the effects of point defects and size effect, and so may be neglected. Likewise, film stress is a negligibly small contributor to resistivity. Size effect contributions to resistivity (32–34) can be quite large, however. Since the mean free path of an electron in a metal is of the order of 100 Å, and since most transparent conductive metal films are thinner than 100 Å, considerable scattering occurs at the film boundaries, and this is further aggravated if the film and/or substrate are rough (nonspecular). Figure 4 shows the theoretical increase in resistivity for films of varying thickness and specularity.

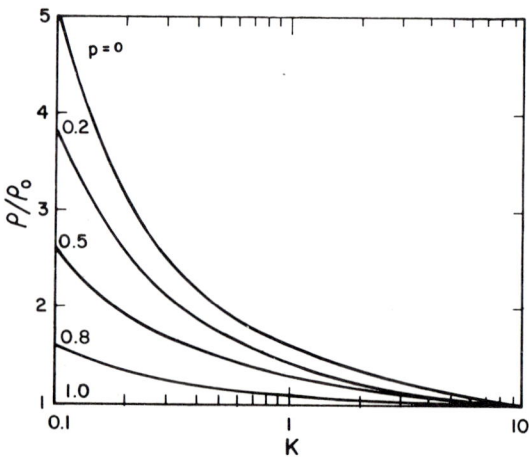

FIG. 4. The calculated variation of the ratio of film to bulk resistivity versus the ratio of film thickness to electron mean free path. The parameter p is related to the film or substrate roughness. Specularity is directly proportional to p. [After Campbell (34).]

The implications of this brief discussion to transparent, conductive metal films are obvious. Pure metals, not alloys should be used, and preferably they should be metals that are not very prone to oxidation (i.e., noble metals). Furthermore, substrate surfaces should be specular and the film deposition technique employed should not result in film surface roughness. Unfortunately, the noble metals are among those that are most prone to agglomeration when deposited from the vapor phase.

3. *Nucleation-Modifying Layers.* Gillham and Preston (*40*, *41*) observed that thin films of Au were much more conductive when a thin layer of Bi_2O_3 was first sputtered on a glass substrate. Later, transmission electron microscope studies showed that the Bi_2O_3 film was amorphous and that the Au deposited over it was continuous at a thickness at which Au films were discontinuous when deposited without the Bi_2O_3 layer. Since this observation, numerous other intermediate layers have proven to reduce the tendency of noble metals to agglomerate. The specifics of these will be discussed later.

There appears to have been no detailed study of the specific nature of this effect. The most probable explanation is that a sputtered, amorphous layer contains many more point defects (possibly charged) than does a glass surface, leading to far more nuclei and early coalescence into a continuous film.

B. Electrical Conduction in Semiconducting Oxide Films

The conduction mechanisms in semiconducting oxide films are, with only one exception, completely different from those of thin metal films. The exception is the size effect, which, of course, includes the specularity of the substrate and film surfaces. Oxide films deposited by whatever means appear to grow on oxide substrates as continuous films from the outset of deposition. This is most probably due to chemical binding at the interface. In any case, one need not be concerned with island structures or coalescence phenomena. Rather, attention is focused on stoichiometric relationships and the incorporation of impurities, either purposeful or inadvertent.

All of the oxide materials that have been used for transparent conductors are known to be *n*-type, so the ensuing discussion will be limited to *n*-type materials. Conductivity mechanisms have been discussed by Mott and Gurney (*42*), Aitchison (*43*), Verwey *et al.* (*44*), Kofstad (*45*), and Wagner (*46*).

1. *Binary Oxides.* If an oxide is completely stoichiometric, it can only be an ionic conductor. Such materials are obviously of no interest as transparent

conductors because of the high activation energy required for ionic conductivity. However, real oxides are hardly, if ever, completely stoichiometric (45). The oxides used for transparent conductors are invariably anion deficient.

Consider the formation of an oxygen vacancy in a perfect crystal. In the process of removing an oxygen atom, two electrons of the oxygen ion are left in the crystal. If both of these electrons are localized at the oxygen vacancy, the charge is the same as in a perfect crystal, and the vacancy has zero effective charge. Such vacancies are neutral. If one or both of the localized electrons are excited and transferred away from the vacancy, the vacancy is left with an effective positive charge with respect to the perfect crystal. The charged vacancy becomes an electron trapping site, but in the process one or more electrons are made available for conduction. If the cation is multivalent (e.g., Sn), the creation of too many oxygen vacancies results in a structure change from SnO_2 to SnO. Somewhere in such a transition, there will occur a compositional range where an excess of oxygen in the SnO structure will exist. Cation vacancies resulting from excess oxygen have the opposite effect to that described above (i.e., they produce holes rather than electrons). Regardless of whether the cation is multivalent or not, for every electron released for conduction, somewhere there is a trap if the cation retains its original valence. Usually, the creation of an anion vacancy results in a cationic valance charge. Clearly, for conduction to be efficient, the volume fraction of traps (and, hence, electrons available for conduction) must be small. Typical free electron concentrations that have been observed range from about 10^{17}–10^{21} cm^{-3}. Even at an electron concentration of 10^{21} cm^{-3}, the number of charged vacancies is small ($<1\%$). Thus, to control conductivity in these materials by controlling the number of oxygen vacancies is difficult at best. The binary oxides used for transparent conductors are all relatively unstable, chemically. They are relatively easy to oxidize and reduce. [At 500 K, the free energies of oxide formation for SnO_2, In_2O_3, and CdO are -114, -123 and -90 kcal/mole, respectively (47).] An approximate correlation between the formation of oxide vacancies and these thermochemical properties has been established (48).

2. *Doped Oxides.* If instead of creating oxygen vacancies by chemical reduction, one incorporates into the host lattice substitutional cations with a valence higher than that of the host, it is electrically the same as creating anion vacancies. Since overall charge neutrality must be preserved, substitution of a higher valent cation requires the addition of an electron. Conversely, incorporation of a lower valent cation produces a hole. For example, if one incorporates Sb^{5+} substitutionally in SnO_2, an additional

electron is added to the lattice. If, instead, In^{3+} is substitutionally added, a hole is produced, which in an n-type semiconductor becomes a trap.

As with oxygen vacancies, not all higher valent dopants incorporated into the lattice produce charge carriers. Some simply remain as neutral point defects. Electrically equivalent effects can occur if anion sites are doped with atoms whose valence is lower than that of oxygen. Of the anion dopants employed, F^- and Cl^- are most often used.

As will be shown, various authors find that equivalent electrical effects are observed in a given host material with widely varying doping levels. In general, deposition processes that involve high substrate temperatures usually have an "optimum" doping level very much lower than processes that use low substrate temperatures. Other deposition process variables also have an effect on the "optimum" doping level. This is a result of the deposition process kinetics, rather than the actual quantity of dopant, determining the ratio of active to inactive dopant sites, and of the number of oxygen vacancies created. (Doped materials are also anion-deficient.) Not all dopants that meet the valency requirements result in increased conductivity. Furthermore, incorporation of some cations and anions with the same valence as that of the host result in conductivity changes, even though they should have no effect based on valency considerations alone. Central to increasing conductivity by doping is the requirement that the dopant ion replace the appropriate host ion substitutionally in the host lattice.

This implies that the ionic radius of the dopant must be the same size as or smaller than the ion it replaces, and that no compounds or solid solutions of dopant oxide with host oxide are formed. For example, Ti^{4+} should have no effect on the conductivity of SnO_2, based on valence considerations alone, but TiO_2 forms a solid solution with SnO_2 (49), resulting in a two-phase material with insulating occlusions. Similarly, Nb should be a donor dopant in CdO, but CdO and Nb_2O_5 combine to form insulating compounds [$Cd_2Nb_2O_7$ and $CdNb_2O_6$ (50)]. Only a few equilibrium phase diagrams for CdO, In_2O_3, and SnO_2 with other oxides have been published (51), so, historically, dopants have been selected empirically.

Even in the absence of solid solution or compound formation, a dopant may not be usable based on its ionic radius. If the dopant ion is too large, an interstitial, rather than a substitutional, site is favored, and the "dopant" will act as a scattering site rather than a source of charge carriers.

All of the host oxides are very hygroscopic. CdO converts to $Cd(OH)_2$ readily (52). The latter cannot be decomposed at temperatures less than about 300°C (53). In_2O_3 partially converts to $In(OH)_3$ at low temperatures (<150°C). This compound is decomposed to InOOH at about 200°C, and single-phase In_2O_3 occurs only above 375°C in the presence of water (54).

SnO_2 readily forms stannic and stannous acids (hydrates of SnO and SnO_2) (53), the most stable of which is β-stannic acid, $SnO_2 \cdot H_2O$ (55). All of these compounds have higher resistivities than the pure oxides. While these chemical effects are not doping, in the strict sense of the word, their effects are equivalent.

3. *The Influence of the Substrate on Oxide Films.* In practice, the most common substrates for transparent conductors are various glasses. From the standpoint of smoothness, glass is a good choice, since size effect problems and scattering of incident light are minimized. However, most glasses provide a virtually inexhaustible supply of ions, which can diffuse into transparent conducting oxides and dope them inadvertently. Unfortunately, the most common and most mobile ions in glasses are acceptors in the transparent conducting oxides most often employed. In particular, the alkali ions Li^+ and Na^+ have very small ionic radii (56) and can fit easily into substitutional sites in CdO, SnO_2, and In_2O_3. The common alkaline earth ions found in glass (Ca^{2+}, Sr^{2+}, and Ba^{2+}) all have ionic radii that are too large for substitutional doping, so presumably they would be incorporated as interstitials (scattering sites). Based on the ionic radii (56) of various glass constituents, the following more or less mobile ions could diffuse into films of oxide transparent conductors in substitutional positions Li, Na, Mg, Ti, Cr, Mn, Fe, Cu, and Zn. All others would be found in interstitial positions.

The quantity of impurities contributed to the film by a given substrate is primarily related to the substrate temperature during film deposition, since it is rare that subsequent operating temperatures exceed the deposition temperature. Thus, high temperature processes are much more prone to substrate influences than lower temperature processes. This is a major concern in hydrolytic processes in which the glass is heated almost to its softening point. Various patents (57–59) describe improved film properties for films deposited on pure SiO_2 rather than on glass, for glass leached with a mineral acid before film deposition, and various other procedures that limit the troublesome impurities at the glass surface. In a more recent patent, Carlson *et al.* (60) have depleted glass surfaces of ions by applying an electric field to the surface at an elevated temperature and cooling the glass with the field applied. Depth profile analysis using ion scattering spectroscopy showed that the first 500 Å of the glass surface was SiO_2, and that the alkali metal ions were depleted to a depth of about 1000 Å. SnO_2 films deposited by spray hydrolysis onto a single piece of glass, half of which was ion-depleted, showed that the resistivity was more than an order of magnitude lower on the depleted half (61).

C. Transparency

The sources of light loss in transparent conducting films are absorption, reflection, and scattering. Charge carrier absorption, absorption by bound charges, and molecular scattering are related to the film materials themselves. The same types of absorption also occur in substrates. However, since nearly all data published on these materials subtract out the effect of substrate absorption, we shall not consider this any further. All of the optical transmission data presented here will follow that convention unless otherwise noted. Reflections at the air–film, film–substrate, and substrate–air interfaces and geometric scattering are not subtracted out in the data presented here.

Since all of these sources of light loss have been reviewed extensively, we shall give only a brief outline of the effects pertinent to transparent conductors.

1. *Absorption.* The ratio of the transmitted light intensity I to the incident light intensity I_0, when light is passed through an absorbing film, is given by the familiar equation (Lambert's law):

$$I/I_0 = e^{-\alpha t} \quad (1)$$

where α is the absorption coefficient, and t is the thickness of the film. This law is not exact for thin films, because it assumes that no reflection phenomena occur. However, for our present purposes, in reviewing absorption per se it is valid.

From the standpoint of transparent conducting films, charge carrier absorption is the most important mechanism. Several mechanisms are known that produce this effect. In metals, the only important mechanism is electron heating by incident photons, and the absorption coefficient in Eq. (1) is related to one of the optical constants of metals (the extinction coefficient k), by the expression

$$\alpha = 4\pi k/\lambda \quad (2)$$

The optical properties of metal films have been reviewed by Abeles (62), and significant compilations of extinction coefficient data on bulk and thin film metals are available (63). Figure 5 shows that the absorption is a slowly varying function of wavelength over a very wide range of wavelengths. Similar data for various noble metals are presented graphically in Fig. 6. The exponential dependence of absorption on film thickness is shown graphically in Fig. 7. The data in Figs. 5–7 are calculated from the extinction coefficients compiled by Hass and Hadley (63). All of these data should be

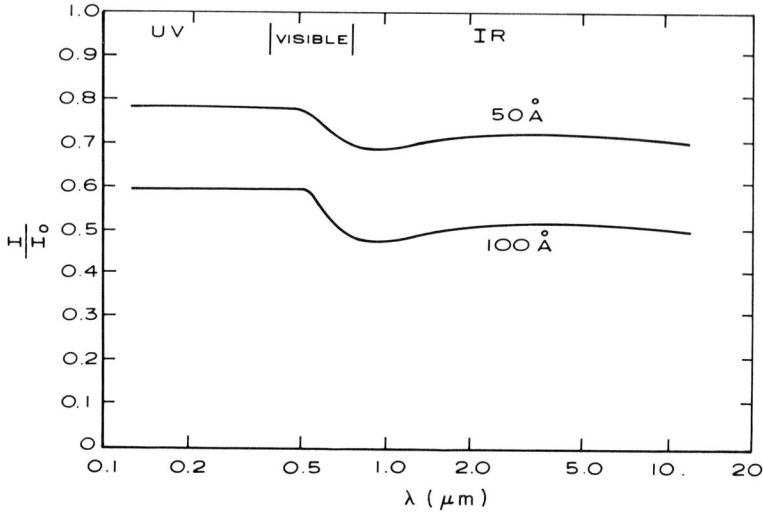

Fig. 5. Calculated optical absorption versus wavelength for Au films of two different thicknesses.

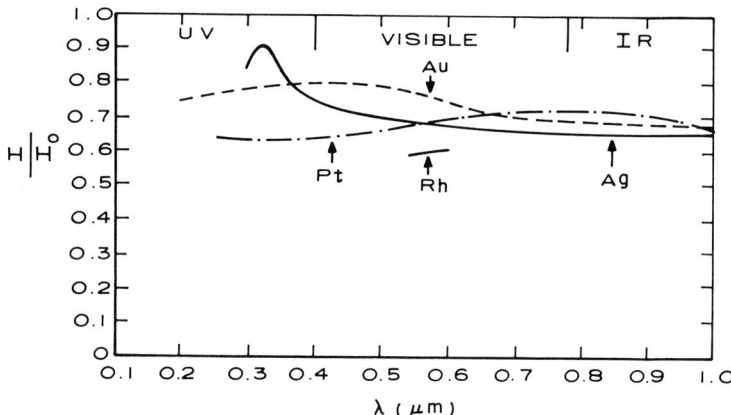

Fig. 6. Calculated optical absorption of selected noble metal films versus wavelength; thickness = 50 Å.

used with caution, especially in the case of very thin films. These data are for "pure" metals, whereas, in the real world of very thin films most of the so-called noble metals are found to exist as oxides—not pure metals. For example, Thomson and Harvey (64) have shown that continuous films of Rh, Pt, Ru, and Pd behave as semiconductors with activation energies for

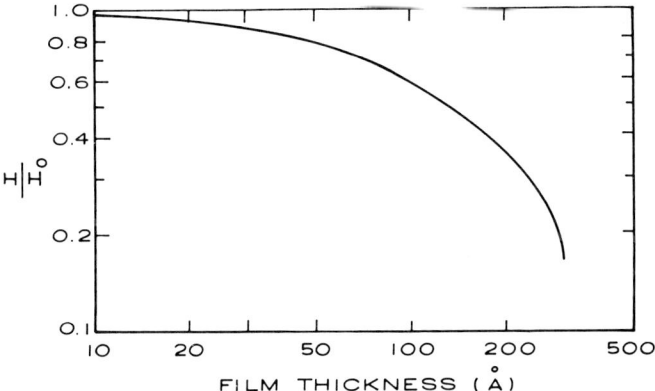

FIG. 7. Calculated optical absorption versus film thickness for evaporated Au films; $\lambda = 5500$ Å.

electrical conductivity between 0.35–12 eV when they were deposited on Al_2O_3 or SiO_2 substrates. Gold is probably the only exception to this. Oxidation of these metals enhances the transmission, but limits conductivity.

In discontinuous thin films, quantum effects in small particles lead to additional light absorption (65–67). In continuous or semicontinuous pure metal films, the problem of charge carrier absorption is much more severe than in semiconducting oxides, since there are far more carriers in the metal case. Nevertheless, it is one source of absorption in metal oxides (68, 69). A related source of absorption is electron scattering by phonons and impurities (70).

The remaining absorption phenomena in metal oxides depend on the energy of the incident radiation. In the far infrared ($\lambda > 20\ \mu m$; $E < 0.05$ eV), only molecular rotation is possible. This gives rise to selective absorption (line absorption) corresponding to quantum numbers. In the near infrared (0.8 $\mu m < \lambda < 20\ \mu m$; 0.05 eV $< E < 1.5$ eV), absorption causes atoms in the molecule to oscillate about their equilibrium positions. The lower energy rotation is superimposed on these oscillations, giving rise to localized absorption bands. In the visible and ultraviolet ($\lambda < 0.8\ \mu m$; $E > 1.5$ eV), absorption consists of displacing an outer electron in the molecule.

As the energy increases, it becomes sufficiently large to overcome the bandgap and drive electrons completely away from the molecule (i.e., result in photoconductivity). Eventually, a sufficiently large number of carriers are liberated to result in an exponential absorption edge. Urbach's rule (71, 72) relates the dependence of the optical absorption coefficient $\alpha(\omega)$ to the photon energy $\hbar\omega$ and temperature T:

$$\alpha(\omega) = A \exp[\sigma(\hbar\omega - \hbar\omega_0)/kT] \qquad (3)$$

where A and ω_0 are constants, σ is a constant of the order of unity and ω is the angular frequency.

2. *Reflection, Refraction, and Interference.* Reflection, refraction, and interference phenomena in thin films have been the subjects of several reviews (73–76). From the standpoint of transparent conducting films, the classical formalism of Airy (77) gives a good intuitive grasp of most of the pertinent effects. This approach is based on suming multiple reflections at the air–film and film–substrate interfaces.

Neglecting absorption for the moment and considering only reflections, if a plane light wave of wavelength λ is incident on a film of refractive index n_f and thickness t, at an angle φ, the reflected light experiences a phase shift δ,

$$\delta = (4\pi/\lambda)n_f t \cos \varphi \tag{4}$$

due to multiple internal reflections at the air–film and film–substrate interfaces. This gives rise to a reflected intensity I_R, given by:

$$I_R = \frac{4R \sin^2 \tfrac{1}{2}\delta}{(1 - R)^2 + 4R \sin^2 \tfrac{1}{2}\delta} \tag{5}$$

and a transmitted intensity I_T:

$$I_T = \frac{T^2}{(1 - R)^2} \left\{ \frac{1}{1 + [4R/(1 - R)^2] \sin^2 \tfrac{1}{2}\delta} \right\} \tag{6}$$

where R is the reflected amount of the incident wave, and T is the transmitted amount ($T + R = 1$). This gives rise to a periodic interference pattern with maxima and minima.

$$I_{max} = T^2/(1 - R)^2 \tag{7}$$

and

$$I_{min} = [(1 - R)/(1 + R)]^2 \tag{8}$$

If some amount A of the incident light is absorbed,

$$T + R + A = 1 \tag{9}$$

the transmitted intensity becomes

$$I = \frac{T^2}{(T + A)^2} \left\{ \frac{1}{1 + [(1 + R)^2/(T + A)^2 - 1] \sin^2 \tfrac{1}{2}\delta} \right\} \tag{10}$$

and the maxima and minima occur at

$$I_{max} = [T/(T + A)]^2 \tag{11}$$

and

$$I_{min} = [T/(1 + R)]^2 \tag{12}$$

respectively.

For detailed definitions of R, T, and A in terms of the optical constants (n and k) of the materials involved, one must turn to electromagnetic theory. These definitions are given for the most general case in numerous optical texts and handbooks. The somewhat simplified versions applicable to light incident from air or vacuum ($n = 1$) on a partially absorbing film of thickness t, refractive index n_f and extinction coefficient k_f in contact with a thick, isotropic, weakly absorbing substrate with refractive index n_s and extinction coefficient k_s are given below.

$$R = \frac{a_1 e^x + a_2 e^{-x} + a_3 \cos\eta + a_4 \sin\eta}{b_1 e^x + b_2 e^{-x} + b_3 \cos\eta + b_4 \sin\eta} \tag{13}$$

$$T = \frac{16 n_s (n_f^2 + k_f^2)}{b_1 e^x + b_2 e^{-x} + b_3 \cos\eta + b_4 \sin\eta} \tag{14}$$

and

$$A = (1 - T - R) \tag{15}$$

where

$a_1 = [(1 - n_f)^2 + k_f^2][(n_f + n_s)^2 + (k_f + k_s)^2]$

$a_2 = [(1 + n_f)^2 + k_f^2][(n_f - n_s)^2 + (k_f - k_s)^2]$

$a_3 = 2[(1 - n_f^2 + k_f^2)(n_f^2 + k_f^2 - n_s^2 - k_s^2) + 4k_f(n_f k_s - n_s k_f)]$

$a_4 = 4[(1 - n_f^2 - k_f^2)(n_f k_s - n_s k_f) - k_f(n_f^2 + k_f^2 - n_s^2 - k_s^2)]$

$b_1 = [(1 + n_f)^2 + k_f^2][(n_f + n_s)^2 + (k_f + k_s)^2]$

$b_2 = [(1 - n_f)^2 + k_f^2][(n_f - n_s)^2 + (k_f - k_s)^2]$

$b_3 = 2[(1 - n_f^2 - k_f^2)(n_f^2 + k_f^2 - n_s^2 - k_s^2) - 4k_f(n_f k_s - n_s k_f)]$

$b_4 = 4[(1 - n_f^2 - k_f^2)(n_f k_s - n_s k_f) + k_f(n_f^2 + k_f^2 - n_s^2 - k_s^2)]$

$x = 4\pi k_f t / \lambda$

and

$\eta = 4\pi n_f t / \lambda.$

Equations (13)–(15) are valid for light incident on a partially absorbing film from the *film side*. When light is incident on the substrate side and the substrate is partially absorbing, the reflectance is lowered, absorption is increased, and the total transmittance remains constant (74). The net effect is that the transmitted intensity [Eq. (10)] is slightly lower when light is incident on the substrate side. In practice, transparent conducting films in devices, such as liquid crystal displays, receive incident light from the *substrate side*, while in almost all cases reported, measurements on these films are for light incident on the *film side*.

The optical constants of both metal and oxide films used for transparent conductors are not constant. They depend strongly on wavelength. This makes the algebra involved in Eqs. (10), (13), and (14) quite cumbersome, but some generalizations may be pointed out. In the case at hand, the index of refraction of most substrates used is in the range 1.4–1.6 and the extinction coefficients are quite small. This usually results in negligible absorption by the substrate. The oxide film materials used have relatively high indices of refraction (~ 2.0) and low extinction coefficients. The metals used have low refractive indices and high extinction coefficients. Both conditions lead to large values of R. In most transparent conductors, reflection is the dominant source of light loss.

However, the high index of refraction of the oxide material results in distinct interference color changes for very small thickness changes. These give a very rapid estimate of the film thickness and uniformity. The thickness of the film is related to the wavelength, index of refraction, and angle of incidence by the well-known relation:

$$t = \tfrac{1}{2}m\lambda/(n_f^2 - \sin^2 \varphi)^{1/2} \tag{16}$$

where m is an integer (the order of interference) and φ is the angle of incidence. Table I is a color chart for films with a refractive index of 2.00. Both SnO_2 and In_2O_3 have indices very close to this value. In some of the very early work reported on these films, the film thickness was very nonuniform, re-

TABLE I

THE COLORS OF THIN FILMS ON REFLECTING SUBSTRATES WHEN VIEWED IN REFLECTED WHITE LIGHT[a]

Color	1st Order t (Å)	2nd Order t (Å)	3rd Order t (Å)	4th Order t (Å)
Gray	75			
Tan	230			
Brown	380			
Blue	620			
Violet	770	2100	3600	5000
Blue	1150	2300	3800	5300
Green	1400	2500	4000	5550
Yellow	1600	2850	4300	5800
Orange	1750	3100	4600	6050
Red	1900	3350	4800	6300

[a] $n = 2.00$.

sulting in multicolored films which were described as "iridescent." Often the films were called "transparent, iridescent coatings" (TIC). In some literature, this terminology still survives.

Finally, the interference phenomena considered above give rise to no little confusion in interpreting transmission data in the literature. In too many cases, transmission data are cited for specific wavelengths, whereas to obtain a true picture of the transmission of a film, the entire spectrum should be shown. Since the optical constants of these materials are functions of wavelength and deposition conditions, the entire interference pattern can shift. It is entirely possible that at one value of λ, the interference pattern could shift from a maximum to a minimum as a result of different film deposition conditions leading to a change in refractive index. Even for very thin films ($t \ll \lambda$), the transmitted light intensity varies considerably with λ, especially near the blue end of the spectrum.

3. *Antireflection Coatings.* Since reflection is the dominant source of light loss in transparent conductors, it would seem advisable to deposit antireflection coatings over them. It is well known (78) that the conditions for zero reflectance in a single-layer film on a substrate are

$$n_f^2 = n_s \tag{17}$$

and

$$n_f t = m\lambda/4 \tag{18}$$

where m is an integer (usually taken as 1) and the remaining symbols have their previous connotation. Thus oxide transparent conductors cannot serve simultaneously as antireflection coatings because high index oxides on low index substrates violate the first condition. In this case, the reflectance of a quarter-wave coating $R_{\lambda/4}$ is a maximum at the chosen wavelength and has a value:

$$R_{\lambda/4} = [(n_f^2 - n_s)/(n_f^2 + n_s)]^2 \tag{19}$$

In principle, it is possible to produce an antireflection coating over the oxide transparent conductors with multiple layers (78), but no attempts to do this appear to have been reported, probably because of the expense involved.

An interesting technique to minimize reflection in oxide transparent conductors has been described by Zaromb (79). The process employed is spray hydrolysis of $SnCl_4$ and $SiCl_4$. Starting at the substrate, 100% SiO_2 is deposited. Gradually, $SnCl_4$ is added to the spray to produce a mixture of SnO_2 and SiO_2. At about halfway through the film, nearly all SnO_2 is deposited and then $SiCl_4$ is again blended in. The result is an SnO_2 film sandwiched between SiO_2 films with a gradual transition in refractive index through the composite film as shown in Fig. 8. This tends to "smear out"

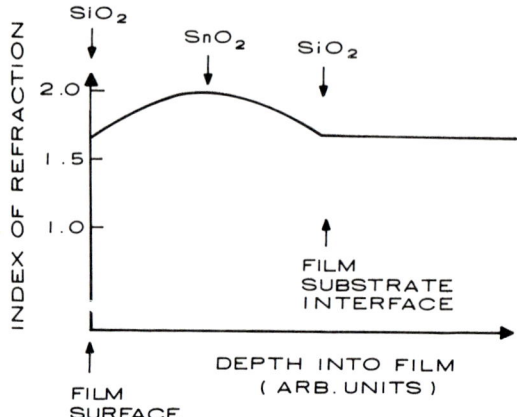

FIG. 8. Variation of refractive index with film depth for SnO_2–SiO_2 films graded by the method of Zaromb (79).

the refractive index change, which, it is claimed, reduces reflection. Apparently there are no compounds formed in the SnO_2–SiO_2 system (51), otherwise the technique probably would not work.

For the case of thin metal films, the usual concern in optics is to *increase* reflectivity with dielectric stacks (80). For the noble metals usually used for transparent conductors, the refractive index is so low that antireflection coatings become impractical (too many layers required). However, it is quite common to overcoat noble metal transparent conductors with insulating films for scratch protection. Clearly, such coatings will have an adverse effect on the optical transmission of the metal film. In addition, dielectric or semiconducting overlayers can affect the electrical properties of the metal films, as will be shown in Section III, B.

Finally, it should be noted that for practical devices such as vidicons and liquid crystal displays, an essential feature is direct contact of the transparent conductor with the active material (photoconductor or liquid crystal). This requirement precludes the use of dielectric coatings for antireflection over the transparent conducting film. Also, since the light is incident from the substrate side in these devices, an antireflection coating over the film has no optical effect. However, antireflection coatings can be applied to the other side of the substrate to reduce reflections by the substrate in such applications (80a).

4. *Scattering and Surface Morphology.* We have already noted the effects of surface roughness on the electrical properties of thin transparent conduc-

tors. In their review of precision optical measurements, Bennett and Bennett (81) have summarized the effects of surface roughness (film and/or substrate) on reflectance. When light is incident on a rough surface, part of the beam is reflected in the specular direction, part is transmitted, part is absorbed, and the remainder is diffusely scattered. If the surface has a Gaussian distribution of heights, the ratio of reflectance in the specular direction R, to the reflectance of a perfectly smooth surface of the same material R_0, is

$$R/R_0 = \exp[-(4\pi r/\lambda)^2] \\ + \{1 - \exp[-(4\pi r/\lambda)^2]\}\{1 - \exp[-2(\pi r\alpha/m\lambda)^2]\} \quad (20)$$

where r is the rms surface roughness, m is the rms slope of the irregularities, and α is the half-acceptance angle of the measuring instrument. Calculated values of R/R_0 are plotted versus surface roughness in Fig. 9.

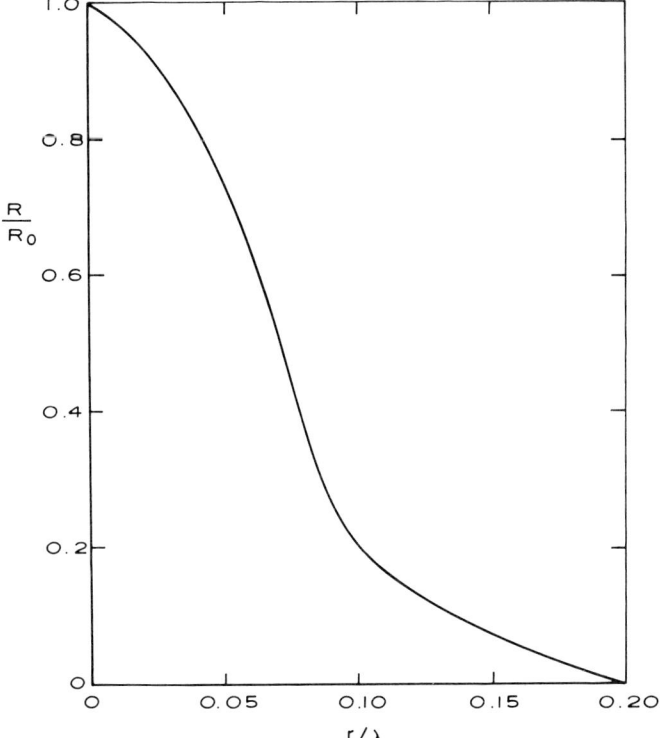

FIG. 9. Calculated relative reflectance versus the ratio of rms surface roughness to wavelength.

Two things should especially be noted. Roughness of only 0.01λ results in a 1% change in reflectance, and the effect is magnified for small values of λ. Thus scattering is more severe at the blue end of the visible spectrum where absorption and reflection are also more severe. To put actual dimensions on the above example, for $\lambda = 4000$ Å, a surface roughness of only 40 Å rms produces a 1% change in reflectance.

While the specularly reflected fraction of the incident light is decreased according to Eq. (20), the total amount of light reflected is increased because of local variations in angle of incidence on a rough surface. Thus, the transmitted intensity is decreased. The virtues of using smooth substrates for transparent conductors are obvious.

D. Bulk Electrical and Optical Properties of Semiconducting Oxides

As a point of reference for the detailed thin film data to follow, the bulk electrical and optical properties of the oxide materials used for transparent conductors are reviewed briefly in this section. Note that many of the phenomena that lead to confusion in oxide thin films apply also to the bulk properties (e.g., stoichiometry, nature of defects, etc.).

1. SnO_2. The electrical and optical properties of doped and undoped SnO_2 single crystals and powders have been the subjects of numerous investigations (82–99). The structure of SnO_2 is tetragonal ($a_0 = 4.73727$ Å; $c_0 = 3.186383$ Å) (100). The structure of SnO is also tetragonal ($a_0 = 3.802$ Å, $c_0 = 4.836$ Å) (100). It is generally agreed that SnO_2 is an extrinsic (defect) semiconductor, and that the pertinent electrical properties depend strongly on stoichiometry. Table II gives the electrical properties of nearly stoichiometric, impurity-free single crystals as reported by various authors. All of the specimens cited in Table II were quite pure and were stoichiometric within analytic error, suggesting that the diversity of properties reported is due to very minor deviations from stoichiometry and/or very minor trace impurities.

Kohnke (83) reported on the properties of naturally occurring SnO_2 crystals (Bolivian cassiterite) which apparently were not very pure and had a fairly high dislocation density. He found a fundamental bandgap of 3.54 eV, a sharp absorption peak at $\lambda = 3.07$ μm (attributed to O—H groups in the crystal), carrier concentrations from 10^{14}–$10^{15}/cm^3$, Hall mobilities from 10–300 cm^2 V^{-1} sec^{-1}, donor densities of about $10^{20}/cm^3$, and a donor activation energy of about 0.7 eV.

TABLE II

ROOM TEMPERATURE ELECTRICAL PROPERTIES OF SnO$_2$ SINGLE CRYSTALS
(NEARLY STOICHIOMETRIC AND IMPURITY-FREE)

ρ (Ω cm)	Carrier concentration (cm^{-3})	Donor concentration (cm^{-3})	Hall mobility (cm^{2}V^{-1} sec^{-1})	Fundamental bandgap (eV)	Donor activation energy (eV)	Ref.
0.2–0.27	1.1–1.5×10^{17}	—	185–240	—	—	(85)
0.4–15	0.7–1.5×10^{17}	0.15–5.7×10^{17}	45–150	—	0.081–0.138	(85)
—	—	—	—	4	0.50–0.74	(87)
10–10^6	—	—	—	—	—	(90)
—	—	—	—	3.7 ($\perp c$ axis)	—	(96)
—	—	—	—	4.1 ($\parallel c$ axis)	—	(96)
—	—	—	—	3.47 ($\perp c$ axis)	—	(97)
—	—	—	—	3.95 ($\parallel c$ axis)	—	(97)
0.5–6.1×10^3	0.11–1.2×10^{18}	—	91–152	—	—	(98)
—	—	—	—	3.57 ($\perp c$ axis)	—	(99)
—	—	—	—	3.93 ($\parallel c$ axis)	—	(99)

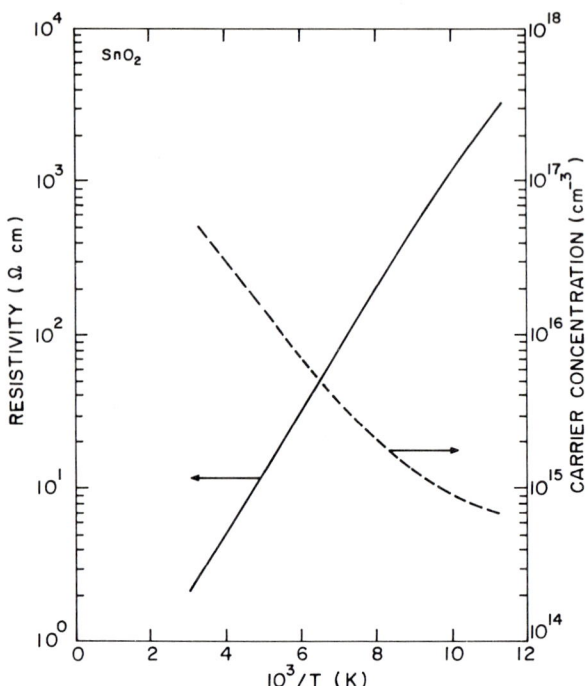

Fig. 10. Resistivity and carrier concentration of a pure SnO_2 single crystal versus reciprocal temperature. Donor concentration = $1.6 \times 10^{17}/cm^3$; donor activation energy = 0.104 eV. [After Marley and Dockerty (86).]

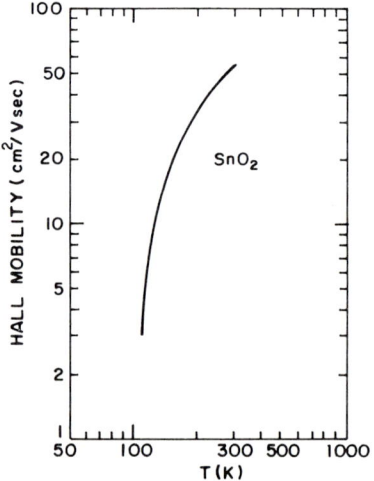

Fig. 11. Hall mobility versus temperature for a pure SnO_2 single crystal. Donor concentration = $1.6 \times 10^{17}/cm^3$, donor activation energy = 0.104 eV. [After Marley and Dockerty (86).]

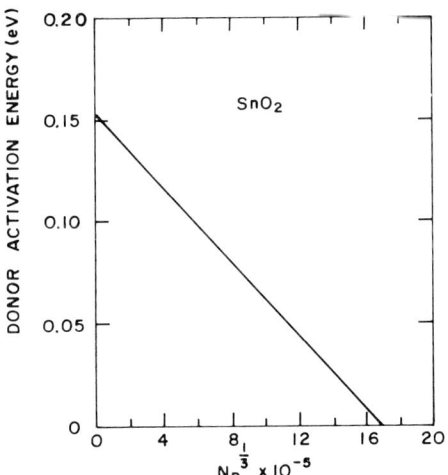

FIG. 12. Donor activation energy versus donor concentration for SnO_2 crystals annealed in O_2 at various temperatures and for various times. [After Marley and Dockerty (86).]

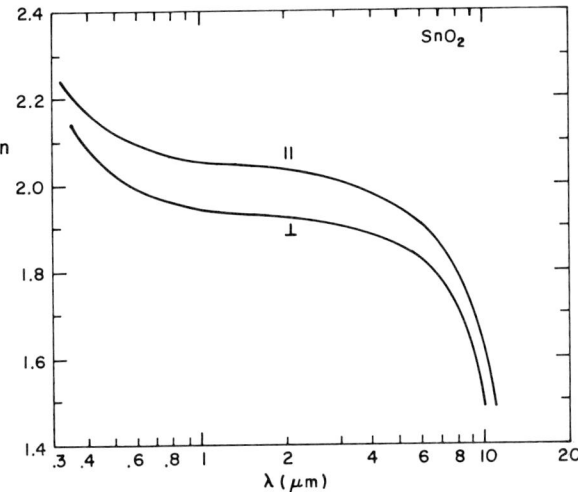

FIG. 13. The refractive index of pure SnO_2 single crystals, perpendicular and parallel to the c axis versus wavelength. [After Reddaway and Wright (96). By permission of The Institute of Physics.]

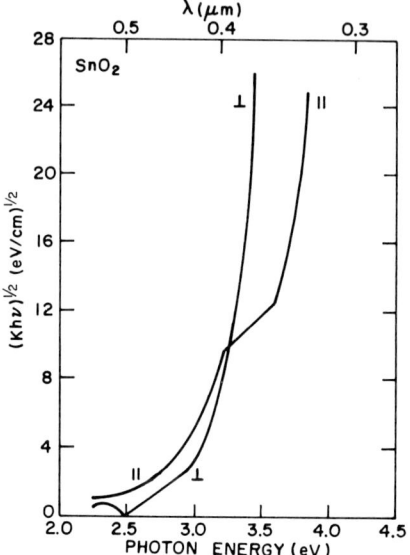

FIG. 14. Average normalized absorption of pure SnO_2 single crystals versus photon energy near the absorption edge, parallel and perpendicular to the c axis. [After Reddaway and Wright (96). By permission of The Institute of Physics.]

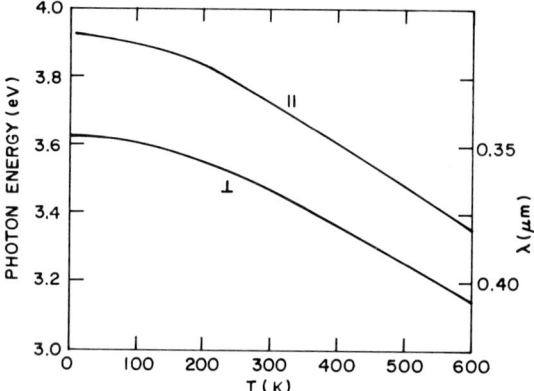

FIG. 15. Temperature dependence of the ultraviolet absorption edge of pure SnO_2 crystals, parallel and perpendicular to the c axis. [After Summitt and Borrelli (97).]

As one might suspect, electrical and optical data on powders and polycrystalline specimens (*82, 84, 89, 91, 92*) are even more scattered than the single-crystal data.

Representative data on the variation of electrical and optical properties of "pure" single crystals are presented in Figs. 10–15. In addition, Cunningham *et al.* (*90*) have concluded from photoconductivity measurements that SnO_2 is a degenerate semiconductor at room temperature when the resistivity is less than 10^4 Ω cm.

Samson and Fonstad (*101*) reported that the electrical conductivity of SnO_2 varies with oxygen pressure (P_{O_2}) as $P_{O_2}^{-1/6.5}$ in the temperature range 1300–1650 K and between $1-10^{-2}$ atm. Mar (*95*) concluded that this oxygen-pressure dependence of conductivity was greatly affected by the presence of Sb ions.

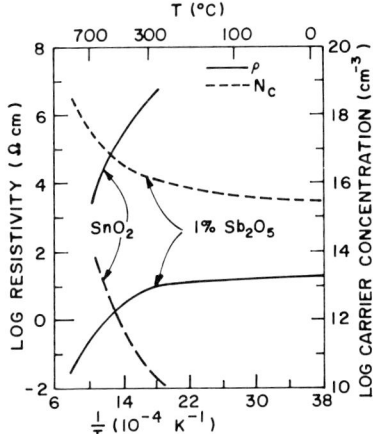

FIG. 16. The resistivity and carrier concentration of pure and Sb-doped SnO_2 sintered powders versus temperature. [After Mar (*95*).]

Morgan and Wright (*88*) studied the effect of Sb doping on SnO_2 single crystals, but did not report the Sb content. Most of the bulk material data on doped SnO_2 have been reported for sintered powder specimens (*84, 92, 95*). Figures 16 and 17 show typical dependencies of electrical properties on Sb doping concentration. There appear to be no reports on the effect of Sb doping on the bulk, optical properties of SnO_2, and other dopants have not been studied in bulk materials.

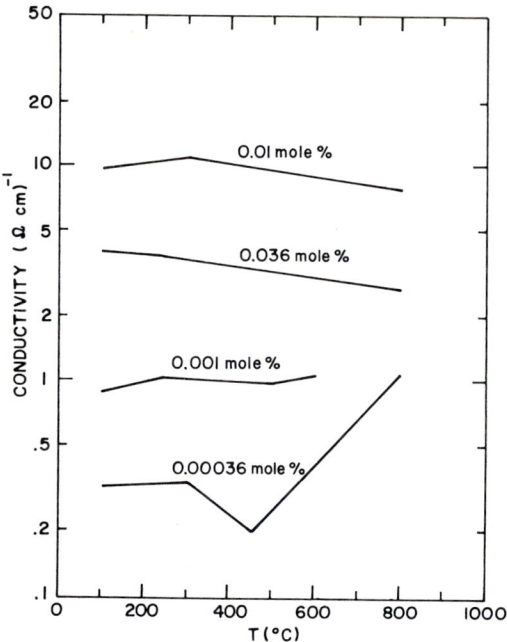

FIG.17. Conductivity versus temperature for various Sb doping levels in SnO_2 sintered powders. [After Loch (*84*).]

2. In_2O_3. The electrical and optical properties of single crystals of "pure" In_2O_3 (*102–105*) and the electrical properties of sintered powder (*106*) have been reported. There do not appear to be any reports on the bulk properties of purposely doped In_2O_3 crystals, but some properties of Sn-doped powders have appeared (*105a*).

The crystal structure is bcc ($a_0 = 10.118$ Å) at room temperature (*100*).

Based on the methods of single-crystal growth used and the resistivities obtained by various workers, the chemical transport method used by DeWit (*105*) probably yields nearly stoichiometric crystals. He reports a room temperature bulk resistivity of 4×10^4 Ω cm in contrast with about 2.5 Ω cm for lead–boron–oxide flux-grown crystals (*103*), and about 20–200 Ω cm for vapor grown crystals from an In–C mixture (*102*). The only extensive electrical and optical data were reported for the vapor-grown crystals (*102, 104*), and these are probably oxygen deficient as a result of the growth technique.

In general, there is agreement that In_2O_3 is a degenerate, extrinsic, n-type semiconductor. Figures 18 and 19 show typical electrical properties of crystals grown from the vapor by Weiher (102). Optical absorption and reflectivity are plotted in Fig. 20. From the optical data, Weiher and Ley (104) found the fundamental energy gap to be 3.75 eV. They also observed an indirect forbidden transition with an energy gap of 2.619 eV.

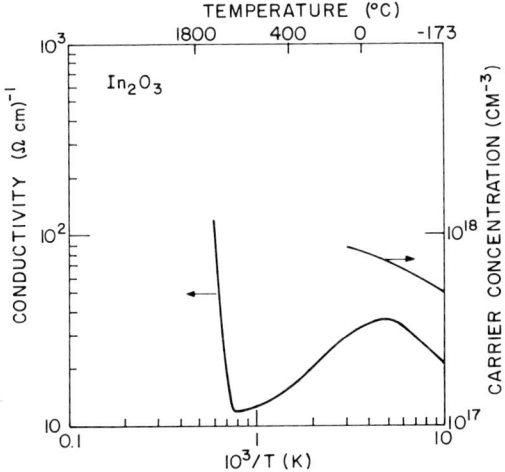

FIG. 18. The temperature dependence of conductivity and charge carrier concentration for typical In_2O_3 single crystals. [After Weiher (102).]

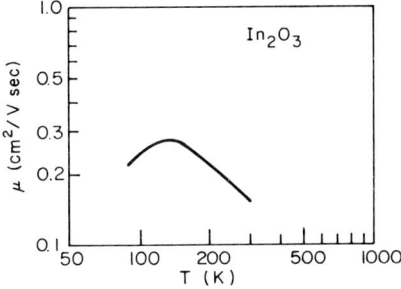

FIG. 19. Carrier mobility versus temperature for a typical In_2O_3 single crystal. [After Weiher (102).]

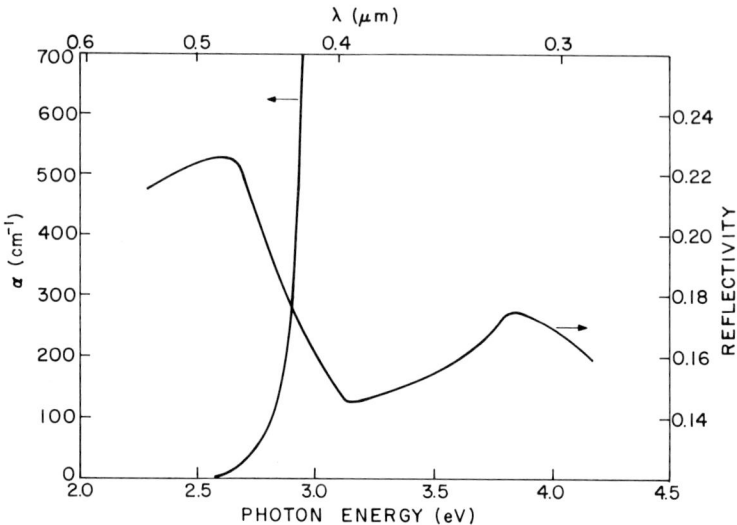

Fig. 20. Absorption coefficient and reflectivity of a typical In_2O_3 single crystal. [After Weiher and Ley (*104*).]

3. CdO. The bulk properties of CdO have been reviewed by Jarzebski (*107*). It is a degenerate, extrinsic semiconductor in which oxygen vacancies are normally doubly ionized. The only studies on single crystals that have appeared are those of Koffyberg (*108, 109*). The bandgap (from optical measurements) is 2.0–2.2 eV (*110*). The structure is cubic (NaCl-type; $a_0 = 4.6953$ Å) (*100*). The electrical resistivity at room temperature is typically $2–10 \times 10^{-3}$ Ω cm (*110*).

The low bandgap of this material results in excessive absorption from about $\lambda = 0.5$ μm down into the uv. Therefore, it is not often used anymore as a transparent conductor.

III. Processes for Film Deposition

In this section we shall consider the numerous methods that have been employed to deposit transparent conducting films. Emphasis is placed on interpretation of the film properties that have been reported as related to film deposition parameters and the physics and chemistry of the materials involved.

A. Chemical Vapor Deposition

It is convenient to subdivide chemical vapor deposition (CVD) into hydrolysis of chlorides and pyrolysis, because, in general, the deposition apparatus and process control parameters are quite different. Reviews of CVD have been done by Powell et al. (*111*), Campbell (*112*), and Feist et al. (*113*).

1. *Hydrolysis of Chlorides.* This method of film preparation depends on the surface hydrolysis of metal chloride on a heated substrate surface via the basic reaction

$$MeCl_x + yH_2O \rightarrow MeO_y + xHCl \qquad (21)$$

in which Me is Sn or In for the usual host materials. Cationic dopants are introduced through the same reaction by mixing the chlorides of Sn or In with dopant chlorides (e.g., $SbCl_5$, PCl_5, etc.). It should be emphasized that the true reaction kinetics are very much more complicated than the basic reaction would indicate. Normally, the simple reagents are mixed with various organic compounds (alcohols or organic acids), and the amount of water introduced from humid ambients is variable (*43*). The decomposition products of the organic materials lead to chemical reduction of the oxide film, and the amount of water present largely determines the amount of free chlorine liberated. The former leads to anion vacancies and the latter leads to doping at anion sites with Cl^-. High substrate temperatures increase the deposition rate and favor reduction (anion vacancies), whereas low temperatures favor Cl^- incorporation in the films (*114, 114a*). The kind and quantity of organic materials mixed with the starting chlorides affects the amount of reduction at any given substrate temperature. A further complication arises in doped systems since the equilibrium constants for the reactions of the host and dopant chlorides are usually different, and they vary with temperature in different ways (*115*). As a result, the ratio of dopant atoms to host cations is not only different from that of the starting reagents, but it varies with deposition temperatures as well. Most reported studies give the starting reagent composition, but few report analyses of the films deposited. All of these reactions compete in a complex manner, the details of which apparently have not been studied. Furthermore, most reported data are given for glass substrates, which, as has been noted, are virtually inexhaustible sources of alkali impurities that are p-type in SnO_2 and In_2O_3.

The two methods of material application that have been used are spraying and dipping. The substrate is preheated to the deposition temperature (typically 400–800°C) and then is sprayed with the starting reagents or dipped into them. Spraying results in a film that is more uniform in thickness,

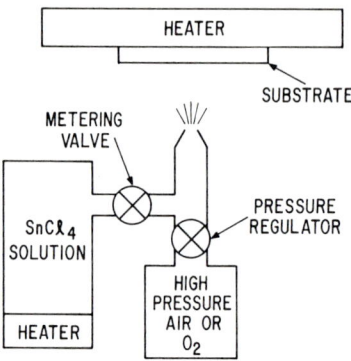

Fig. 21. Schematic of an apparatus for spray hydrolysis.

but has several disadvantages. It is more affected by atmospheric impurities. It is wasteful of starting reagents (a major problem with expensive materials, such as $InCl_3$). Since glass substrates are often heated to temperatures near their softening point, stresses in the deposited films tend to warp or bend large substrates unless the coating is sprayed onto both sides simultaneously (*116*) or unless the back side of the substrate is cooled at the same rate as the spray cools the front surface. Dipping is less affected by atmospheric contaminants, coats both sides of substrates simultaneously, and is less wasteful of expensive reagents, but has the disadvantage that the film thickness is inherently nonuniform. The leading edge of the substrate upon entering the solution is the trailing edge upon leaving the solution. This, coupled with surface tension effects of the solution on the substrate result in very nonuniform film thickness, especially for large substrates. Finally, since the reagent temperature is much lower than the substrate temperature, dipping results in a greater thermal shock to the substrate than does spraying. For these reasons, the most common hydrolytic technique is spraying. Figure 21 illustrates the essentials of the equipment required.

While the thermal shock involved in spray hydrolysis is less than that for dipping processes, it is a major problem—not so much from the viewpoint of substrate cracking, but rather because of its effect on the instantaneous substrate temperature during deposition. The spray cools the substrate, and so the instantaneous substrate temperature decreases with deposition time. This effect is usually minimized by using the highest practical substrate temperature (which implies a very high deposition rate) so that the cooling effect of the spray is negligible during the short deposition time. This problem is overcome at the expense of greater dependence of film properties on impurities introduced from the substrate. For small substrate areas, the required substrate temperature can be lowered by the use of a "fogging" apparatus (*116a*).

The interplay of these process-related effects with the formation of semiconducting oxides has resulted in an enormous array of spraying and dipping procedures devised on an empirical basis for different substrates and different applications. Starting reagents and procedures have been described for films of SnO_2 (*116–127*), In_2O_3 (*124–129*), CdO (*130*), SnO_2: Sb (*43, 131–135*), SnO_2:In (*43*), SnO_2:Cd (*131*), SnO_2:Bi (*136*), SnO_2:Mo (*136*), SnO_2:B (*137*), SnO_2:P (*138, 139*), SnO_2:Te (*140*), SnO_2:W (*140*), In_2O_3:Sn (*141*), In_2O_3:Ti (*141*), In_2O_3:Sb (*141*), SnO_2:F (*142–146*), In_2O_3:F (*144, 147*), SnO_2:Sb, F (*148*), and In_2O_3:Sn, F (*149*). It would be futile to review any or all of these in any detail in view of the rather specific reasons for their original formulation.

About the only generalizations that one can make about these processes are related to substrate surface treatment and the ambient humidity during film deposition. Since the quantity of water available largely determines the way in which the reactions proceed, one must control the amount of water in the starting reagents and the ambient humidity (*150*). For optimum film properties, the substrate surface should be alkali-free. In the case of glass, this can be accomplished by ion depletion (*60, 151, 152*), by precoating the glass with SiO_2 (e.g., by hydrolysis of $SiCl_4$) (*58, 59*) or by selectively etching alkali ions (e.g., with HNO_3) from the surface to leave an SiO_2 surface layer (*153*). These surface treatments are, in fact, beneficial for all film deposition processes, but are especially important in processes that use high substrate temperatures either during or subsequent to film deposition. Aside from the obvious process control requirements (starting reagent composition, substrate temperature, etc.), it has been shown that the film conductivity and transmission of sprayed films is a maximum when the angle of incidence of the spray is normal to the substrate (*154*); that soda-lime glasses are not attacked as much by the HCl liberated in the reaction if an air jet is directed at the surface (*155*); and that spray nozzles, substrate fixtures, etc. should not contain Fe, which can be etched by the HCl and included in the film as an insulating impurity (*57*). Finally, postdeposition annealing at 300°C has been shown to decrease the resistivity and temperature coefficient of resistivity of SnO_2 films prepared by spray hydrolysis at 350–400°C (*156*). The decrease in resistivity is inversely proportional to film thickness. For films 300 Å thick, the decrease is about tenfold, while for 3500 Å-thick films, the decrease is about 30%.

The electrical and optical properties of films prepared by various hydrolytic techniques have been studied in detail, but the details of the deposition procedures have rarely been given. Therefore, the properties to be cited

TABLE III

ELECTRICAL PROPERTIES OF "UNDOPED" SnO_2 FILMS PREPARED BY HYDROLYSIS

Film thickness (Å)	Resistivity at 25°C (Ω cm)	Temperature coefficient of resistivity (ppm/°C)	Hall coefficient (cm^3/C)	Hall mobility (cm^2 V^{-1} sec^{-1})	Carrier concentration (cm^{-3})	Carrier mean free path (Å)	Energy gap (eV)	Ref.
1100	0.00233	~0	0.075	32	8.4×10^{19}	29	—	(157)
820	0.0065	<0	0.11	17	5.8×10^{19}	13	—	(157)
420	0.016	<0	0.145	—	—	—	—	(157)
5000	0.00207	+25	—	—	—	—	—	(158)
3500	0.006	−400	—	—	—	—	—	(156)
1800	0.013	−900	—	—	—	—	—	(156)
300	0.019	~0	—	—	—	—	—	(156)
—	—	—	—	—	—	—	3.45	(159)
—	—	—	—	—	8×10^{17}	—	4.0	(160)
—	—	—	—	—	7.3×10^{20}	—	4.6	(160)
—	—	—	—	—	—	—	3.82	(161)

below should be taken as representative and subject to considerable variation depending on the exact deposition technique. The electrical and optical properties of hydrolytic SnO_2 films which have not been doped purposely have been studied extensively. Typical electrical properties are given in Table III. In addition, the secondary electron emission ratio has been reported (*162*). The maximum ratio is 1.11 at a primary bombarding energy of 600 eV. Most of the optical studies (*157, 159–161, 163*) have concentrated on the ultraviolet and infrared portions of the spectrum. The index of refraction has been reported as 1.96 at $\lambda = 5000$ Å and 1.85 at $\lambda = 7000$ Å (*157*). The extinction coefficient was given as 0.00 at both of these wavelengths (*157*). Arai (*161*) observed a very gradual decrease in refractive index from about 2.2 at $\lambda = 3000$ Å to about 2.0 at $\lambda = 7000$ Å. He also measured the transmission and reflectivity of the films, but the resistivity and thickness were not given. SnO_2 films prepared from $SnCl_2$ (*163a*) and $SnCl_4$ (*163b*) have been employed in SnO_2–Si heterojunctions to form solar cells.

No detailed reports of the properties of undoped In_2O_3 and CdO prepared by hydrolysis have appeared.

The electrical and optical properties of doped SnO_2 films have been studied by several authors (*43, 114, 131, 132, 135, 161, 164, 165*). Representative electrical data on doped SnO_2 films are given in Table IV. Table V gives the Sb doping level that was found to yield the lowest resistivity in various studies. Rohatgi *et al.* (*135*) also studied SnO_2:P and found that the P-doping level corresponding to minimum resistivity was 5 mole %. Haayman *et al.* (*138*) found that the optimum P-content varied from 1–3% depending on the P salt used in the starting reagents.

Light loss in the visible is mainly a result of reflection (*43*). The data of Peaker and Horsley (*132*) are typical of the average optical transmission in the visible for various values of sheet resistivity (Fig. 22).

In view of what has been discussed previously, the scatter in all the electrical and optical data on these films should not be surprising.

Groth (*141*) studied the electrical and optical properties of spray-hydrolyzed In_2O_3:Sn, In_2O_3:Ti, and In_2O_3:Sb. Ti and Sb are acceptors in In_2O_3 and are of no interest for transparent conductors. He found the optimum Sn doping level to be 2 at.% (relative to In). For substrate deposition temperatures between 450–550°C, the carrier concentration and mobility averaged about 5×10^{20} cm^{-3} and 45 cm^2 V^{-1} sec^{-1}, respectively, leading to a resistivity of about 2.3×10^{-4} Ω cm. For films with sheet resistivity of 20–25 Ω/□, the average transmission in the visible was about 85%, and the average reflectivity was about 15%. The absorption edge was found at about $\lambda = 3500$ Å, and the reflectivity increased to about 90% in the near infrared.

TABLE IV

ELECTRICAL PROPERTIES OF DOPED SnO_2 FILMS PREPARED BY HYDROLYSIS

Dopant/Concentration (mole %)	Spray temperature (°C)	Resistivity (Ω cm)	Carrier concentration (cm^{-3})	Hall mobility ($cm^2 V^{-1} sec^{-1}$)	Bandgap (eV)	Ref.
Sb/0.4	650	9×10^{-4}	2×10^{21}	35	—	(113)
Sb/1.0	650	7×10^{-4}	3×10^{21}	30	—	(113)
Sb/7.0	650	1.5×10^{-3}	8×10^{21}	8	—	(113)
Sb/11.0	650	8×10^{-3}	4×10^{20}	3	—	(113)
Sb/1.5	700	1×10^{-3}	1×10^{21}	20	4.1	(131)
Sb/1.0	550	3.3×10^{-3}	—	—	—	(160)
Sb/0.2	600	2.1×10^{-3}	2.6×10^{20}	—	—	(165)
Sb/0.6	600	9.7×10^{-4}	3.4×10^{20}	—	—	(165)
Sb/1.7	600	1.23×10^{-3}	4.8×10^{20}	—	—	(165)
Sb/5.4	600	1.62×10^{-3}	1.3×10^{21}	—	—	(165)
Sb/13.0	600	1.32×10^{-2}	2.9×10^{21}	—	—	(165)
In/3.0	550	5×10^{-1}	—	—	—	(164)
In/5.0	600	$\sim 8 \times 10^{-2}$	—	—	—	(43)

TABLE V

Sb DOPING LEVELS IN
SnO$_2$ YIELDING MINIMUM RESISTIVITY
(HYDROLYSIS)

Deposition temperature (°C)	Sb concentration (mole %)	Ref.
500	0.4	(132)
550	1.0	(164)
600	0.6	(165)
650	3.0	(113)
700	3.0	(135)

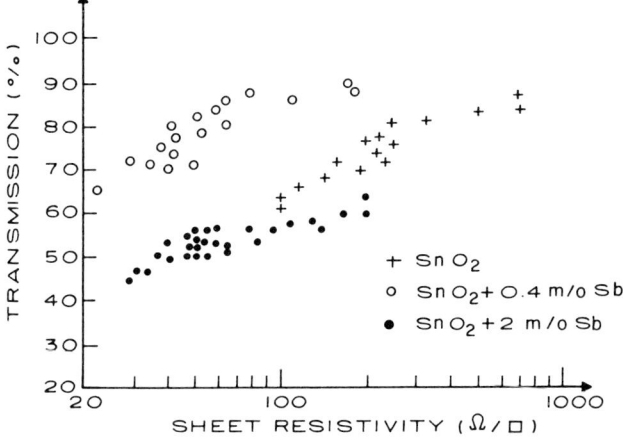

FIG. 22. The variation of average optical transmission in the visible with film sheet resistivity for SnO$_2$ and SnO$_2$:Sb films. [After Peaker and Horsley (132).]

Fischer (131) studied the electrical properties of In$_2$O$_3$:Cd (as 1.5 mole % CdO), prepared at 700°C. He found the resistivity of these films was 3.3×10^{-4} Ω cm, the electron concentration was 3×10^{20} cm^{-3}, the mobility was 6 cm^2 V^{-1} sec^{-1}, and the bandgap was 5.1 eV. The quantum mechanical aspects of the optical and electrical properties of hydrolyzed In$_2$O$_3$:Sn have been treated by Köstlin et al. (165a). The best film properties that have been reported for hydrolyzed films are the In$_2$O$_3$:Sn films of Groth (141). However, these have not been applied extensively, probably because of the high cost of the In halides used as starting reagents.

The main use of hydrolytic processes seems to be for large substrate applications because of relatively low cost. However, the film properties are difficult to reproduce and usually have too rough a surface for most electronic applications. Hydrolytic techniques are rarely used for sophisticated applications.

2. *Pyrolysis.* Pyrolysis may be defined as the thermal decomposition of a metal-organic compound at a heated substrate surface. The requirements on the organometallic starting reagent include relatively low decomposition temperature (to minimize substrate interactions with the growing film), relatively high vapor pressure at temperatures well below the decomposition temperature (for ease of vapor transport), and chemical stability at room temperature (for long reagent life). In addition, it is helpful if the reagent is inexpensive and if it does not present major safety hazards in storage or use.

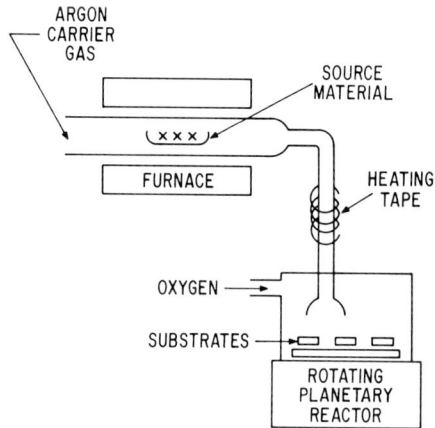

FIG. 23. Diagram of an apparatus for pyrolytic deposition.

The main process control parameters in pyrolytic deposition are the gas flow, gas composition (including carrier gases), substrate temperature, and the geometry of the deposition apparatus. Figure 23 is a schematic illustration of a typical pyrolytic deposition system. Some of the same problems encountered in hydrolysis occur in pyrolytic deposition as well. The substrate temperature and gas flow rate largely determine the deposition rate. If the substrate temperature is low, carbon occlusions are found in the films as a result of incomplete oxidation of the organic material. If the substrate temperature is too high, excessive diffusion of substrate impurities during film growth can occur. A further complication arises if the substrate temperature and reagent gas flow rate are too high; namely, radiation from the substrate heater preheats the gas in the reaction chamber, resulting in decomposition of the organometallic in the gas phase rather than at the substrate surface. In the limiting case, large particles are formed in the gas phase which produce a powder deposit rather than a smooth film.

The gas flow and system geometry determine the uniformity of film deposition over large areas. Thus, it is important to admit the gas into the reaction chamber in a controlled manner both in terms of gas inlet location and flow rate. Most reaction chambers contain baffles and/or planetary rotating substrate holders to ensure that the gas flow in the vicinity of the substrates is as uniform as possible. A very large variety of reactor designs have been described in the literature (*113*).

To obtain optimum quality films, all of these parameters must be adjusted on an empirical basis and then controlled.

As compared to other film deposition techniques, pyrolytic deposition of transparent conducting oxides is relatively inexpensive in terms of equipment cost. In a properly designed reactor and after suitable adjustment of process parameters, it is quite reproducible. Dopants can be added with relative ease. The major limitations of the process are related to the relatively small areas of uniform film deposition that have been achieved and, in some cases, to the cost of the organometallic starting reagents.

Korzo and co-workers (*166–169*) have mainly studied the conductivity mechanisms of In_2O_3 films prepared by the pyrolysis of indium acetylacetonate, $In(C_5H_7O_2)$. Kane *et al.* have studied the electrical and optical properties of SnO_2 prepared from dibutyl tin diacetate, $(C_4H_9)_2Sn(OOCCH_3)_2$ (*170*), SnO_2:Sb prepared from dibutyl tin diacetate and $SbCl_5$ (*171*) and In_2O_3:Sn prepared from an In chelate derived from dipivaloyl methane (2,2,6,6,-tetramethylheptane-3,5-dione) and dibutyl tin diacetate (*172*). All of these reagents meet the requirements outlined above, except the In acetylacetonate which required relatively high power uv radiation to precondition the reagent for complete decomposition at substrate temperatures between 270–520°C (*169*). However, Blandenet *et al.* (*172a*) did not require uv with an aerosol pyrolysis process using indium acetylacetonate at 480°C. The aerosol process uses a mist of the starting reagents derived from ultrasonic agitation of the reagents rather than thermal vaporization. Tin and/or fluorine doping of the films could be accomplished by adding dibutyl tin diacetate or trifluoracetic acid in known quantities. The carrier gas used was air. Tabata (*172b*) deposited SnO_2 films from $SnCl_4$ by pyrolytic, rather than hydrolytic techniques. The substrate temperature range found most useful was 600–800°C. Controlled amounts of O_2 and an Ar carrier gas were used.

The intent of the experiments of Korzo and co-workers (*166–169*) was to produce high resistivity films, not transparent conductors, and to study the dependence of film resistivity on electric field and film structure. The low field resistivity of these films was of the order of 10^{12} Ω cm, which is much too high for the present application. In their studies of various transparent conducting films, Kane et al. (*170–172*) used a reactor described earlier by Kern (*173*). In the undoped SnO_2 work (*170*), oxygen was used as the reactant gas, nitrogen was used as the carrier gas; and it was found that the addition of some water vapor considerably improved the film properties. The substrates used were mainly fused silica and low-alkali glasses. It was found that the mere presence of soda-lime glass in the reaction chamber not only resulted in poor films on the soda-lime glass substrate, but also on other alkali-free substrates in the same run. Furthermore, subsequent runs were spoiled and the reactor had to be thoroughly cleaned before good quality films could be deposited.

Optimum film properties were produced with the Sn source at 98°C, substrate temperature of 420°C, and a large N_2 flow (to minimize gas phase nucleation of particles). The flow rates of O_2 and H_2O were not found to be critical. The optimum conditions were found to be a function of time. That is, for a given set of temperature-flow conditions, the sheet resistivity of the films decreased with deposition time, reached a minimum, and then increased as more material was deposited. This behavior was ascribed to competition between increasing film thickness and the combined effects of film oxidation (annihilation of anion vacancies) and diffusion of acceptors from the substrate.

Similar processing procedures were used for SnO_2:Sb films (*171*). The $SbCl_5$ source temperature was adjusted to yield the optimum doping concentration in the films (0.6–2.7 at. % Sb relative to Sn).

For In_2O_3:Sn films (*172*), the source temperature for the In chelate was 220°C (vapor pressure ≈ 10 torr) and the connections between the vapor source and the reactor were heated to prevent condensation. The optimum substrate temperature was found to be 500°C. Otherwise the conditions were much the same as those used for SnO_2 (*170*). The electrical and optical properties of various films prepared by pyrolysis under optimum conditions are given in Table VI. All of the films could be deposited with very smooth surfaces as evidenced by scanning electron microscopy at 10,000 ×.

The main applications of pyrolytically deposited transparent conductors appear to be those that require relatively small substrates due to gas flow limitations. The SnO_2-based systems are generally preferred to the In_2O_3-based systems because of the high cost of the organo-In compounds. The SnO_2 films prepared by Tabata (*172b*) have been used to fabricate SnO_2–Si heterojunction solar cells.

TABLE VI

TYPICAL PROPERTIES OF OXIDE FILMS PREPARED BY PYROLYSIS

Film material	Resistivity (Ω cm)	Carrier concentration (cm^{-3})	Mobility ($cm^2 V^{-1} sec^{-1}$)	Optical transmission (%)	Refractive index	Ref.
SnO_2	$0.5-9 \times 10^{-2}$	9×10^{18}	10	$90-95^b$	$1.75-2.20^c$	(170)
SnO_2:Sb (0.6–2.7 at. %^{a})	$1.5-3.2 \times 10^{-3}$	1.2×10^{29}	23	$85-91^d$	1.89	(171)
In_2O_3	$1-5 \times 10^{-2}$	—	—	$\geq 90^b$	—	(174)
In_2O_3: 8 at. % Sn^a	$2.2-7.1 \times 10^{-4}$	—	—	$67-89^b$	$1.68-2.48$	(172)

^{a} Relative to the host cation.
^{b} Average transmission between $\lambda = 4000-6000$ Å; actual value depends on film thickness.
^{c} Inversely proportional to film thickness.
^{d} Average transmission between $\lambda = 4000-7200$ Å; actual value depends on film thickness.

B. Evaporation and Sputtering of Thin Metal Films

In this section, we shall cover the evaporation and sputtering of thin metal films, with and without nucleation modifying layers, and deposited metal films that are subsequently heat treated to form oxides. A recent review of evaporation processes was done by Glang (*175*). Sputtering processes have been reviewed recently by Maissel (*176*) and Vossen (*177*).

1. Metal Films on Amorphous Substrates. We have already discussed the problems of film agglomeration that occur during the early stages of metal film nucleation and growth (Section II,A) in relation to electrical conductivity and transmission. There appear to have been only two reported uses of noble metal films directly deposited on glass, and both of these utilize a semiconductive layer over the metal (*178, 179*). In both cases a semicontinuous, noble metal film is evaporated and then, in a separate chamber, a thin, semiconducting oxide is reactively sputtered over the metal film. In both cases, the metal films were deposited very rapidly on cooled substrates to promote early film continuity. The semiconducting oxide overcoats were quite thin (350 Å) and the metal films were about 100 Å average thickness. The metal-oxide combinations used were $Ag-In_2O_3$, $Ag-ZnO$, and $Au-Bi_2O_3$. The effect of the oxide overcoat was to increase the sheet resistivity by about 20–30%. Although it was not stated, the thickness of the oxide layers were apparently adjusted to yield a maximum transmission in the interference pattern at certain wavelengths. At other wavelengths, the transmission was very much degraded due to the effects discussed earlier (Section II,C,3). The indices of refraction of these oxides are very much higher than those of the metal films used, thus promoting reflective losses.

The increase in film resistivity that occurs when noble metals are overcoated with another material have been studied extensively (*180–183*). It is generally agreed that any film deposited over a noble metal will result in microscopic roughening of the original film surface which leads to more diffuse scattering. Meyer (*183*) ascribes this to oxygen incorporation in the original film surface layer. This same effect is observed if more Au is deposited onto a Au film that has been exposed to the atmosphere.

Such films are, therefore, not useful as broad-band transparent conductors; but only as wavelength-peaked conducting filters. The oxide overcoat does provide scratch protection to the soft noble metals, but only at the expense of decreased conductivity and transmission.

2. Metal Films Deposited over Nucleation-Modifying Layers. In nearly all of the reported work on thin metal film transparent conductors, some kind of nucleation-modifying layer was deposited prior to metal deposition. The work of Gillham and Preston (40, 41) on Bi_2O_3 layers for modifying the nucleation of Au films has already been noted. The Bi_2O_3–Au sandwich appears to be the most prevalent combination used for transparent conductors, but numerous other materials have been used as well. In general, there have been three classes of materials reported that have been shown to promote early continuity in subsequently deposited noble metal films. These are semiconducting compounds, other metals, and adsorbed oxygen.

The semiconducting compounds used have been deposited by direct evaporation, evaporation with subsequent oxidation, and reactive sputtering. There does not appear to be a great difference in the results produced by these various methods. Improved conductivity and transmission of Au films have been reported for nucleation-modifying compound layers of the following; Bi_2O_3 (40, 41, 184–188), Au_2O (189), Cu_2O (189, 190), Ag_2O (189), NiO (189), Al_2O_3 (189), PbS (189), Sb_2S_3 (189), $PbSO_4$ (189), MgF_2 (187, 191), Sb_2O_3 (185, 188, 191), In_2O_3 (185), PbO (185, 188), PbO_2 (185, 188), CdO (188), TeO_2 (188), and various metal sulfides, tellurides, and selenides (192). In a similar manner, Ag, Cu, and Pt have been shown to coalesce at an early stage of film growth when deposited on films of Bi_2O_3 (184), Sb_2O_3, or MgF_2 (191). Several other compound/thin metal combinations have also been described (189); (ZnO/Ag, Al_2O_3/Ag, PbO/Ag, Au_2O/Ag, Ag_2O/Ag, Au_2O/Cu, Cu_2O/Cu, FeO/Fe, and Cu_2O/Ni). In addition, sandwich structures of various kinds have been reported: Bi_2O_3/Au/Bi_2O_3 (41, 182, 183, 186), Al_2O_3/Au/Al_2O_3 (189), and Al_2O_3/Au/Al_2O_3 + SiO_2 (189). No complete transmission–wavelength curves were given for any of these sandwiches, but it can be expected that they would increase the reflectivity in some portions of the visible spectrum for the reasons outlined earlier. Typical films deposited over these compounds yield average optical transmissions of about 75% for sheet resistivities of about 50 $\Omega/\square$.

The metal films that have been used to modify the nucleation of Au have been Cr (25, 193, 194) and Pt (195). In the case of Cr, it was found that the best properties were obtained when the Cr thickness was minimized. Hoffman (193) utilized an estimated 5-Å-thick Cr film followed by 30–80 Å of Au deposited by evaporation from resistance heated filaments. In that range of film thicknesses, the optical transmission did not vary appreciably (60–70%), but the sheet resistivity decreased from about 10K $\Omega/\square$ at 30 Å of Au to about 100 $\Omega/\square$ at 80 Å. The 5 Å of Cr was, in all likelihood, converted to oxide. Figure 24 illustrates the degree of film continuity obtained at relatively small average thicknesses. This may be compared to the completely discontinuous Au film deposited on an amorphous substrate (Fig. 2). Changes in reflectivity with time are likely because of interdiffusion in the system (194).

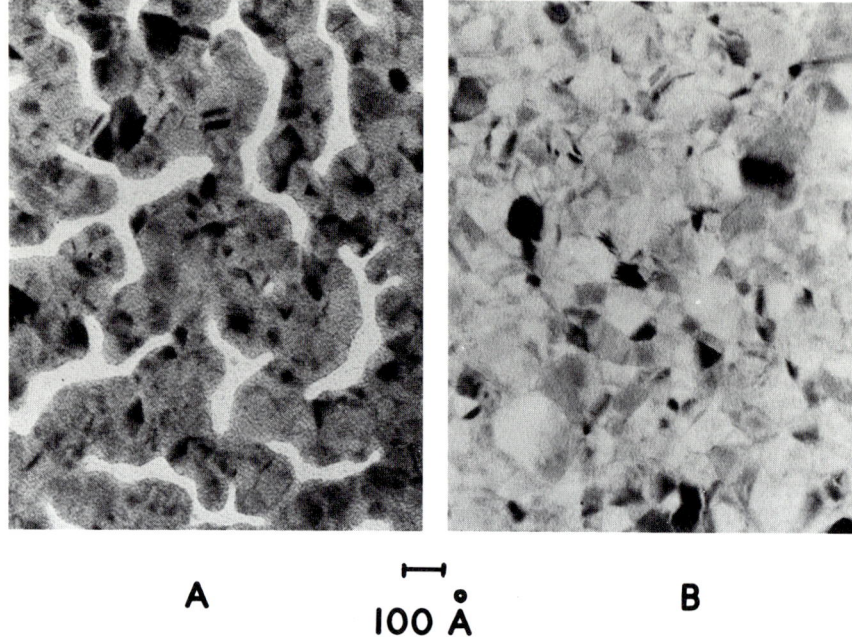

FIG. 24. Transmission electron micrographs of (A) ~5-Å Cr + 60-Å Au, and (B) ~5-Å Cr + 80-Å Au. (Courtesy of D. M. Hoffman and M. D. Coutts, RCA Laboratories, Princeton, N.J.)

Chaurasia and Voss (*195*) evaporated 3–10-Å-thick Pt, followed by various thicknesses of Au, and found that the onset of conductivity in the films occurred at a sandwich thickness of only 11 Å. At this thickness the sheet resistivity was 2000–3000 $\Omega/\square$. With ~40 Å Au over the thin Pt layer, they found a resistivity of ~40 $\Omega/\square$. Annealing the films decreased their resistivity up to about 210°C. Above that temperature the resistivity increased. This annealing behavior suggests a two-step process: grain growth followed by interdiffusion.

Hoffman (*196, 197*) found that a layer of adsorbed oxygen promoted the early coalescence of Rh films. The films were prepared by first heating the substrates in vacuum to drive off water vapor, then cooling them while admitting a small partial pressure of O_2. When the substrates were cooled nearly to room temperature, the O_2 was pumped out and Rh was evaporated either from a plated filament (*196*) or from an electron beam source (*197*). Figure 25 shows the degree of film continuity as a function of film thickness. The oxygen partial pressure during substrate cooling was 5×10^{-5} torr. Increasing the oxygen pressure increased film resistivity, transmission, adhe-

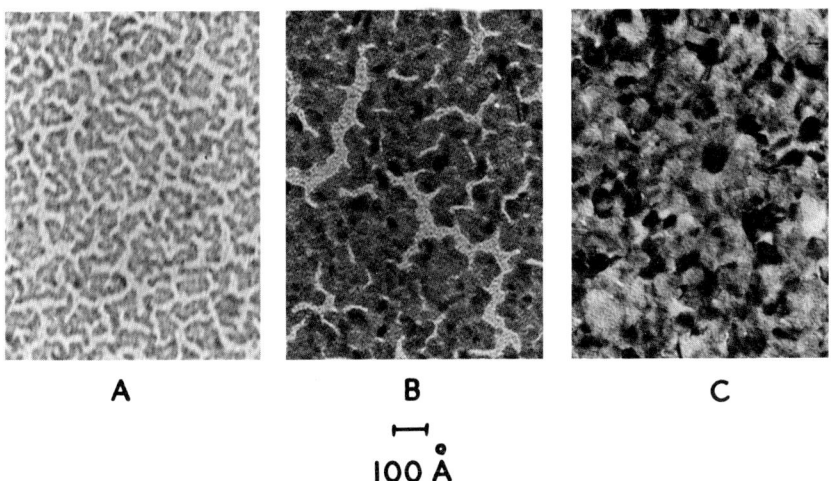

FIG. 25. Transmission electron micrographs of Rh films deposited on an adsorbed layer of O_2 (see text). (A) 395 kΩ/□, 80% transmission. (B) 745 Ω/□, 66% transmission. (C) 25 Ω/□, 25% transmission. (Courtesy of D. M. Hoffman and M. D. Coutts, RCA Laboratories, Princeton, N.J.)

sion, and hardness. Cox (198) confirmed the durability of Rh films prepared by rf sputtering using an oxygen interlayer.

Thin metal films with or without nucleation-modifying layers are not extensively used as transparent conductors. Their electrical and optical properties are not, in general, as good as those of the semiconducting oxides, and they are not very resistant to abrasion and other forms of mechanical damage. They are, however, widely applied as infrared reflectors.

One can only speculate as to the physical reasons for early film coalescence when nucleation-modifying layers are used. As has been shown, there are many different materials used, and these apparently can be deposited by many different deposition processes with essentially the same results. Only one common thread seems evident. During both nucleation-modifying-layer and metal-film deposition, the substrates are subjected to a flux of electrons which could charge the substrates negatively. In sputtering, the source of electrons is the sputtering target (secondary electrons), which are accelerated across the target dark space toward the substrates. In electron beam evaporation, stray electrons from the beam are known to charge surfaces in the vacuum chamber (199–201). In evaporation from resistance-heated filaments, one has, in addition to thermionic electrons, electron emission upon melting of the evaporant (202) and upon alloying with the filament material (203).

A fraction of the metal vapor arriving at the substrate would be expected to be ionized. In sputtering processes, positive ions cannot escape from the target because of the high negative potential at that surface, but they can be ionized by electron impact and/or by the Penning mechanism in transit from target to substrate (*177*). In electron beam evaporation, the evaporant must pass through the electron beam as it leaves the source, resulting in some ionization (*199–201*). In evaporation from resistance-heated filaments, surface ionization (sometimes known as Saha–Langmuir ionization) is known to occur (*204–208*). Levine (*209*) has presented a generalized, theoretical treatment of this subject.

Therefore, it is clear that in all cases the substrates can be negatively charged and the metal can be partially ionized. It is entirely possible that the sole function of the nucleation-modifying layer is to provide a fresh, dry surface that can be uniformly charged. The charged sites can then act as nucleation sites for incoming positive ions of the metal film material.

A recent experiment in this laboratory (*210*) tends to support this view. Au films were deposited on reactively sputtered Bi_2O_3 by dc sputtering with and without rf-induced substrate bias (*211*). In the zero bias mode,

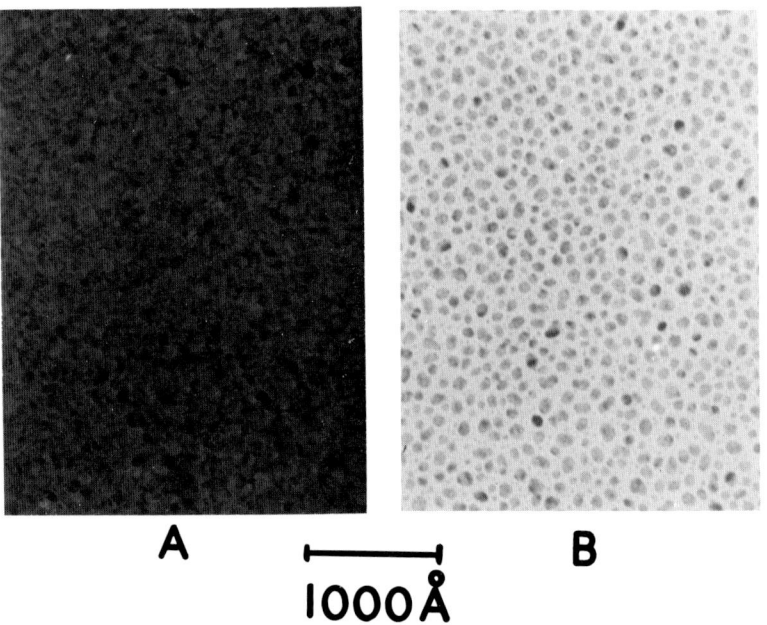

FIG. 26. Transmission electron micrographs of ~50-Å-thick Au films sputtered on Bi_2O_3-coated substrates. (A) No applied substrate bias, (B) −50-V substrate bias.

the substrates receive a relatively high flux of electrons, whereas with the negative bias, they receive a positive ion flux, but few electrons. Figure 26 shows that the bias-sputtered films are discontinuous, while the zero-bias films are continuous.

3. *Postoxidation of Thin Metal Films.* Mainly because of the disruption of the film–substrate bond which results from volumetric changes during oxidation of metal films (*212*), such films are not often used for transparent conductors. Studies of these films are only mentioned here for the purpose of completeness. There have been reports on transparent conducting films produced by postoxidation of evaporated In (*213*), sputtered Sn and In (*214*), evaporated Sn (*215*), and evaporation of Sn–Sb mixtures (*216*).

C. REACTIVE EVAPORATION

Several authors (*197, 217–221*) have reported on transparent conductors prepared by reactive evaporation. The main control parameters involved in reactive evaporation are the metal evaporation rate, the oxygen background pressure, the substrate temperature, and the source–substrate separation. A careful balance must be maintained between metal vapor and oxygen arriving at the substrate at any given substrate temperature to ensure that repeatable stoichiometry in the films is achieved. Of these parameters, perhaps the most difficult to control is the evaporation rate because of the steep dependence of vapor pressure on source temperature (*222*). This problem is especially severe when doped films are to be deposited, since two evaporation sources are ordinarily used (*221*), requiring fairly elaborate controls. Oxidation of the evaporant in the source must be minimized, or at least controlled, as this can affect the film stoichiometry also. One method of controlling source oxidation is to use oxides as the evaporation source (*222a, 222b*). When In_2O_3 or In_2O_3–SnO_2 mixtures are evaporated, they reduce, forming opaque films, which can be rendered transparent by oxidation during and/or after deposition.

The substrate temperature must be maintained high enough to ensure the desired degree of reaction, but not so high as to result in the diffusion of p-type impurities from glass substrates into the film during deposition. Glass-walled vacuum chambers represent another source of p-type impurities when high temperature processes are used (*223, 224*). This latter complication applies not only to reactive evaporation, but to all vacuum deposition processes involving heat and/or particle bombardment (e.g., sputtering).

Table VII gives the process control parameters and electrical and optical results reported by various workers.

TABLE VII

Process Parameters and Properties of Reactively Evaporated Transparent Conductors

Film material	O_2 Pressure (torr)	Metal deposition rate (Å/min)	Source-substrate separation (cm)	Substrate temperature (°C)	Film resistivity (Ω cm)	Average transmission (%)	Thickness (Å)	Ref.
In_2O_3	1.8×10^{-3}	—	—	25	—	—	—	(217)
In_2O_3	$4-7 \times 10^{-3}$	60–180	13	25	$0.1-1 \times 10^{-2}$	~70	900	(218)
SnO_2	$0.2-2 \times 10^{-4}$	60–300	—	300	—	—	—	(219)
In_2O_3:Sn	1×10^{-3}	180 (In) 12 (Sn)	35	400[a]	5×10^{-3}	~80	800	(221)
Rh:O	5×10^{-5}	~300	30	125	2.4×10^{-4}	54	125	(197)
Rh:O	8×10^{-5}	~300	30	125	6×10^{-4}	58	125	(197)
Rh:O	2×10^{-4}	~300	30	125	2×10^{-3}	68	125	(197)

[a] Temperature for optimum film properties.

The films prepared by evaporating Rh in O_2 (197) exhibited more diffuse electron diffraction patterns as the oxygen content was increased. No oxides of Rh were detected by electron diffraction or infrared spectroscopy, but they may be present as small quantities of amorphous rhodium oxide. Figure 27 illustrates the structural dependence of these films on oxygen concentration.

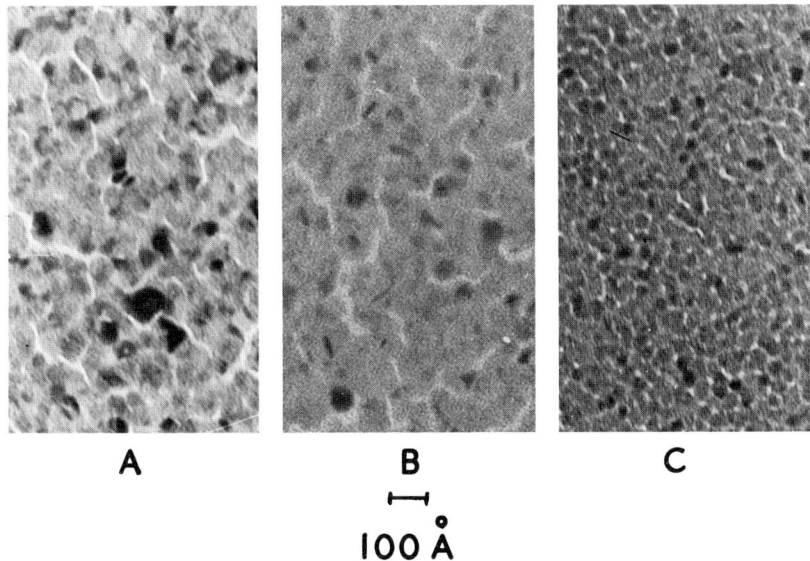

FIG. 27. Transmission electron micrographs of 50-Å-thick Rh–O films prepared by reactive evaporation at different O_2 pressures. (A) 5×10^{-5} torr, (B) 8×10^{-5} torr, (C) 2×10^{-4} torr. (Courtesy of D. M. Hoffman and M. D. Coutts, RCA Laboratories, Princeton, N.J.)

Reactively evaporated In_2O_3 has been deposited on p-type Si to form solar cells with efficiencies comparable to Si p–n junction devices (224a).

D. REACTIVE SPUTTERING

The control parameters in reactive sputtering are similar to those in reactive evaporation, except that sputtering rates are more easily controlled than are evaporation rates. Since sputtering is more directional than evaporation, there is less waste of expensive source material. Also, the sputtering of alloy targets is easier to control, thus making the production of doped films easier. However, these advantages are often balanced by certain disadvantages.

Sputtering is slower and much more complicated than evaporation; the equipment cost is higher, especially for large substrate areas; and, in the case of alloy targets, special care must be exercised to ensure that no diffusion occurs in the target which can change the film composition in an uncontrolled manner (*177*). The problem of diffusion in low-melting alloy targets [e.g., In–Sn or Sn–Sb (*225*)] is especially severe (*226*).

The mechanism of compound synthesis in reactive sputtering is mainly related to the reactive gas pressure (or partial pressure) during deposition. Compound decomposition is favored at the target (*227*) because of the high bombarding energy at that surface. However, at high reactive gas pressure, the buildup of compound material on the target surface is faster than the bombardment-induced decomposition (*228*). When low reactive gas pressures are used, the opposite is true. Thus, in the low pressure regime, one is sputtering metal, while in the high pressure regime, a compound is being sputtered. In the latter case, the material actually ejected from the target surface is usually not the stoichiometric compound, but rather some fragment of the compound.

From the standpoint of transparent conducting oxides, the high pressure regime is more difficult to control and usually results in nearly stoichiometric films of high resistivity. It is far more controllable to use a low oxygen pressure and synthesize the anion-deficient oxide completely at the substrate.

The usual sputtering gas mixture for reactive sputtering of oxides is Ar–O, but some workers have added H to reduce the target and/or the growing film (*229*). Better control of the film stoichiometry was claimed for the H additions when high gas pressures were used.

Most of the early work in reactively sputtered transparent conductors centered on films of CdO (*186, 230–236*). Because of the low bandgap of this material, the absorption edge occurs at wavelengths between 5000–6000 Å. Thus, CdO is limited as a transparent conductor to the red end of the visible spectrum. CdO–Si heterojunctions have been employed as solar cells (*236a*). SnO_2 and doped SnO_2 films prepared by reactive sputtering have not received too much attention in the literature. Using a pure, cast tin cathode, Holland and Siddall (*186*) obtained films with sheet resistivities of 10 kΩ/□ and 80% optical transmission. Their films had to be annealed in air at 450°C after deposition. As deposited, the films probably contained metallic tin occlusions. Lieberman and Medrud (*237*) measured the refractive index and extinction coefficient of SnO_2 films, but no data on electrical properties were presented. The refractive index varied slowly from about 1.95 at $\lambda = 3750$ Å to about 1.88 at $\lambda = 5500$ Å. The extinction coefficient was zero for wavelengths greater than 4500 Å, and then slowly increased to 0.005 at $\lambda = 4000$ Å. Hecq and Portier (*238*) studied the reactive sputtering process for the Sn–O system in detail and reported on film stoichiometry and crystal structure for various sputtering conditions, but they did not give the electrical and optical properties of their films.

The Sb-doped films deposited by Sinclair et al. (239) were amorphous, as deposited, and were probably anion deficient. Upon heating in air at 300°C or higher, the films first crystallized and then oxidized as a function of time. Films containing 10% Sb gave a minimum resistivity ($\sim 7-8 \times 10^{-3}$ Ω cm). Vaynshteyn (240) found the minimum resistivity when using a Sn cathode with 10 at.% Sb, target voltage and current density of 3.6 kV and 0.4–0.5 mA/cm^2, respectively, 60% O_2 in Ar, a total pressure of 40–80 mtorr and substrate temperature of 300°C. He reports a resistivity for unannealed films of 10^{-2} Ω cm. Based on the results of Sinclair et al. (239), the resistivity of these films might have been lowered by mild annealing. Lehmann and Widmer (241) studied rf-reactive sputtering of SnO_2:Sb. They found that the minimum resistivity (3×10^{-3} Ω cm) occurred for a doping level of 10 mole % Sb_2O_3 in SnO_2. It should be pointed out that in rf sputtering, there can occur substantial amounts of resputtering, including preferential resputtering of the growing film (177). The amount of resputtering is a complex function of sputtering conditions and sputtering chamber geometry (242). The optimum doping level determined by Lehmann and Widmer refers to the film material. Because the sputtering yield of Sb is about 2.5 times higher than that of Sn, the target composition should ordinarily contain excess Sb to compensate for resputtering at the substrate.

There have been a greater number of reports on reactively sputtered In and In–Sn mixtures. The reports on pure In_2O_3 (186, 237, 243–247) are for nearly stoichiometric films of high resistivity and are of no interest here. In_2O_3:Sn films deposited by reactive dc or rf sputtering have been described by several authors (236, 241, 248–253). All of the dc sputtered films apparently had metallic occlusions in them as deposited and required some kind of postdeposition treatment to increase the transmission. Sihvonen et al. (248, 249) deposited their films from a 82–18% In–Sn target in pure O_2. After deposition, they either annealed their films in air at 200°C for a few minutes or sputter-etched them in a high pressure (150 mtorr) glow discharge. In both cases, they reported the film resistivity as 5×10^{-4} Ω cm. For 500-Å-thick films, the sheet resistivity was 100–150 $\Omega/\square$ and the average transmittance was about 80% from the blue end of the visible spectrum to the near infrared. Williams (250) used a 20% Sn cathode and sputtered in a similar manner, but with an Ar–O_2 (70–30) mixture. Based on an assumed thickness of 50 μm, he estimated a film resistivity of 10^{-1} Ω cm. Judging from the sputtering conditions and time, the thickness estimate is high by a factor of at least 100, leading to a probable resistivity of something less than 10^{-3} Ω cm. Bonnet and Marchal (*252*) used a triode sputtering system and targets of In, 80 In–20 Sn, and 90 In–10 Sn and sputtered in various Ar–O_2 mixtures. They found that the resistivity of the films was directly related to the ratio of oxygen pressure to the ion current density at the target. The best films were prepared with the 20 at. % Sn target with an oxygen to current ratio of about 7×10^{-3} torr/A. The films were amorphous if the substrates were cooled (20°C), and highly crystalline if they were heated to 400°C during deposition. The resistivity of the best films was about $6–7 \times 10^{-4}$ Ω cm. Films with sheet resistivity of 4 Ω/□ averaged about 80% transmission in the visible spectrum.

Mehta and Vogel (*236*), using an In–Sn alloy target (18 at. % Sn), deposited films using rf sputtering in a pure oxygen discharge at a rate of about 35 Å/min. Ar–O_2 and Ar–N_2 discharges yielded higher resistivities. They found that a substrate temperature of 300–350°C and a postdeposition anneal in air at 340°C for 1 hr yielded optimum film properties. The resistivity of the best films was about 3×10^{-3} Ω cm. Films with a sheet resistivity of about 90 Ω/□ yielded an average transmission of about 85% in the visible.

Lehmann and Widmer (*241*) found the optimum Sn doping level to be 10 mole % SnO_2 in their rf-reactively sputtered films. Again, this result cannot be universally translated into a target composition due to differing amounts of preferential resputtering in different sputtering systems. Under optimum conditions, they found that the film resistivity was 1.5×10^{-3} Ω cm, and that the films were highly structured with a strong (111) texture, and they were extremely smooth as viewed in a scanning electron microscope.

Pankratz (*251*) studied a wide range of process variations in his rf-reactively sputtered films. He found the optimum In:Sn ratio to be 82:18 by weight. The films were deposited at room temperature in various Ar–O_2 mixtures. Annealing in O_2, N_2, and $N_2 + 10\%$ H_2 lowered the resistivity of the films to a saturation value that was a factor of about 2–200 lower than the as-deposited value, depending upon the film thickness, annealing atmosphere, and temperature. Prolonged annealing in O_2 reversed this trend, indicating that two mechanisms were operative: recrystallization (removal of scattering centers) and oxidation (removal of anion vacancies). Films with a sheet resistivity after annealing of 150–200 Ω/□ exhibited average transmission values of 90–95%.

Molzen (*253*) rf-sputtered In–Sn (5% Sn) targets in pure oxygen so as to produce nearly stoichiometric, high-resistivity films after deposition. Anion vacancies were introduced by postdeposition annealing in Ar at atmospheric pressure. Typical annealing temperatures to achieve minimum resistivity ranged from 500–600°C. Depending upon the annealing temperature, the resistivity, extinction coefficient, refractive index, and transmission could be varied widely. Interference effects were subtracted out of the transmission data, thus yielding only peak transmission. The true, average transmission for films with sheet resistivity of 2–3 $\Omega/\square$ was about 75–80%.

Murayama (253a) prepared In_2O_3 films by a modified ion plating process (253b). Ion plating is a hybrid process in which material is evaporated thermally or from an electron beam source through a glow discharge onto a substrate that is the cathode of the glow discharge. Thus, the growing film is ion bombarded (sputter etched). In-metal was evaporated up the axis of an rf coil operating at 150 W onto a substrate attached to a dc sputtering target. The glow discharge pressure was about 8×10^{-4} torr (O_2). Oxidation was enhanced by the use of the rf coil. The dc substrate bias most often used was -500 V. The average optical transmission in the visible was about 85% for a 1500-Å-thick film. The resistivity was found to decrease dramatically with increasing deposition rate.

In nearly every case, reactively sputtered transparent conductors had to be annealed after deposition to optimize the film properties. This negates one of the principal advantages of sputtering, namely, the ability of the process to deposit films at relatively low substrate temperatures. The requirement for high temperature annealing puts a limit on the types of substrates that can be used. For example, glass with a high alkali content obviously would result in degraded film properties. Nevertheless, excellent film properties can be achieved by this technique.

E. Sputtering of Oxide Targets

Sputtering from oxide targets to form semiconducting oxide films is significantly different than reactive sputtering of metal targets. The control of film stoichiometry has been found to be much easier with oxide targets, thus obviating the need for high temperature, postdeposition annealing. Furthermore, there are no problems with diffusion in mixed oxide targets of SnO_2–Sb_2O_3 or In_2O_3–SnO_2. However, a new set of problems are encountered with oxide targets. Target fabrication is quite difficult. Hot-pressed targets of these materials are rarely more than 90% of theoretical density. Porous targets are known to be virtually inexhaustible sources of contamination (254). With most oxide materials, this is not a major problem, because the principal contaminant is oxygen; but In_2O_3 and SnO_2 are quite hygroscopic, leading to contamination by water vapor and other waterborne contaminants. In sputtering systems that must be vented to the atmosphere between runs, this implies the need for extensive presputtering onto a shutter before film deposition. Ideally, one should use a sputtering system with a load lock arranged so that the target is not exposed to the atmosphere between deposition runs.

Another problem relates to target heating. All of the materials involved have relatively high secondary-electron-emission yields, leading to very high currents in the glow discharge. This is further aggravated by the electrical properties of the sputtering target. When dc sputtering is used, the target is simply Joule-heated (i^2R). When rf sputtering is used, the power loss in the target per unit volume is:

$$P_L = \varepsilon_0 \omega \mathscr{E}^2 K \tan \delta \qquad (22)$$

where ε_0 is the permittivity of free space, ω is the angular frequency, $\mathscr{E}$ is the electric field across the volume, K is the relative dielectric constant, and $\tan \delta$ is the dissipation factor (227). Under rf conditions, semiconducting oxides behave as lossy insulators (i.e., K and $\tan \delta$ are high). Furthermore, both K and $\tan \delta$ have positive temperature coefficients. All of these factors lead to a large amount of target heating, thus requiring a very well cooled target assembly. To minimize all of these problems, the target should be hot-pressed at the highest practical temperature to ensure both high density and high electrical conductivity. Further, the targets should be relatively thin to minimize the electrical losses.

The control of film stoichiometry when sputtering from mixed oxide targets is virtually automatic in dc sputtering, whereas preferential resputtering of deposited film material in rf sputtering usually requires that the target composition be suitably adjusted to compensate for the material lost at the substrate (177, 242).

Yamanaka and Oohashi (255) dc-sputtered a sintered SnO_2 target in Ar, O_2, N_2, and air. All of their films had too high a resistivity (<5 Ω cm) to be used as transparent conductors. The films sputtered in Ar apparently were reduced, since heat treatment in air resulted in nearly insulating films (10^5 Ω cm). The absorption edge varied from 3.7 to 3.8 eV depending on the sputtering gas. Deitch et al. (256) deposited SnO_2 films by rf sputtering from a hot-pressed target. Their intent was to deposit nearly stoichiometric films for integrated optic light guides. Films sputtered in pure Ar and Ar + 28% O_2 gave indices of refraction of 2.07 and 1.95, respectively. Sb-doped films have been studied by Vossen and co-workers (257, 258). These films were deposited by rf sputtering SnO_2 + 10 mole % Sb_2O_3 targets in pure O_2. As deposited, the films had a thin (~50 Å) surface layer of 5 $SnO \cdot 2H_2O$, which could be converted to a more conducting SnO_2 phase by annealing at any temperature above about 240°C. The bulk of the films were very highly oriented, tetragonal SnO_2 with the c axis normal to the substrate surface. No Sb compounds were found in the films, indicating that complete substitution of Sb in the SnO_2 lattice was achieved during deposition. The annealing of the films reduced the resistivity of the films (Fig. 28). As the film thickness is decreased, the surface layer becomes a larger fraction of the film volume. For film thicknesses greater than 400 Å, the resistivity of the films is 6×10^{-3} Ω cm. The resistivity increases from that thickness

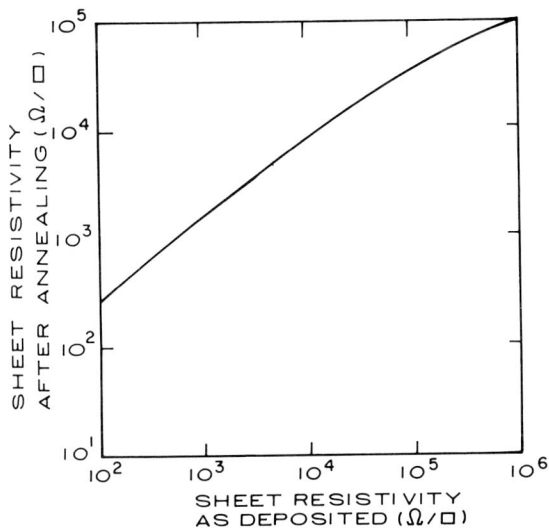

FIG. 28. Sheet resistivity of rf-sputtered SnO_2:Sb after annealing at 300°C in air versus as-deposited sheet resistivity. [After Vossen (257).]

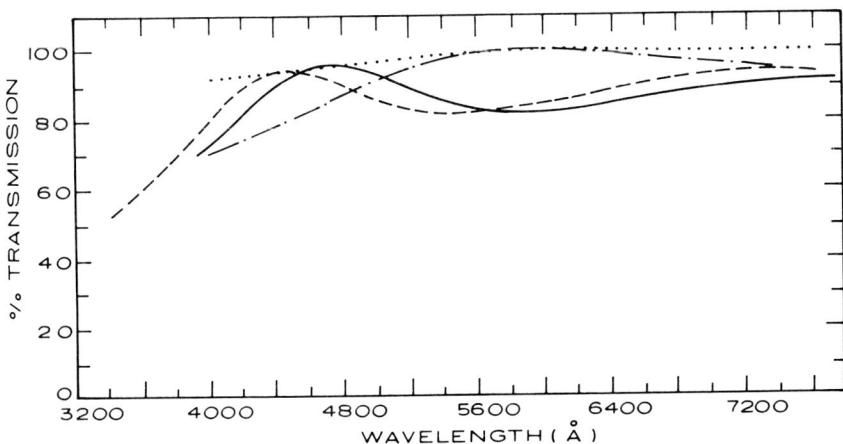

FIG. 29. Transmission versus wavelength for various thicknesses of rf-sputtered SnO_2:Sb films. (———) 500 Ω/□, (– – –) 1500 Ω/□, (–·–·–) 2500 Ω/□, (·····) 20 kΩ/□.

to $5 \times 10^{-2}\,\Omega\,cm$ at a thickness of 50 Å. Except for very thin films, annealing had little effect on the optical transmission of the films. The transmission was increased slightly by annealing. Figure 29 shows representative transmission–wavelength characteristics of these films. The film surfaces were extremely smooth when viewed at 10,000× in a scanning electron microscope.

Several authors have reported on the properties of In_2O_3:Sn sputtered from oxide targets (258–264). Vossen (258, 259, 263) rf-sputtered targets of In_2O_3 + 20 mole % SnO_2 in pure Ar. Under the optimum sputtering conditions, the films were anion deficient and very highly conducting as deposited. Annealing the films in air resulted in the formation of a stoichiometric passivating surface layer, the thickness of which was a function of the annealing temperature. Subsequent heat treatments at or below the original annealing temperature did not affect the film resistivity or transmission. The initial anneal increased the resistivity of the films (Fig. 30), did not change the average transmission, but shifted the interference pattern as a result of a change in the refractive index. Since glass substrates were used, it is highly probable that impurities were diffused into the films during high temperature annealing. In films thinner than about 1000 Å, the resistivity was very high and uncontrollable in annealed films, presumably because the stoichiometric surface layer extended through most of the film. For film thicknesses greater than 2000 Å, the resistivity was $6.25 \times 10^{-4}\,\Omega\,cm$. Typical film transmission characteristics are shown in Fig. 31. Films with sheet resistivity of 2–3 $\Omega/\square$ were produced with 80% average transmission in the visible. The films were highly crystalline as deposited and exceptionally smooth when viewed in a scanning electron microscope at 10,000×. The surface morphology and crystallinity were unaffected by annealing. The lattice parameter of the cubic structure was higher than that of bulk In_2O_3.

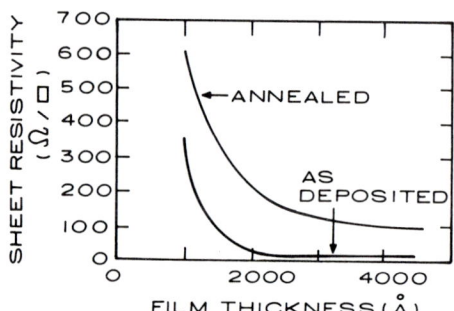

FIG. 30. Sheet resistivity versus film thickness for as-deposited and annealed (550°C, 2hr) In_2O_3:Sn films prepared by rf-sputtering. [After Vossen (259).]

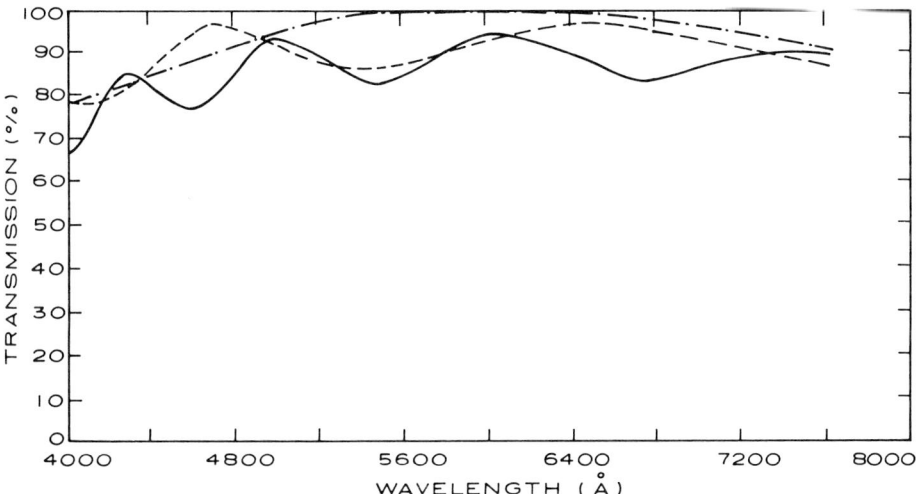

FIG. 31. Transmission versus wavelength for various rf-sputtered In_2O_3:Sn films (———) 14 $\Omega/\square$, (-----) 28 $\Omega/\square$, (-·-·-) 320 $\Omega/\square$. [After Vossen (259).]

Using similar sputtering techniques, Bosnell and Waghorne (261) studied the structure and chemical binding of these films using electron diffraction and ESCA. Their results tended to confirm that annealing produces a passivating layer. They also found that Sn is incorporated, not substitutionally but as Sn_3O_4.

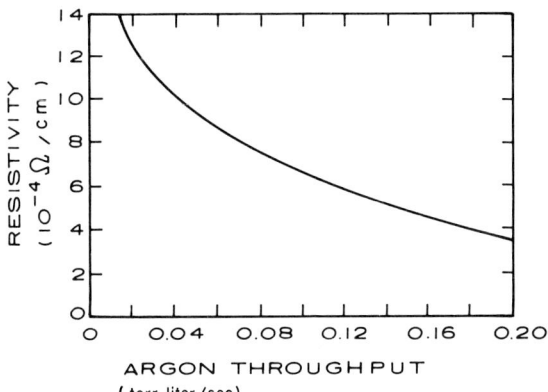

FIG. 32. Resistivity of In_2O_3:Sn films dc-sputtered from a porous target versus sputtering gas throughput. [After Fraser and Cook (260).]

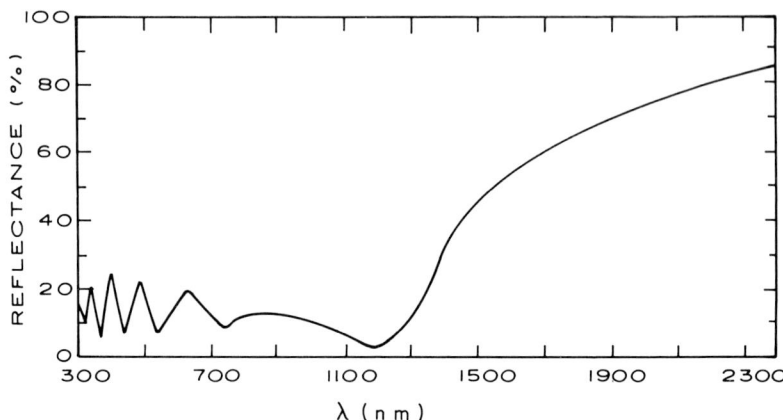

FIG. 33. Reflection characteristics of dc-sputtered In_2O_3:Sn films. [After Fraser and Cook (260).]

Fraser and Cook (260) studied dc sputtering of In_2O_3 + 9 mole % SnO_2 in pure Ar. Other sputtering gases or gas mixtures resulted in inferior films. Aside from target composition and density, they found a marked influence of Ar throughput on film resistivity (Fig. 32). Dense targets were not as sensitive to Ar thoughput. They found that increasing the substrate temperature during film deposition lowered the resistivity. The lowest resistivity was found for a substrate temperature of 500°C (1.77×10^{-4} Ω cm). Films with resistivities of 2–3 Ω/□ yielded average transmission values of about 80%. The dominant cause of light loss was reflection (Fig. 33). The films were found to be good infrared reflectors ($\sim 90\%$ reflectance) for wavelengths from 2.5–15 μm. The films were quite rough as deposited and had to be polished to reduce scattering of light.

Hecq et al. (262, 264) used an rf hollow-cathode source to sputter In_2O_3 + 20 mole % Sn in Ar. With the hollow cathode source, there is no substrate heating as is the case in diode sputtering. Films deposited on substrates at room temperature were amorphous. When the substrates were heated to 400°C during deposition, deposited films were highly crystalline. Post-deposition annealing in air decreased the resistivity of amorphous films and increased the resistivity of crystalline films. Their results seemed to show that the Sn is incorporated into the film as SnO and that its presence stabilized anion vacancies in In_2O_3.

Nozik (265) and Haacke (265a) studied the electrical and optical properties of Cd_2SnO_4 rf-sputtered from pressed oxide targets in Ar and Ar + O_2 mixtures. Nearly stoichiometric films had a low bandgap (2.06 eV) leading to an absorption edge in the middle of the visible spectrum. However, when films are deposited in an Ar-rich Ar–O_2 mixture, the conductivity increases and there is a large Burstein shift of the absorption edge to about 2.8–2.9 eV. Films with average transmission and conductivity similar to the best sputtered In_2O_3:Sn films have been produced (265a) and the cost of starting materials are considerably less than for In_2O_3:Sn.

F. MISCELLANEOUS PROCESSES

Several other processes have been described for the deposition of transparent conducting films. None of these appear to be in wide use, so they will not be described in detail. Some of these processes are borrowed from ceramic fabrication and metallization. There have been reports of screen-printed (266) or doctor-bladed (267) powders of In_2O_3 and In_2O_3:Sn for

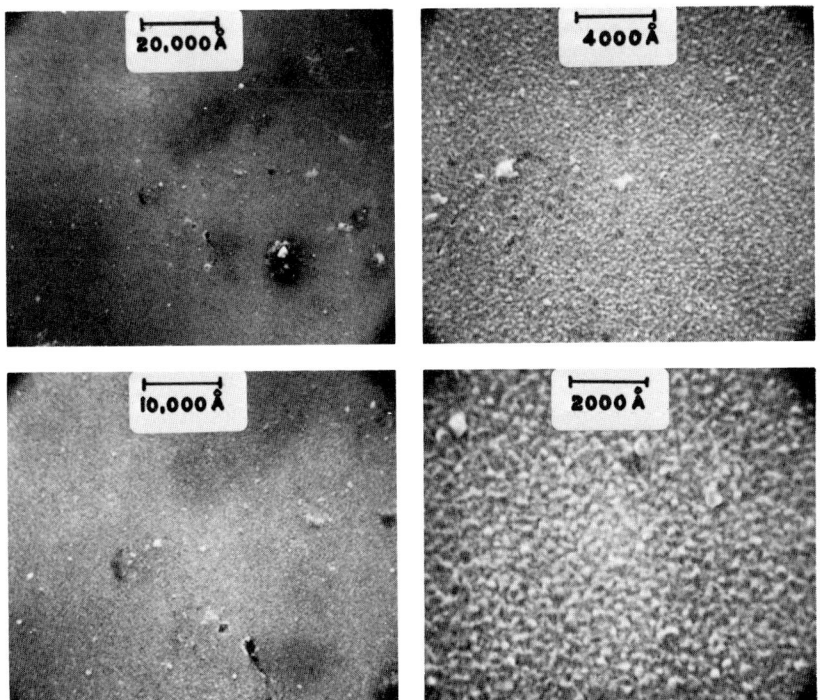

FIG. 34. Scanning electron micrographs of screen-printed and fired CdO films prepared from a metal resinate. (Courtesy of J. J. O'Neill, Jr., RCA Laboratories, Princeton, N.J.)

flexible electroluminescent cells, and screen-printed and fired metal-resinates to form In_2O_3 (*268*) or CdO (*269*). The major problem with these processes is the roughness of the surface produced (Fig. 34).

Thin Au films have been produced by a modified electroless-plating technique (*270*). A water-soluble Au-salt and a reducing agent are subjected to uv radiation which decomposes the salt to a thin Au film. Single-crystal films of CdS have been deposited onto various II–VI and III–V compound substrates by a vapor transport technique (*271*). Above the absorption edge, the transmittance of these films was 60–80% with a sheet resistivity of 1–2 $\Omega/\square$. SnO_2 films have been prepared by glow-discharge decomposition of $SnCl_2$ (*272*), and $Sn(CH_3)_4$ or $SnCl_4$ (*273*) in oxygen. The best films were obtained with 0.06 at. % Sb in the films (from $SbCl_5$) and ~ 0.33 at. % Cl. This yielded a film resistivity of 1.7×10^{-3} Ω cm and average visible light transmission of $\sim 90\%$ for a film with sheet resistivity of ~ 1000 $\Omega/\square$ (*273*).

IV. Application of Materials and Processes

Selection of the appropriate material and process for a given application is quite complex. Both technical and economic factors must be considered. The ratio of average optical transmission to sheet resistivity is by no means the only criterion to be used. There are many applications in which transparency is many times more important than electrical conductivity. For example, the photoconductors used in television camera vidicons are rather inefficient, especially at the blue end of the spectrum. Light loss must be minimized in the transparent conductor used as a field plate as far as possible, even if this results in a sheet resistivity as high as 20 k$\Omega/\square$. Surface imperfections too small to be observed in optical microscopes cause enough scattering to result in large, visible defects in the television pictures produced. For this application, SnO_2 is usually employed because its absorption edge occurs further into the uv than other materials, resulting in higher transmission at the blue end of the spectrum.

In other classes of devices, transparency must be sacrificed for maximum conductivity. An example of this category is the liquid crystal display. For small alphanumeric displays, the transparent conductor must be etched into a pattern of relatively fine lines using photolithographic techniques. Excessive resistance in these lines necessitates the use of higher input power to operate the display. For these devices, In_2O_3:Sn is ordinarily used because it yields the highest conductivity and because it can be etched easily (Fig. 35). The

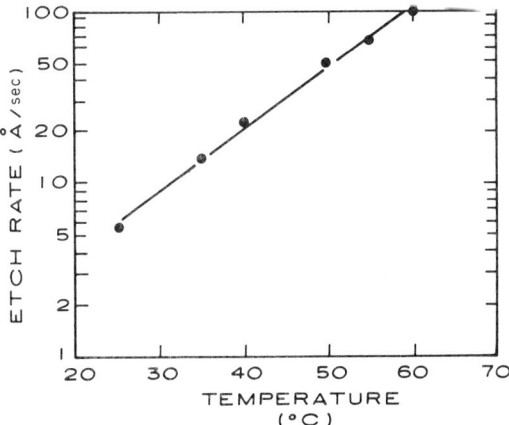

FIG. 35. The etch rate of rf-sputtered In_2O_3:Sn films in 36.7% HCl versus temperature.

average optical transmission can be as low as 70–75% and still be acceptable. In_2O_3:Sn is almost always used in preference to SnO_2 or SnO_2:Sb when the film must be etched because the chemical etching of SnO_2 is very difficult. Sputter etching is normally employed when SnO_2 films must be patterned. However, the resistance to chemical etching implies that the SnO_2 films are more resistant to harsh environments, which is important for other applications (e.g., front surface infrared reflectors). There are other applications that require conductivity and transparency which are intermediate to the applications described above.

Table VIII is a compilation of the best electrical and optical properties of transparent conductors produced by various deposition techniques. These are subdivided into low and high transmission ranges. Table IX represents an attempt to rate objectively the more commonly used deposition processes on a scale of 0–10 on twelve different technical and economic factors of importance. No weight is given to any of these factors in the table, but, clearly, weighing factors should be applied when selecting a process for a particular application. (Some of the factors may be of paramount importance or of no importance whatever.)

Taken together, Tables VIII and IX graphically illustrate that there are no panaceas in this field and that there is no "optimum" material or process, since the differences among the various processes, on balance, are negligibly small. Thus, not only the material, but the deposition process must be chosen to suit the exact application at hand.

TABLE VIII

Lowest Sheet Resistivity of Films Produced by Various Deposition Techniques for Average Visible Transmission of 80% and 95%[a]

Process	Average transmission = 80%			Average transmission = 95%		
	Material	σ ($\Omega/\square$)	Ref.	Material	σ ($\Omega/\square$)	Ref.
Spray hydrolysis	In_2O_3:Sn	15–20	(141)	SnO_2:Sb	250	(132)
Pyrolysis	In_2O_3:Sn	10–15	(172)	SnO_2:Sb	100	(171)
Evaporation of metals over nucleation-modifying layers	Au/Bi_2O_3	10–15	(40, 41)	—	—	—
Reactive evaporation	In_2O_3:Sn	40–60	(221)	—	—	—
Sputtering noble metals over nucleation-modifying layers	Au/Bi_2O_3	10–15	(186)	—	—	—
Reactive sputter	In_2O_3:Sn	3–5	(251–253)	—	—	—
dc Sputter oxides	In_2O_3:Sn	2–3	(260)	—	—	—
rf Sputter oxides	In_2O_3:Sn	2–3	(259, 262–264)	SnO_2:Sb	2500	(257, 258)
	Cd_2SnO_4	2–3	(265a)			

[a] Values are approximate.

TABLE IX: The Quality of Various Parameters of Transparent Conductors Prepared by Different Processes[a]

Parameter \ Process	Hydrolysis (spray)	Pyrolysis	Evaporation noble metal + nucleation modifier	Reactive evaporation	dc Sputter noble metals + nucleation modifier	Reactive dc sputter	Reactive rf sputter	dc Sputter oxides	rf Sputter oxides
General film properties	3	8	6	7	6	6	7	8	8
Control of uniformity (small areas)	7	9	9	9	9	9	9	9	9
Control of uniformity (large areas)	5	2	5	5	8	7	7	7	7
Reproducibility	2	8	6	5	6	7	7	8	7
Film surface morphology	2	9	7	7	7	7	9	3	9
Substrate temperature[b]	2	4	10	5	7	5	5	5	7
Ability to coat large areas	10	2	7	7	6	6	5	6	5
Ability to coat nonflat surfaces	9	2	6	6	5	5	5	5	5
Capital equipment cost (small area)	9	8	4	4	3	3	1	3	1
Capital equipment cost (large area)	10	4	3	3	2	2	1	2	1
Process cost/area coated	5	6	7	7	8	8	7	8	7
Recurring materials cost	5	6	1	4	4	8	8	8	8
Total	69	68	71	69	71	73	71	72	74

[a] Scale: 0–10 (10, best).
[b] Lower substrate temperature required either during or after deposition is assumed better, since there is less influence on film properties of material diffusing into the film from the substrate.

V. Conclusion

In this review, an attempt has been made to put into scientific perspective, insofar as is possible, an area of thin film technology that is ordinarily viewed as an art. It is clear that all of the details of the physics and chemistry of these films are not completely understood. The problems of the complex interplay of the deposition processes with film properties have been addressed by many investigators, especially in the past 5–10 years. Despite this intensified study, there are still many unexplained subtleties apparent in all the deposition processes.

The importance of this area of thin film technology is ever increasing. The recent accelerated pace of activity has been generated largely by the needs of the electronic industry for better display devices of a variety of kinds (274). Further acceleration can be expected as the use of these same materials and processes expands into the field of thin films related to solar energy. Thin metal films are now used as infrared reflectors on architectural glass to minimize heating and air-conditioning costs. Because of their superior durability, some of the oxides used for transparent conductors may find application in that area. At present, the infrared reflectance of the oxides, which is related to the electrical conductivity, is not as high as that of the thin noble metals, but a combination of advances in oxide deposition technology and increasing cost of noble metals could easily open the possibility of applications in various kinds of coated glass for thermal insulation.

ACKNOWLEDGMENTS

I should like to express my gratitude to many of my colleagues at RCA Laboratories for very fruitful discussions on this and related subjects. I am especially indebted to D. M. Hoffman, M. D. Coutts, and J. J. O'Neill, Jr. for permitting me to use hitherto unpublished material. Thanks are also due to G. L. Schnable, W. Kern, and D. M. Hoffman for their critical appraisal of the manuscript.

References

1. K. Bädeker, *Ann. Phys. (Leipzig)* **22**, 749 (1907).
2. L. Holland, "Vacuum Deposition of Thin Films," pp. 492–509. Wiley, New York, 1958.
3. L. S. Palatnik and Y. F. Komnik, *Sov. Phys. -Dokl.* **4**, 663 (1959).
4. L. A. Goodman, *J. Vac. Sci. Technol.* **10**, 804 (1973).
5. L. I. Maissel, in "Handbook of Thin Film Technology" (L. I. Maissel and R. Glang, eds.), Ch. 13. McGraw-Hill, New York, 1970.

6. C. A. Neugebauer, in "Handbook of Thin Film Technology" (L. I Maissel and R. Glang, eds.), Ch. 8. McGraw-Hill, New York, 1970.
7. C. A. Neugebauer, in "Physics of Thin Films" (G. Hass and R. E. Thun, eds.), Vol. 2, p. 1. Academic Press, New York, 1964.
8. D. W. Pashley, "Thin Films," p. 59. Am. Soc. Metals, Metals Park, Ohio, 1964.
9. G. A. Bassett, in "Condensation and Evaporation of Solids" (E. Rutner et al., eds.), p. 521. Gordon & Breach, New York, 1964.
10. G. A. Bassett, Proc. Eur. Conf. Electron Microsc., Delft p. 270 (1960).
11. H. Poppa, J. Vac. Sci. Technol. **2**, 42 (1965).
12. K. L. Chopra, J. Appl. Phys. **37**, 3405 (1966).
13. K. L. Chopra and M. R. Randlett, J. Appl. Phys. **39**, 1874 (1968).
14. L. Kasprzak, R. Laibowitz, S. Herd, and M. Ohring, Thin Solid Films **22**, 189 (1974).
15. K. L. Chopra, Appl. Phys. Lett. **7**, 140 (1965).
16. K. L. Chopra, J. Appl. Phys. **37**, 2249 (1966).
17. D. I. Kennedy, R. E. Hayes, and R. W. Alsford, J. Appl. Phys. **38**, 1986 (1967).
18. J. Le Bas, C. R. Acad. Sci., Ser. B **268**, 1393 (1969).
19. E. Ahilea and A. A. Hirsch, J. Appl. Phys. **42**, 5601 (1971).
20. L. E. Murr and H. P. Singh, Appl. Phys. Lett. **20**, 512 (1972).
21. M. Koedam, Philips Res. Rep. **16**, 266 (1961).
22. E. A. Allen, G. D. Scott, K. T. Thompson, and F. Veas, J. Opt. Soc. Am. **64**, 1190 (1974).
23. K. L. Chopra and R. Randlett, Appl. Phys. Lett. **8**, 241 (1966).
24. S. L. Chow, N. E. Hedgecock, M. Schlesinger, and J. Rezek, J. Electrochem. Soc. **119**, 1614 (1972).
25. R. Cornely and N. Fuschillo, Bull. Am. Phys. Soc. **18**, 16 (1973).
26. S. S. Minn, J. Rech. Cent. Natl. Rech. Sci. Lab. Bellevue **51**, 131 (1960).
27. R. M. Hill, Thin Solid Films **1**, 39 (1967).
28. R. M. Hill, Nature (London) **204**, 35 (1964).
29. C. A. Neugebauer and M. B. Webb, J. Appl. Phys. **33**, 74 (1962).
30. E. S. Herman and T. N. Rhodin, J. Appl. Phys. **37**, 1594 (1966).
31. T. E. Hartman, J. Appl. Phys. **34**, 943 (1963).
32. D. C. Larson, in "Physics of Thin Films" (M. H. Francombe and R. W. Hoffman, eds.), Vol. 6, p. 81. Academic Press, New York, 1971.
33. E. H. Sondheimer, Adv. Phys. **1**, 1 (1952).
34. D. S. Campbell, in "The Use of Thin Films in Physical Investigations" (J. C. Anderson, ed.), p. 299. Academic Press, New York, 1966.
35. S. Mader, in "The Use of Thin Films in Physical Investigations" (J. C. Anderson, ed.), p. 433. Academic Press, New York, 1966.
36. N. F. Mott and H. Jones, "The Theory of the Properties of Metals and Alloys," Dover, New York, 1958.
37. A. Malliaris and D. T. Turner, J. Appl. Phys. **42**, 614 (1971).
38. S. M. Aharoni, J. Appl. Phys. **43**, 2463 (1972).
39. F. Bueche, J. Appl. Phys. **43**, 4837 (1972).
40. E. J. Gillham and J. S. Preston, Proc. Phys. Soc., London, Sect. B **65**, 649 (1952).
41. E. J. Gillham, J. S. Preston, and B. E. Williams, Phil. Mag. **46**, 1051 (1955).
42. N. F. Mott and R. W. Gurney, "Electronic Processes in Ionic Crystals," Oxford Univ. Press, London and New York, 1948.
43. R. E. Aitchison, Aust. J. Appl. Sci. **5**, 10 (1954).
44. E. J. Verwey, P. W. Haaijman, F. C. Romeijn, and G. W. van Oosterhout, Philips Res. Rep. **5**, 173 (1950).

45. P. Kofstad, "Nonstoichiometry, Diffusion, and Electrical Conductivity in Binary Metal Oxides," Wiley, New York, 1972.
46. C. Wagner, *J. Phys. Chem. Solids* **33**, 1051 (1972).
47. T. B. Reed, "Free Energy of Formation of Binary Compounds," MIT Press, Cambridge, Massachusetts, 1971.
48. A. K. Vijh, *J. Mater. Sci.* **9**, 334 (1974).
49. E. M. Levin, C. R. Robbins, and H. F. McMurdie, "Phase Diagrams for Ceramists," p. 142. Am. Ceram. Soc. Columbus, Ohio, 1964.
50. E. M. Levin, C. R. Robbins, and H. F. McMurdie, "Phase Diagrams for Ceramists," p. 109. Am. Ceram. Soc., Columbus, Ohio, 1964.
51. E. M. Levin, C. R. Robbins, and H. F. McMurdie, "Phase Diagrams for Ceramists—1969 Supplement." Am. Ceram. Soc., Columbus, Ohio, 1969.
52. L. S. Dent Glasser and R. Roy, *J. Inorg. Nucl. Chem.* **17**, 100 (1961).
53. R. C. Weast, ed., "Handbook of Chemistry and Physics," 47th Ed. Chem. Rubber Publ. Co., Cleveland, Ohio, 1967.
54. R. Roy and M. W. Shafer, *J. Phys. Chem.* **58**, 372 (1954).
55. H. F. Mark, J. J. McKetta, Jr., and D. F. Othmer, eds., "Encyclopedia of Chemical Technology," Vol. 20, p. 317. Wiley (Interscience), New York, 1969.
56. J. T. Waber and D. T. Cromer, *J. Chem. Phys.* **42**, 4116 (1965).
57. Libbey-Owens-Ford Glass Co., Brit. Patent 671,767 (1952).
58. R. F. Raymond and B. J. Dennison, U.S. Patent 2,617,745 (1952).
59. W. O. Lytle, U.S. Patent 2,617,741 (1952).
60. D. E. Carlson, K. W. Hang, and G. F. Stockdale, U.S. Patent 3,811,855 (1974).
61. D. E. Carlson, personal communication.
62. F. Abelès, *in* "Physics of Thin Films" (M. H. Francombe and R. W. Hoffman, eds.), Vol. 6, p. 151. Academic Press, New York, 1971.
63. G. Hass and L. Hadley, *in* "American Institute of Physics Handbook" (D. E. Gray, ed.), 2nd Ed., pp. 6–103. McGraw-Hill, New York, 1963.
64. S. J. Thomson and G. A. Harvey, *J. Catal.* **22**, 359 (1971).
65. G. Rasigni and P. Rouard, *J. Opt. Soc. Am.* **53**, 604 (1963).
66. R. H. Doremus, *J. Appl. Phys.* **37**, 2775 (1966).
67. R. H. Doremus, *Thin Solid Films* **1**, 379 (1967/68).
68. B. S. Krishnamurthy and V. V. Paranjape, *Phys. Rev.* **181**, 1153 (1969).
69. P. Kaw, *J. Appl. Phys.* **40**, 793 (1969).
70. V. A. Pazdzerskii, *Sov. Phys.—Semicond.* **6**, 658 (1972).
71. F. Urbach, *Phys. Rev.* **92**, 1324 (1953).
72. F. Moser and F. Urbach, *Phys Rev.* **102**, 1519 (1956).
73. S. Tolansky, "Multiple Beam Interferometry." Oxford Univ. Press, London and New York, 1948.
74. O. S. Heavens, "Optical Properties of Thin Solid Films." Butterworth, London, 1955.
75. P. H. Berning, *in* "Physics of Thin Films" (G. Hass, ed.), Vol. 1, p. 69. Academic Press, New York, 1963.
76. O. S. Heavens, *in* "Physics of Thin Films" (G. Hass and R. E. Thun, eds.), Vol. 2, p. 193. Academic Press, New York, 1964.
77. G. B. Airy, *Trans. Cambridge Phil. Soc.* **4**, 409 (1832).
78. J. T. Cox and G. Hass, *in* "Physics of Thin Films" (G. Hass and R. E. Thun, eds.), Vol. 2, p. 239. Academic Press, New York, 1964.
79. S. Zaromb, U.S. Patent 3,378,396 (1968).
80. H. A. McCloud, "Thin Film Optical Filters." Am. Elsevier, New York, 1969.

80a. G. Kienel and G. Gallus, *Jpn. J. Appl. Phys.*, Suppl. 2, Pt. 1, p. 479 (1974).
81. H. E. Bennett and J. M. Bennett, *in* "Physics of Thin Films" (G. Hass and R. E. Thun, eds.), Vol. 4, p. 1. Academic Press, New York, 1967.
82. B. P. Kryzhanovskii, *Sov. Phys.—Tech. Phys.* **3**, 1378 (1958).
83. E. E. Kohnke, *J. Phys. Chem. Solids* **23**, 1557 (1962).
84. L. D. Loch, *J. Electrochem. Soc.* **110**, 1081 (1963).
85. M. Nagasawa, S. Shionoya, and S. Makishima, *Jpn. J. Appl. Phys.* **4**, 195 (1965).
86. J. A. Marley and R. C. Dockerty, *Phys. Rev. A* **140**, 304 (1965).
87. J. E. Houston and E. E. Kohnke, *J. Appl. Phys.* **36**, 3931 (1965).
88. D. F. Morgan and D. A. Wright, *Brit. J. Appl. Phys.* **17**, 337 (1966).
89. A. Peterson, Ph.D. Thesis, Rutgers Univ., New Brunswick, New Jersey, 1968.
90. R. D. Cunningham, J. P. Marton, and M. Schlesinger, *J. Appl. Phys.* **40**, 4664 (1969).
91. C. A. Vincent, *J. Electrochem. Soc.* **119**, 515 (1972).
92. C. A. Vincent and D. G. C. Weston, *J. Electrochem. Soc.* **119**, 518 (1972).
93. J. L. Jacquemin, C. Alibert and G. Bordure, *Solid State Commun.* **10**, 1295 (1972).
94. F. P. Koffyberg, *J. Appl. Phys.* **36**, 844 (1965).
95. R. W. Mar, *J. Phys. Chem. Solids* **33**, 220 (1972).
96. S. F. Reddaway and D. A. Wright, *Brit. J. Appl. Phys.* **16**, 195 (1965).
97. R. Summitt and N. F. Borrelli, *J. Appl. Phys.* **37**, 2200 (1966).
98. R. Summitt and N. F. Borrelli, *J. Phys. Chem. Solids* **26**, 921 (1965).
99. R. Summitt, J. A. Marley, and N. F. Borrelli, *J. Phys. Chem. Solids* **25**, 1465 (1964).
100. R. W. G. Wyckoff, "Crystal Structures," 2nd Ed., Vols. 1–3. Wiley (Interscience), New York, 1963.
101. S. Samson and C. G. Fonstad, *142nd Natl. Meet., Electrochem. Soc., Miami Beach, Fla.* Recent News Pap. No. 372 (1972).
102. R. L. Weiher, *J. Appl. Phys.* **33**, 2834 (1962).
103. J. P. Remeika and E. G. Spencer, *J. Appl. Phys.* **35**, 2803 (1964).
104. R. L. Weiher and R. P. Ley, *J. Appl. Phys.* **37**, 299 (1966).
105. J. H. W. DeWit, *J. Cryst. Growth* **12**, 183 (1972).
105a. T. Vojnovich and R. J. Bratton, *Am. Ceram. Soc., Bull.* **54**, 216 (1975).
106. M. J. Arvin, *J. Phys. Chem. Solids* **23**, 1681 (1962).
107. Z. M. Jarzebski, "Oxide Semiconductors," pp. 239–242. Pergamon, Oxford, 1973.
108. F. P. Koffyberg, *Phys. Lett. A* **30**, 37 (1969).
109. F. P. Koffyberg, *J. Solid State Chem.* **2**, 176 (1970).
110. N. B. Hannay, "Semiconductors," pp. 52, 590. Reinhold, New York, 1959.
111. C. F. Powell, J. H. Oxley, and J. M. Blocher, "Vapor Deposition." Wiley, New York, 1966.
112. D. S. Campbell, *in* "Handbook of Thin Film Technology" (L. I. Maissel and R. Glang, eds.), Ch. 5. McGraw-Hill, New York, 1970.
113. W. M. Feist, S. R. Steele, and D. W. Ready, *in* "Physics of Thin Films" (G. Hass and R. E. Thun, eds.), Vol. 5, p. 237. Academic Press, New York, 1969.
114. J. A. Aboaf, V. C. Marcotte, and N. J. Chou, *J. Electrochem. Soc.* **120**, 701 (1973).
114a. H. Kim and H. A. Laitinen, *J. Am. Ceram. Soc.* **58**, 23 (1975).
115. R. W. Gress, J. A. Murphy, and A. T. Talwalker, *Proc. Electron. Components Conf. IEEE, New York* p. 164 (1968).
116. J. G. Marriott, R. M. Felt, J. H. Boicy, and J. D. Ryan, Ger. Patent 1,045,612 (1958).
116a. T. R. Viverito, E. W. Rilee, and L. H. Slack, *Am. Ceram. Soc., Bull.* **54**, 217 (1975).
117. R. G. Livesay, E. Lyford, and H. Moore, *J. Phys. E* **1**, 947 (1968).
118. E. W. Wartenburg and P. W. Ackerman, *Glastech. Ber.* **41**, 55 (1968).

119. V. M. Novikov, *Steklo Keram.* **25**, 13 (1968).
120. R. F. Bartholomew and H. M. Garfinkel, *J. Electrochem. Soc.* **116**, 1205 (1969).
121. H. A. McMaster, U.S. Patent 2,429,420 (1947).
122. R. A. Gaiser, U.S. Patent, 2,602,032 (1951).
123. L. D. Thomas, Ger. Patent, 1,045,613 (1958).
124. K. H. Reiss, Ger. Patent 971,072 (1958).
125. G. Wendel, DDR Patent 14,536 (1958).
126. Glaverbel, Fr. Patent 1,479,586 (1967).
127. T. Suzukawa and Y. Yamane, Jpn. Patent 18,747 (1969).
128. D. W. Roe, Fr. Patent 1,568,948 (1970).
129. M. J. Zunick, U.S. Patent 2,516,663 (1950).
130. R. F. Raymond and B. J. Dennison, U.S. Patent 2,592,601 (1952).
131. A. Fischer, *Z. Naturforsch. A* **9**, 508 (1954).
132. A. R. Peaker and B. Horsley, *Rev. Sci. Instrum.* **42**, 1825 (1971).
133. J. M. Mochel, U.S. Patent 2,564,707 (1951).
134. J. A. Lely and J. G. Bos, U.S. Patent 3,014,815 (1961).
135. A. Rohatgi, T. R. Viverito, and L. H. Slack, *J. Am. Ceram. Soc.* **57**, 278 (1974).
136. H. Ladwig, *Silikattechnik* **15**, 182 (1964).
137. I. Golovcenco, G. I. Rusu, V. Stefan, and M. Rusu, *Iasi, Sect. Ib Fiz.* **11**, 77 (1965).
138. P. W. Haayman, P. C. van der Linden, D. Veeneman, and G. H. Janssen, U.S. Patent 2,772,190 (1956).
139. Philips Electrical Industries, Ltd., Br. Patent 732,566 (1955).
140. J. W. McAuley, U.S. Patent 2,692,836 (1954).
141. R. Groth, *Phys. Status Solidi* **14**, 69 (1966).
142. A. Y. Kuznetsov, A. V. Kruglova, and B. P. Kryzhanovskii, *Zh. Prikl. Khim.* **32**, 1161 (1959).
143. B. P. Kryzhanovskii, A. V. Kruglova, and A. Y. Kuznetsov, *Zavod. Lab.* **31**, 1366 (1965).
144. B. P. Kryzhanovskii and M. A. Okatov, *Zh. Prikl. Khim.* **39**, 2832 (1966).
145. W. O. Lytle and A. E. Junge, U.S. Patent 2,566,346 (1951).
146. A. E. Saunders and W. E. Wagner, U.S. Patent 3,107,177 (1963).
147. W. O. Lytle and A. E. Junge, U.S. Patent 2,740,731 (1956).
148. Union des Verreries Mecaniques Belges, Brit. Patent 892,708 (1962).
149. M. S. Tarnopol, U.S. Patent 2,694,649 (1954).
150. W. O. Lytle and A. E. Junge, Ger. Patent 971,957 (1959).
151. D. E. Carlson, *J. Am. Ceram. Soc.* **57**, 291 (1974).
152. D. E. Carlson, K. W. Hang, and G. F. Stockdale, *J. Am. Ceram. Soc.* **57**, 295 (1974).
153. A. Y. Kuznetsov, *Zavod. Lab.* **23**, 90 (1957).
154. O. V. Vorob'eva and V. V. Vonogradova, *Steklo Keram.* **20**, 13 (1963).
155. Libbey-Owens-Ford Glass Company, Brit. Patent 682,342 (1952).
156. I. Viscrian and V. Georgescu, *Thin Solid Films* **3**, R17 (1969).
157. K. Ishiguro, T. Sasaki, T. Arai, and I. Imai, *J. Phys. Soc. Jpn.* **13**, 296 (1958).
158. R. Gomer, *Rev. Sci. Instrum.* **24**, 993 (1953).
159. H. Koch, *Phys. Status Solidi* **7**, 263 (1964).
160. S. P. Lyashenko and V. K. Miloslavskii, *Opt. Spectros. (USSR)* **19**, 55 (1965).
161. T. Arai, *J. Phys. Soc. Jpn.* **15**, 916 (1960).
162. H. E. Mendenhall, *Phys. Rev.* **72**, 532 (1947).
163. H. Koch, *Phys. Status Solidi* **3**, 1619 (1963).
163a. K. Kajiyama and Y. Furukawa, *Jpn. J. Appl. Phys.* **6**, 905 (1967).
163b. T. Nishino and Y. Hamakawa, *Jpn. J. Appl. Phys.* **9**, 1085 (1970).

164. I. Imai, *J. Phys. Soc. Jpn.* **15**, 937 (1960).
165. D. Elliott, D. L. Zellmer, and H. A. Laitinen, *J. Electrochem. Soc.* **117**, 1343 (1970).
165a. H. Köstlin, R. Jost, and W. Lems, *Phys. Status Solidi* **A 29**, 87 (1975).
166. V. F. Korzo, *Izv. Vyssh. Uchebn. Zaved., Fiz.* **10**, 86 (1967).
167. V. F. Korzo and L. A. Ryabova, *Sov. Phys.—Solid State* **9**, 745 (1967).
168. L. A. Ryabova and Y. S. Savitskaya, *J. Vac. Sci. Technol.* **6**, 934 (1969).
169. V. F. Korzo and V. N. Chernyaev, *Phys. Status Solidi* **A 20**, 695 (1973).
170. J. Kane, H. P. Schweizer, and W. Kern, *J. Electrochem. Soc.* **122**, 1144 (1975).
171. J. Kane, H. P. Schweizer, and W. Kern, *J. Electrochem. Soc.* **123**, 270 (1976).
172. J. Kane, H. P. Schweizer, and W. Kern, *Thin Solid Films* **29**, 155 (1975).
172a. G. Blandenet, Y. Lagarde, and J. Spitz, in "Proceedings of the Fifth International Conference on Chemical Vapor Deposition" (J. M. Blocher, H. E. Hintermann, and L. H. Hall eds.), p. 190. Electrochem. Soc., Princeton, New Jersey, 1975.
172b. O. Tabata, in "Proceedings of the Fifth International Conference on Chemical Vapor Deposition" (J. M. Blocher, H. E. Hintermann, and L. H. Hall, eds.), p. 681. Electrochem. Soc., Princeton, New Jersey, 1975.
173. W. Kern, *RCA Rev.* **29**, 525 (1968).
174. J. Kane, U.S. Patent 3,854,992 (1974).
175. R. Glang, in "Handbook of Thin Film Technology" (L. I. Maissel and R. Glang, eds.), Ch. 1. McGraw-Hill, New York, 1970.
176. L. I. Maissel, in "Handbook of Thin Film Technology" (L. I. Maissel and R. Glang, eds.), Ch. 4. McGraw-Hill, New York, 1970.
177. J. L. Vossen, *J. Vac. Sci. Technol.* **8**, S12 (1971).
178. Society of European Research Associates S. A., Belg. Patent 555,277 (1957).
179. J. L. Van Cakenberghe and J. F. Gilles, U.S. Patent 3,039,896 (1962).
180. M. S. P. Lucas, *Appl. Phys. Lett.* **4**, 73 (1964).
181. K. L. Chopra and M. R. Randlett, *J. Appl. Phys.* **38**, 3144 (1967).
182. M. S. P. Lucas, *Thin Solid Films* **2**, 337 (1968).
183. D. T. Meyer, *Thin Solid Films* **2**, 27 (1968).
184. S. Minn and S. Offret, *C. R. Acad. Sci.* **242**, 2117 (1956).
185. A. E. Ennos, *Brit. J. Appl. Phys.* **8**, 113 (1957).
186. L. Holland and G. Siddall, *Vacuum* **3**, 375 (1953).
187. S. Minn and S. Offret, *C. R. Acad. Sci.* **244**, 1624 (1957).
188. J. S. Preston and E. J. Gillham, U.S. Patent 2,825,687 (1958).
189. W. H. Colbert, A. R. Weinrich, and W. L. Morgan, U.S. Patent 2,628,927 (1953).
190. W. L. Morgan, U.S. Patent 2,750,832 (1956).
191. S. Minn, *J. Rech. Cent. Natl. Rech. Sci., Lab. Bellevue* **51**, 131 (1960).
192. W. H. Colbert and W. L. Morgan, U.S. Patent 2,676,117 (1954).
193. D. M. Hoffman, personal communication.
194. J. R. Rairden, C. A. Neugebauer, and R. A. Sigsbee, *Metall. Trans.* **2**, 719 (1971).
195. H. K. Chaurasia and W. A. G. Voss, *Nature (London)* **249**, 28 (1974).
196. D. M. Hoffman, *Natl. Vac. Symp. 12th, Am. Vac. Soc., New York, 1965.*
197. D. M. Hoffman, *J. Vac. Sci. Technol.* **13**, 122 (1976).
198. R. E. L. Cox, *Thin Solid Films* **11**, 323 (1972).
199. W. R. Chase and F. L. Shuermeyer, *Bull. Am. Phys. Soc.* **16**, 836 (1971).
200. F. L. Shuermeyer, *J. Appl. Phys.* **42**, 5856 (1971).
201. D. Hoffman and D. Leibowitz, *J. Vac. Sci. Technol.* **9**, 326 (1972).
202. Y. K. Shalabutov, *Sov. Phys.—Semicond.* **7**, 322 (1973).
203. S. A. Hoenig and R. A. Pope, *Appl. Phys. Lett.* **14**, 271 (1969).

204. J. F. Wendt and A. B. Cambel, *J. Appl. Phys.* **34**, 176 (1963).
205. F. L. Reynolds, *J. Chem. Phys.* **39**, 1107 (1963).
206. R. F. Tinder, G. A. Antypas, and E. E. Donaldson, *J. Appl. Phys.* **35**, 3452 (1964).
207. D. Lichtman, *J. Vac. Sci. Technol.* **2**, 91 (1965).
208. M. D. Scheer and J. Fine, *J. Chem. Phys.* **42**, 3645 (1965)
209. J. D. Levine, Ph.D. Thesis, pp. 107–114. Mass. Inst. of Technol., Massachusetts, Cambridge, 1963.
210. J. L. Vossen, J. J. O'Neill, Jr., and M. D. Coutts, unpublished data.
211. J. L. Vossen and J. J. O'Neill, Jr., *RCA Rev.* **29**, 566 (1968).
212. L. Holland, "Vacuum Deposition of Thin Films," pp. 450–455. Wiley, New York, 1958.
213. G. Rupprecht, *Z. Phys.* **139**, 504 (1954).
214. J. S. Preston, U.S. Patent 2,769,778 (1956).
215. H. Watanabe, *Jpn. J. Appl. Phys.* **9**, 1551 (1970).
216. E. Jedlicka and J. Jandus, *Trans. Czech. Conf. Electron. Vac. Phys., 5th, Brno, 1972.*
217. General Electric Company, Neth. Patent 6,509,917 (1966).
218. Y. Hori, K. Doi, and I. Ikuzo, *Oyo Butsuri* **34**, 507 (1965).
219. W. Spence, *J. Appl. Phys.* **38**, 3767 (1967).
220. J. Naegele, Ger. Patent 1,907,394 (1970).
221. D. Furuuchi, *Microelectron. Soc. Jpn., Meet., 1973.*
222. R. E. Honig and D. A. Kramer, *RCA Rev.* **30**, 285 (1969).
222a. S. Sobajima, H. Okaniwa, N. Takagi, I. Sugiyama, and K. Chiba, *Jpn. J. Appl. Phys.*, Suppl. 2, Pt. 1, p. 475 (1974).
222b. B. P. Kryzhanovskii and E. N. Orel, *Sov. J. Opt. Technol. (Engl. Transl.)* **41**, 479 (1974).
223. F. G. Allen, T. M. Buck, and J. T. Law, *J. Appl. Phys.* **31**, 979 (1960).
224. E. G. Bylander, J. R. Piedmont, L. D. Shubin, and R. C. Smith, *J. Appl. Phys.* **34**, 3407 (1963).
224a. H. Matsunami, K. Oo, H. Ito, and T. Tanaka, *Jpn. J. Appl. Phys.* **14**, 915 (1975).
225. F. H. Huang and H. B. Huntington, *Phys. Rev. B* **9**, 1479 (1974).
226. G. S. Anderson, *J. Appl. Phys.* **40**, 2884 (1969).
227. J. L. Vossen and J. J. O'Neill, Jr., *RCA Rev.* **29**, 149 (1968).
228. J. Sosniak and F. B. Alexander, *Proc. Symp. Deposition Thin Films Sputtering, 3rd, Bendix Corp., Rochester, N.Y.* p. 114, (1969).
229. F. H. Gillery and J. P. Pressau, Ger. Patents 1,909,869 and 1,909,910 (1969).
230. J. S. Preston, *Proc. Roy. Soc. London, Ser. A* **202**, 449 (1950).
231. G. Helwig, *Z. Phys.* **132**, 621 (1952).
232. H. Dunstädter, *Z. Phys.* **137**, 383 (1954).
233. F. Lappe, *Z. Phys.* **137**, 380 (1954).
234. J. Stuke, *Z. Phys.* **137**, 401 (1954).
235. T. K. Lakshmanan, *J. Electrochem. Soc.* **110**, 548 (1963).
236. R. R. Mehta and S. F. Vogel, *J. Electrochem. Soc.* **119**, 752 (1972).
236a. A. Kunioka and Y. Sakai, *Jpn. J. Appl. Phys.* **7**, 1138 (1968).
237. M. L. Lieberman and R. C. Medrud, *J. Electrochem. Soc.* **116**, 242 (1969).
238. M. Hecq and E. Portier, *Thin Solid Films* **9**, 341 (1972).
239. W. R. Sinclair, F. G. Peters, D. W. Stillinger, and S. E. Koonce, *J. Electrochem. Soc.* **112**, 1096 (1965).
240. V. M. Vaynshetyn, *Sov. J. Opt. Technol.* **34**, 45 (1967).
241. H. W. Lehmann and R. Widmer, *Thin Solid Films* **27**, 359 (1975).
242. J. J. Cuomo and R. J. Gambino, *J. Vac. Sci. Technol.* **12**, 79 (1975).
243. V. M. Vainshtein and V. I. Fistul, *Sov. Phys.—Semicond.* **1**, 104 (1967).

244. V. I. Fistul and V. M. Vainshtein, *Sov. Phys.—Solid State* **8**, 2769 (1967).
245. H. K. Muller, *Phys. Status Solidi* **27**, 723 (1968).
246. H. K. Muller, *Phys. Status Solidi* **27**, 733 (1968).
247. V. M. Vainshtein, L. Gerasimova, and I. N. Nikolaeva, *Izv. Akad. Nauk SSSR, Neorg. Mater.* **4**, 357 (1968).
248. Y. T. Sihvonen and D. R. Boyd, *Rev. Sci. Instrum.* **31**, 992 (1960).
249. D. R. Boyd, Y. T. Sihvonen, and C. D. Woelke, U.S. Patent 3,235,476 (1966).
250. V. A. Williams, *J. Electrochem. Soc.* **113**, 234 (1966).
251. J. M. Pankratz, *J. Electron Mater.* **1**, 182 (1972).
252. M. Bonnet and M. Marchal, *Proc. Colloq. Int. Pulverisation Cathodique, 1st, Soc. Fr. Vide, Montpellier* p. 157 (1973).
253. W. W. Molzen, *J. Vac. Sci. Technol.* **12**, 99 (1975).
253a. Y. Murayama, *J. Vac. Sci. Technol.* **12**, 818 (1975).
253b. D. M. Mattox, *Electrochem. Technol.* **2**, 295 (1964).
254. J. L. Vossen, *J. Vac. Sci. Technol.* **8**, 751 (1971).
255. S. Yamanaka and T. Oohashi, *Jpn. J. Appl. Phys.* **8**, 1058 (1969).
256. R. H. Deitch, E. J. West, T. G. Giallorenzi, and J. F. Weller, *Appl. Opt.* **13**, 712 (1974).
257. J. L. Vossen, *Proc. Symp. Deposition Thin Films Sputtering, 3rd, Bendix Corp., Rochester, N.Y.* p. 80 (1969).
258. J. L. Vossen and E. S. Poloniak, *Thin Solid Films* **13**, 281 (1972).
259. J. L. Vossen, *RCA Rev.* **32**, 289 (1971).
260. D. B. Fraser and H. D. Cook, *J. Electrochem. Soc.* **119**, 1368 (1972).
261. J. R. Bosnell and R. Waghorne, *Thin Solid Films* **15**, 141 (1973).
262. M. Hecq, A. DuBois, and J. Van Cakenberghe, *Thin Solid Films* **18**, 117 (1973).
263. J. L. Vossen, U.S. Patent 3,749,658 (1973).
264. M. Hecq, A. DuBois, and J. Van Cakenberghe, *C. R. Colloq. Int. Pulverisation Cathodique, 1st, Soc. Fr. Vide, Montpellier* p. 151 (1973).
265. A. J. Nozik, *Phys. Rev. B* **6**, 453 (1972).
265a. G. Haacke, *Appl. Phys. Lett.* **28**, 622 (1976).
266. R. L. Amans, U.S. Patent 3,295,002 (1966).
267. M. S. Jaffe and E. G. Fridrich, U.S. Patent 3,315,111 (1967).
268. M. S. Jaffe, Ger. Patent 934,848 (1955).
269. C. Y. Kuo, *Solid State Technol.* **17**, 49 (1974).
270. A. Kushihashi and K. Fujiwara, U.S. Patent 3,484,263 (1969).
271. A. Yoshikawa, R. Kondo, and Y. Sakai, *Jpn. J. Appl. Phys.* **12**, 1096 (1973).
272. J. Goodman, U.S. Patent 3,239,368 (1966).
273. D. E. Carlson, *J. Electrochem. Soc.* **122**, 1334 (1975).
274. J. L. Vossen, ed., *Proc. Symp. Display Mater. Devices, Am. Inst. Phys.*, New York, 1973.

Metal–Dielectric Interference Filters*

I. Introduction 74
II. Basic Theory 75
   1. Definitions of Admittance and Amplitude Reflection 75
   2. The Characteristic Matrix 77
   3. Radiant Reflectance and Transmittance 78
   4. The Net Flux Ratio 79
   5. The Radiant Absorptance 81
   6. The Net Flux Ratio for a Single Layer 81
   7. The Net Flux Ratio for an Assembly of Layers 82
   8. The Maximum Value of the Net Flux Ratio 86
   9. A Recapitulation of the Properties of the Net Flux Ratio 87
   10. Fabry–Perot Type Filters 88
   11. The Transmittance of a Symmetrical Multilayer 90
III. General Considerations in Bandpass Filter Design 93
   1. Summary of the Attributes of Bandpass Filters 93
   2. The Choice of Metals for Bandpass Filters 95
IV. Single-Cavity Filter Design 102
   1. The Design of an MDM Filter 102
   2. The Design of Wedged Filters 105
   3. Filters with Augmented Spacers 106
   4. Filters with Equivalent Layer Spacers 110
V. One-M Filter Design 112
   1. Three-Layer Symmetrical Systems 112
   2. Procedure for One-M Filter Design 116
   3. Examples of One-M Filters 117
VI. Multiple Cavity and Other Designs 124
   1. An Example of Multiple-Cavity Filter Design 124
   2. Augmented MDM Filters 125
   3. Augmented Double-Cavity Filter 127
VII. Reflection Filters 128
VIII. The Production of Filters 132
   1. Deposition Techniques for Metals 132
   2. Reflection or Transmission Monitoring 134
   3. Examples of Optical Monitoring 136
   4. Cementing and the Addition of Blocking Filters 140
IX. Future Developments 142
   References 143

* This manuscript was compiled from the publications of many physicists actively engaged in research in this field. Please refer to the references at the end of the article.

I. Introduction

Multilayer interference filters are widely used today to enhance the reflectance of surfaces, to antireflect optical components, and to filter radiant flux. Most of these coatings contain dielectric, that is, nonabsorbing, films. Although these dielectric multilayers may be slightly absorbing, this is usually not a welcomed effect and considerable effort is usually made to eliminate the absorption. This article discusses coatings in which one or more of the layers is absorbing. This is usually achieved with metal films; the absorptance in these films produces such remarkable effects as very low offband transmittance in multiple-layer bandpass filters.

The Fabry–Perot interferometer was developed early in this century and consists of a single optical resonant cavity that is formed by placing two semireflecting coatings on either side of a spacer. Geffcken (1) was one of the pioneers who in 1939 conceived of the idea of evaporating the entire single-cavity filter. As shown in Fig. 1, the semitransparent metal films and the magnesium fluoride spacer layer are all evaporated. The coatings contour the substrate and thus it need not be optically flat. The design was further embellished by Hadley and Dennison (2). In the 1950s Hermansen (3) and Wolter (4) developed the theory of the three-cavity filter. Dufour (5) and Schroeder (6) also described some triple-cavity filters. A reflection filter, alias a *dark mirror*, was also produced by Hass et al. (7). Its spectral reflectance depended more on the dispersion of the optical constants than on optical interference effects. Berning and Turner (8) developed the theory of the filter that contains only a single metal layer surrounded by dielectric stacks—the *induced transmission filter*, as they called it. They also produced some prototype filters. Later workers (9, 10) in the 1960s refined the concept of *admittance matching* and applied it to the design of filters for the ultraviolet part of the spectrum.

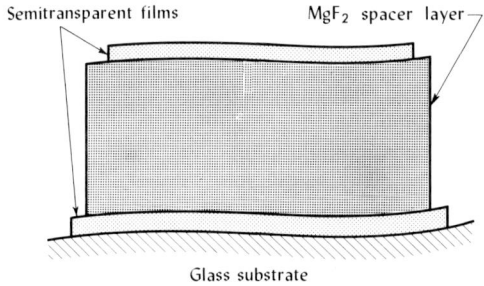

FIG. 1. A cross section of an MDM bandpass filter. The lack of planeness of the substrate is exaggerated for the purposes of illustration.

II. Basic Theory

1. Definitions of Admittance and Amplitude Reflection

Space does not permit us to develop the complete theory for the propagation of an electromagnetic wave into a multilayer stack. Thus we will rely on the results of other authors, such as Berning (*11*) and Abelès (*12*). The physical model for the multilayer assumes that a plane wave is incident on the stack of homogeneous films deposited on a substrate of optical constant $\hat{n}_s = n_s - jk_s$, as shown in Fig. 2. At any point in space it is possible to decompose the electric field into

$$E^+ = E_0^+ \exp[j(\omega t - \ell z)] \tag{1}$$

and

$$E^- = E_0^- \exp[j(\omega t + \ell z)] \tag{2}$$

which are interpreted physically as the wave components propagating in the $+z$ and $-z$ directions, respectively. ℓ is the propagation constant

$$\ell = 2\pi\sigma\hat{n} \tag{3}$$

where $\sigma = 1/\lambda$ is the vacuum wavenumber of the incident flux. The foregoing equations pertain to normal incidence. Similar equations could be written when $\theta_0 \neq 0$.

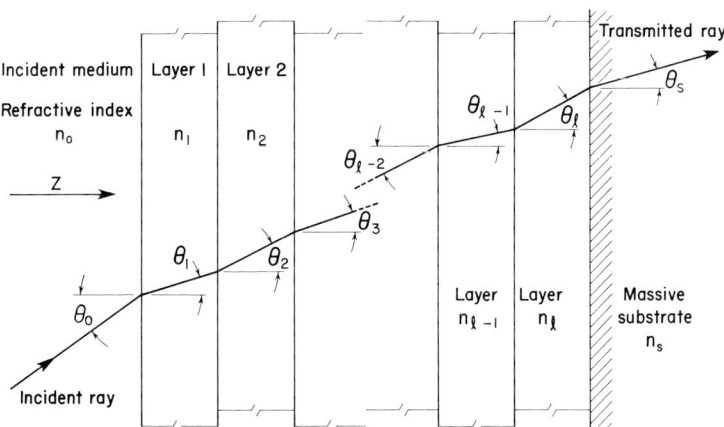

FIG. 2. The propagation of a ray through a multilayer stack, in which layer 1 is adjacent to the incident medium and layer *l* is adjacent to the substrate. The reflections at each interface are not shown.

At a given point in the stack, the reflection coefficient for the electric vector ρ_0, alias amplitude reflection coefficient, is

$$\rho_0 = E_0^-/E_0^+ = |\rho_0| \exp(j\delta_0) \tag{4}$$

If ρ_0 is measured at the surface of a multilayer stack, then δ_0 is the phase shift upon reflection and determines the position of the node of the standing wave.

The foregoing equations refer only to the electric field E. Standard texts on electromagnetic theory show that a magnetic field H accompanies E. For a plane wave propagating in the $+z$ direction E_0 and H_0 are related by

$$E_0 = Z_0(\mu/\varepsilon)^{1/2} H_0 \tag{5}$$

where ε and μ are the dielectric permittivity and magnetic permeability of the medium, respectively. Z_0 is the impedance of free space and is 377 Ω in MKS units. It is a nuisance to carry the Z_0 in the equations related to multilayer design, and it is omitted at the risk of being dimensionally inconsistent. Experiments have shown that $\mu = 1$ at optical frequencies. Starting from the definition of the refractive index, Eq. (5) reduces to

$$E_0 = nH_0 \tag{6}$$

The foregoing discussion considers only a wave propagating in the $+z$ direction. In the more general case where there are both incident and reflected components, the optical admittance Y is

$$Y = H/E \tag{7}$$

where H and E are the total fields at a given point in space. At nonnormal incidence, the E and H in the foregoing equation are the components tangential to a surface, in which case Y is termed the surface admittance. Standard texts (*11*, *12*) show that the relationship between Y and ρ is

$$\rho = (n_0 - Y)/(n_0 + Y) \tag{8}$$

where n_0 is the refractive index of the medium in which Y is measured. This is also written as

$$\rho = (1 - y)(1 + y)^{-1} \tag{9}$$

where the reduced admittance y is

$$y = Y/n_0 \tag{10}$$

The standing wave ratio is another useful concept that is used in multilayer design. This is introduced by considering the total field E which is the sum of the incident wave and reflected waves

$$E = E_0[\exp(-jkz) + |\rho| \exp(j\delta_0) \exp(jkz)] \tag{11}$$

where ℓ is defined in Eq. (3) and the time dependence has been omitted. It is assumed that the origin of the coordinate system is at the interface between the top layer of the multilayer and the incident medium. The E field is a maximum when the term $\exp[j(\delta_0 + 2\beta)]$ is real and has a value of $+1$. The magnitude of the field is then proportional to $1 + |\rho|$. At another position in space, a distance of $\lambda/4$ away, the field is a minimum. In this case the term $\exp[j(\delta_0 + 2\beta')]$ is real and has a value of -1. The total E field has a value proportional to $1 - |\rho|$. The standing wave ratio V is defined as

$$V = \frac{\text{maximum value of total } E \text{ field}}{\text{minimum value of total } E \text{ field}} \quad (12)$$

or

$$V = \frac{1 + |\rho|}{1 - |\rho|} \quad (13)$$

and is always greater than one:

$$1 \leq V \leq \text{VLN} \quad (14)$$

where VLN stands for very large number—of the order of 10^5 to 10^6 for optical coatings. As an example, suppose that the radiant reflectance of a coating is 36%. $|\rho|$ is 0.6 and V is 4. The standing wave ratio is discussed in greater detail in texts on electrical transmission lines (13).

Before leaving this section, we establish some terminology that is used in later sections. In general, both the admittance and the amplitude reflectance are complex quantities. The real and imaginary parts are defined as

$$\hat{Y} = x + jz \quad (15)$$

and

$$\hat{\rho} = u + jw \quad (16)$$

2. The Characteristic Matrix

We next briefly discuss the characteristic matrix method of computing the properties of a multilayer stack. This matrix expresses a linear relationship between the fields E_0, H_0 on the left side of a layer to the fields E_1, H_1 on its emergent side

$$\begin{bmatrix} E_0 \\ H_0 \end{bmatrix} = \begin{bmatrix} \cos \beta_1 & jn_1^{-1} \sin \beta_1 \\ jn_1 \sin \beta_1 & \cos \beta_1 \end{bmatrix} \begin{bmatrix} E_1 \\ H_1 \end{bmatrix} \quad (17)$$

where n_1 is its refractive index and the phase of retardation β_1 is

$$\beta_1 = 2\pi\sigma n_1 h_1 \quad (18)$$

where h_1 is its metric thickness.

The admittance and other attributes of a stack are computed as follows. The matrix in Eq. (17) $\mathcal{M}_i$ is computed for each layer. The product $\mathcal{M}$ of these matrices is

$$\mathcal{M} = \prod_{i=1}^{l} \mathcal{M}_i \qquad (19)$$

$\mathcal{M}$ is called the characteristic matrix of the stack and is defined in terms of the matrix elements:

$$\mathcal{M} = \begin{bmatrix} \hat{a}_1 & \hat{a}_3 \\ \hat{a}_2 & \hat{a}_4 \end{bmatrix} \qquad (20)$$

where the caret denotes a complex quantity. The real and imaginary parts of the matrix elements are $\hat{a}_1 = a_1 + jb_1$, $\hat{a}_2 = a_2 + jb_2$, and so on. In the special case of a stack of nonabsorbing layers, the diagonal elements of $\mathcal{M}$ are pure real and the off-diagonal elements pure imaginary:

$$\mathcal{M} = \begin{bmatrix} a_1 & jb_3 \\ jb_2 & a_4 \end{bmatrix} \qquad (21)$$

We next show how to compute the reflectance. The multilayer described by Eq. (20) is contiguous to an admittance $\hat{Y}_s$ on its emergent side. $\hat{Y}_s$ can represent either the optical constant $\hat{n}$ of the substrate or $\hat{Y}$ of another multilayer stack. The admittance is

$$\hat{Y} = (\hat{a}_2 + \hat{a}_4 \hat{Y}_s)(\hat{a}_1 + \hat{a}_3 \hat{Y}_s)^{-1} \qquad (22)$$

Standard texts (*11, 12*) show how the characteristic matrix is used to develop other properties of a multilayer stack, such as its radiant transmittance. The amplitude reflection can be found from Eq. (8) and its radiant reflectance $R = \rho\rho^*$.

3. Radiant Reflectance and Transmittance

Assume that a beam of finite width impinges upon a multilayer stack and emerges into the substrate. A detector is inserted into the incident and emergent beams so that it collects the total flux in each beam. The definition of the radiant transmittance T is

$$T = \frac{\text{response of the detector inserted into the emergent beam}}{\text{response of the detector inserted into the incident beam}} \qquad (23)$$

Before we derive an equation for T, we first consider how the Poynting vector S is obtained from the complex representation of the E and H fields. Standard texts on electromagnetic theory show that the cross product of E

and H gives the Poynting vector S which represents the power per unit area transported in a direction collinear with S. At a given point in space the time varying E and H fields are represented as

$$E(t) = \text{Re} \left| \hat{E}_0 \right| \exp(j\alpha_1) \exp(j\omega t) \right| \tag{24}$$

$$H(t) = \text{Re} \left| \hat{H}_0 \right| \exp(j\alpha_2) \exp(j\omega t) \right| \tag{25}$$

where the Re means the *real part of*.

At microwave frequencies it is possible to measure the instantaneous E field, but at higher frequencies we measure an average of S over a time $\mathcal{T}$ that is long compared to the period of oscillation $2\pi/\omega$. We denote this time average by $\langle S \rangle$

$$\langle S \rangle = \mathcal{T}^{-1} \int_0^{\mathcal{T}} S(t)\, dt \tag{26}$$

Substituting the product of Eqs. (24) and (25) for $S(t)$ in the foregoing equation yields

$$S = \tfrac{1}{2} |\hat{E}_0| |\hat{H}_0| \cos(\alpha_1 - \alpha_2) \tag{27}$$

Another way of writing Eq. (27) is

$$S = \tfrac{1}{2} \text{Re}[\hat{E}_0 \hat{H}_0^*] \tag{28}$$

where the asterisk denotes a complex conjugate.

In order to find the radiant transmittance, we first compute the magnitude of the Poynting vector S_0^+ of the incident wave and also the magnitude S_s^+ in the substrate. Then T is

$$T = \frac{S_s^+ \cos \theta_s}{S_0^+ \cos \theta_0} \tag{29}$$

where θ_0 and θ_s are the angles of refraction of the incident medium and substrate, as shown in Fig. 2. If S_0^- is the magnitude of the Poynting vector of the reflected component of the wave, then

$$R = S_0^- / S_0^+ \tag{30}$$

is the radiant reflectance.

4. THE NET FLUX RATIO

The term net flux ratio is defined below and has appeared in various aliases in the literature, such as potential transmittance (*8*) or radiant power flow ratio (*10*). It is an extremely useful concept and is used extensively in the design of filters and absorbing multilayers. The net flux ratio ψ is the fraction of the net flux incident upon a stack that emerges

$$\psi = \frac{S_1^+ - S_1^-}{S_0^+ - S_0^-} \tag{31}$$

where S_0^+ and S_0^- are, respectively, the incident and reflected components of the magnitude of the Poynting vector on its incident side, and S_1^+ and S_1^- are the corresponding quantities on its emergent side. In the situation where there is no reflected component of the wave on the emergent side, $S_1^- = 0$ and Eq. (31) reduces to

$$\psi = \frac{T}{1 - R} \tag{32}$$

Since the optical medium is passive, i.e., nonamplifying, ψ is in the range $0 < \psi < 1$. The lower limit occurs when the stack absorbs all of the incident flux, and the upper limit when the stack is nonabsorbing.

ψ can be defined for any subassembly of layers. As an example, consider a three-layer stack, as shown in Fig. 3. For purposes of illustration, the layers are separated so that the net flux $S_i^+ - S_i^-$ on the emergent side of each layer can be shown. In the emergent medium, $S_3^- = 0$. Then, starting with

$$\psi_{ij} = (S_i^+ - S_i^-)/(S_j^+ - S_j^-) \tag{33}$$

it follows that the ψ's have a multiplicative property. For example, ψ_{03} for the entire stack is

$$\psi_{03} = \psi_{01}\psi_{12}\psi_{23} \tag{34}$$

or

$$\psi_{13} = \psi_{12}\psi_{23} \tag{35}$$

represents the net flux ratio for the two layers that are adjacent to the substrate.

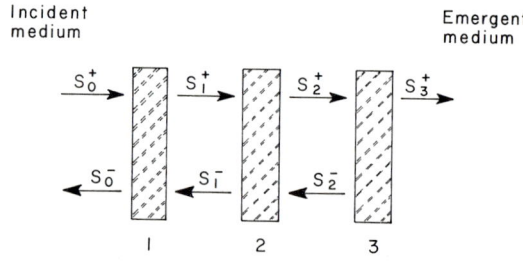

FIG. 3. The incident components (superscript $+$) and reflected components (superscript $-$) of the Poynting vector in a three-layer stack.

5. The Radiant Absorptance

The radiant absorptance A is that fraction of the incident flux that is converted into Joule heat. A is that fraction of the incident flux that remains after the reflected and transmitted components are deducted:

$$A = (S_0^+ - S_0^- - S^+)/S_0^+ = 1 - R - T \tag{36}$$

The absorption in a layer or a subassembly of layers in a stack is computed from

$$A = \frac{\text{net flux incident} - \text{net flux emergent}}{\text{flux incident on the stack}} \tag{37}$$

Using the notation of Fig. 3, the absorption A_1 in layer one is

$$A_1 = [S_0^+ - S_0^- - (S_1^+ - S_1^-)]/S_0^+ \tag{38}$$

and for layer two

$$A_2 = [S_1^+ - S_1^- - (S_2^+ - S_2^-)]/S_0^+ \tag{39}$$

and so on. From the definition of ψ,

$$A_1 = (1 - R)(1 - \psi_{01}) \tag{40}$$

$$A_2 = (1 - R)\psi_{01}(1 - \psi_{12}) \tag{41}$$

$$A_3 = (1 - R)\psi_{01}\psi_{12}(1 - \psi_{23}) \tag{42}$$

Thus the absorptance in the ith layer is proportional to (1) one minus the reflectance of the stack; (2) the product of the ψ's, $\psi_{01}\psi_{12} \cdots \psi_{i-2,i-1}$ of the layers on the incident side; (3) one minus $\psi_{i-1,i}$ of the layer itself.

6. The Net Flux Ratio for a Single Layer

It is necessary to define some terminology before we launch into this section. We define optical constant $\hat{n}$ for the ith layer as

$$\hat{n}_i = n_i - jk_i \tag{43}$$

and the real and imaginary parts of its complex propagation constant as

$$\hat{\beta} = \beta_r - j\beta_j = 2\pi\sigma n_i h_i - j2\pi\sigma k_i h_i \tag{44}$$

where h_i is its metric thickness. In the foregoing equations, n is the refractive index of the layer and the real part of β is its phase of retardation defined in Eq. (18).

Berning and Turner (8) were the first to develop an explicit equation for the ψ of a single layer. As a specific example, suppose the ψ of layer of 1 as shown in Fig. 3 is computed. This procedure (6) is followed:

(a) Use the characteristic matrix method to find the admittance $\hat{Y}$ of the two layers on the emergent side of layer 1. This is discussed in Section II,B.

(b) Compute the amplitude reflection coefficient at the interface between this layer and the stack on its emergent side. From Eq. (8) this is

$$\hat{\rho} = \frac{\hat{n}_1 - \hat{Y}}{\hat{n}_1 - \hat{Y}} = |\rho|e^{j\delta} \tag{45}$$

where $\hat{n}_1$ is the optical constant of layer 1.

(c) The ψ is

$$\psi = \frac{1 - |\rho|^2 - 2(k/n)|\rho|\sin\delta}{\exp(2\beta_j) - |\rho|^2 \exp(-2\beta_j) - 2_n^k|\rho|\sin(\delta - 2\beta_r)} \tag{46}$$

where β_j and β_r are computed from Eq. (43) for layer 1. Costich (14) has provided a derivation of the foregoing equation that is considerably more detailed than in Berning and Turner (8).

In the computation of $|\rho|$ via Eq. (45), we should not be dismayed if this is greater than one. As discussed by Berning (15), this does not violate any laws of physics nor does it mean that we have necessarily made a computational error. The Poynting vector also depends upon the magnitude of the H field, and even if the E field increases, as is implied by $|\rho| > 1$, there is a corresponding decrease in H so that the magnitude of the Poynting vector that emerges is still less than that entering the medium.

7. The Net Flux Ratio for an Assembly of Layers

There is a simple equation (10) for the ψ of an assembly of layers in terms of the elements of its characteristic matrix. This is bounded on its emergent side by either a substrate or stack of admittance $\hat{Y}$. From the definitions of the Poynting vector in Eq. (28) and of ψ, it follows that

$$\psi = x\{\text{Re}[(\hat{a}_1 + \hat{Y}\hat{a}_3)(\hat{a}_2 + \hat{Y}\hat{a}_4)^*]\}^{-1} \tag{47}$$

The following form is more convenient for computational purposes:

$$\psi = x(P_{11}P_{12} + P_{22}P_{21})^{-1} \tag{48}$$

where

$$P_{11} = a_1 + a_3 x - b_3 z$$
$$P_{22} = b_1 + b_3 x + a_3 z$$
$$P_{12} = a_2 + a_4 x - b_4 z \qquad (49)$$
$$P_{21} = b_2 + b_4 x + a_4 z$$

The foregoing can be expressed as a rational polynomial

$$\psi = x[D_{10} + D'_{10}x + D_{11}z + D_{12}(x^2 + z^2)]^{-1} \qquad (50)$$

where x and z are defined in Eq. (15). The coefficients are expressed in terms of the matrix elements in Table I. In the special case of a nonabsorbing stack, Eq. (21) is valid and ψ is identically 1. This also follows from the conservation of energy.

TABLE I

The Coefficients That Appear in Eq. (50) in Terms of the Matrix Elements Defined in Eq. (20)

Coefficient	Term
D_{10}	$a_1 a_2 + b_1 b_2$
D'_{10}	$a_2 a_3 + a_1 a_4 + b_1 b_4 + b_2 b_3$
D_{11}	$-a_2 b_3 - a_1 b_4 + a_4 b_1 + a_3 b_2$
D_{12}	$a_3 a_4 + b_3 b_4$

We consider the case where ψ is constant and x and z are variables. Equation (50) represents a circle in the admittance plane. The coordinates of the center of a circle are

$$x_c = -D_2/(2D_{12})$$
$$z_c = -D_{11}/(2D_{12}) \qquad (51)$$

where

$$D_2 = D'_{10} - \psi^{-1} \qquad (52)$$

and its radius is

$$r = \left[\left(\frac{D_2}{2D_{12}}\right)^2 + \left(\frac{D_{11}}{2D_{12}}\right)^2 - \frac{D_{10}}{D_{12}}\right]^{1/2} \qquad (53)$$

When plotted for various parametric values of ψ, the circles lie along a constant ordinate in the Y plane and are nested as shown schematically in Fig. 4. These are not computed for a particular absorbing film or multilayer because it is preferable to plot the circles directly in the amplitude reflection plane, as described below.

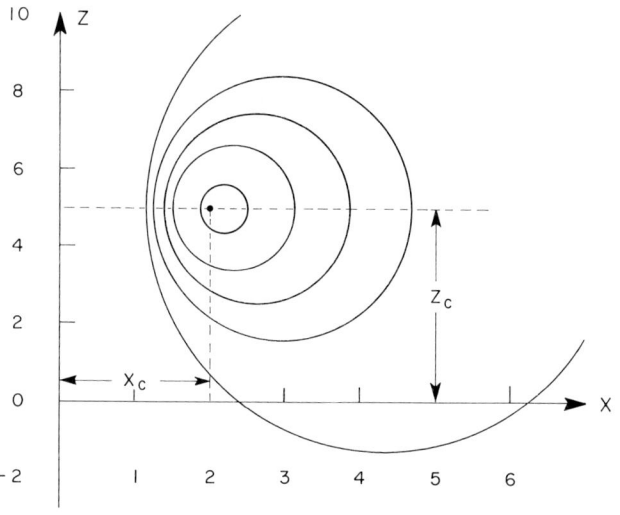

FIG. 4. Circles of constant net flux ratio ψ as a function of the real part x and imaginary part z of the admittance on its emergent side. x_c and z_c are the coordinates of ψ_{max}.

The admittance is transformed into the amplitude reflection plane (with $n_0 = 1$) via Eq. (9). This may be regarded as a conformal transformation that maps points in the complex y plane into the ρ plane. Equation (9) is a bilinear transformation that maps circles into circles (16) and is the basis for the Smith chart (13). The transformation of Eq. (50) into the ρ plane is accomplished via the substitutions

$$x = (1 - u^2 - w^2)/D \tag{54}$$

$$z = -2w/D \tag{55}$$

where

$$D = (1 + u)^2 + w^2 \tag{56}$$

This equation becomes

$$\Psi = (1 - \rho^2)[e_2\rho^2 + e_{10}u + e_{01}w + e_0]^{-1} \tag{57}$$

where the coefficients e_2, e_{10}, e_{01}, and e_0 are given in Table II, along with the identity

$$|\rho|^2 = u^2 + w^2 \tag{58}$$

The centers of the circles are

$$u_c = -\tfrac{1}{2} e_{10}/f_2 \tag{59}$$

$$w_c = -\tfrac{1}{2} e_{01}/f_2 \tag{60}$$

and have a radius

$$r = \left[\frac{f_1}{f_2} - \frac{1}{2}\left(\frac{e_{10}}{f_2}\right)^2 - \frac{1}{2}\left(\frac{e_{01}}{f_2}\right)^2 \right]^{1/2}$$

where

$$f_1 = e_0 - \psi^{-1} \tag{61}$$

and

$$f_2 = e_2 - \psi^{-1} \tag{62}$$

Since both u_c and w_c depend on f_2, which is a function of ψ, the centers lie on a radial line in the ρ plane.

TABLE II

The Coefficients That Appear in Eq. (57) in Terms of the Matrix Elements in Eq. (20)

Coefficient	Term
e_2	$a_2 a_1 - a_4 a_1 - a_3 a_2 + b_2 b_1 - b_4 b_1 - b_3 b_2 + a_4 a_3 + b_3 b_4$
e_{01}	$2(a_2 a_1 + b_2 b_1 - a_4 a_3 - b_3 b_4)$
e_{10}	$2(b_4 a_1 + b_3 a_2 - a_4 b_1 - a_3 b_2)$
e_0	$a_2 a_1 + a_4 a_1 + a_3 a_2 + b_2 b_1 + b_4 b_1 + b_3 b_2 + a_4 a_3 + b_3 b_4$

As an example, Fig. 5 shows the circles of constant ψ in the ρ plane for a single layer of palladium at $\lambda = 546$ nm. The radial line on which the centers lie is inclined at an angle of 193°. The radii of the circles decrease monotonically as ψ increases from zero. This implies that ψ attains a maximum value, which is shown as a dot in Fig. 5. The significance of this maximum value of ψ is discussed in the next section. These circles of constant ψ depend on the matrix elements and will change if the wavelength, film thickness, or optical constants are altered.

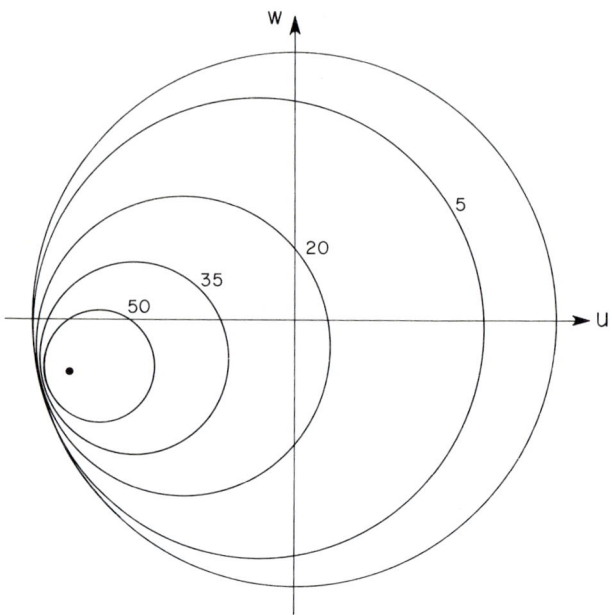

Fig. 5. Circles of constant net flux ratio ψ as a function of the amplitude reflection coefficient of the medium on its emergent side. The film is 0.037 wavelength thick and has an optical constant $\hat{n} = 2.3 - j2.7$.

8. The Maximum Value of the Net Flux Ratio

The equations for ψ show that it is a single-valued function of either x, z or u, w in the admittance or reflectance plane, respectively. There are several ways its maximum value can be found. One way is to find the coordinates x_c, z_c in the limit as the radius in Eq. (53) goes to zero. It is also possible to solve the simultaneous equations resulting from $\partial \psi / \partial x = 0$ and $\partial \psi / \partial z = 0$. This shows (10) that ψ attains a maximum value of

$$\psi_{\max} = (D'_{10} + 2F)^{-1} \tag{63}$$

where

$$F = (D_{12}D_{10} - 0.25 D_{11}^{2})^{1/2} \tag{64}$$

and the D's are defined in Table I. The foregoing equations can be used to find the $\psi_{\max}$ of either a single layer or any multilayer that contains one or more absorbing films. In the special case of a single layer, this reduces to Eq (8)

$$\psi_{max} = \mu - (\mu^2 - 1)^{1/2} \tag{65}$$

where

$$\mu = \frac{\cosh(2\beta_j) + (k^2/n^2)\cos(2\beta_r)}{1 + (k^2/n^2)} \tag{66}$$

The admittance Y_{max} at which ψ is a maximum is

$$x_{max} = FD_{12}^{-1} \tag{67}$$

and

$$z_{max} = -0.5 D_{11} D_{12}^{-1} \tag{68}$$

The coordinates in the ρ plane u_{max} and w_{max} are found by substituting the foregoing equations into Eq. (8).

The fact that ψ attains a maximum value greatly facilitates the design of bandpass filters. The process of designing a stack that has the admittance in Eqs. (67) and (68) is called *matching*. This is discussed in Section V,2.

9. A Recapitulation of the Properties of the Net Flux Ratio

It is useful to summarize the properties of ψ as theorems so that we can refer to them in other sections. In most cases the proof is in Section II,7.

Theorem I: ψ is multiplicative for layers or any subassemblies of multi-layers that are cascaded.

$$\psi_l = \prod_{i=1}^{l} \psi_i \tag{69}$$

Theorem II: ψ is 1 for a nonabsorbing layer. From Theorem I it follows that a stack of nonabsorbing films has a ψ of unity.

Theorem III: ψ depends on only the characteristic matrix elements of the stack and the admittance of the emergent medium. In other words, ψ is independent of the refractive index of the incident medium.

Corollary: The ratio A/T for a stack depends only on its matrix elements and the admittance of the emergent medium. From the definition of A,

$$\psi = 1/(1 + A/T) \tag{70}$$

Theorem IV: If an absorbing stack has a net flux ratio of ψ_{12}, then its transmittance is ψ_{12}, provided a subassembly of nonabsorbing layers (to antireflect it) is added to its incident side.

Proof: From Theorem III, ψ_{12} is unaltered by the addition of the subassembly because this does not affect the admittance on the emergent side of the absorbing stack. Since the ψ_{01} of the subassembly is 1 (Theorem II), it follows from Eqs. (69) and (32) that $T = \psi_{12}$.

Theorem V: $\psi_{\max}$ is the same for a stack of absorbing films [represented by $\mathscr{M}$ in Eq. (20)] in which the layers are ordered 1, 2, 3, ... l, and a stack $\mathscr{M}'$ in which the order of the layers is reversed: $l, l-1, \ldots 3, 2, 1$.

Proof: The elements of the matrix $\mathscr{M}$ are identical to those of $\mathscr{M}'$, with the exception that the diagonal elements are exchanged, i.e., $a_1 = a'_4$, $a_4 = a'_1$, $b_1 = b'_4$, $b_4 = b'_1$. Equation (50) is unaltered by such an exchange of matrix elements.

Theorem VI: The admittance of a symmetrical absorbing stack is $\hat{Y}^*_{\max}$ provided the admittance on its emergent side is $\hat{Y}_{\max}$. The asterisk denotes a complex conjugate and $\hat{Y}_{\max}$ is obtained from Eqs. (67) and (68). In other words, a symmetrical multilayer acts like an admittance transformer that converts $\hat{Y}_{\max}$ on its emergent side into $\hat{Y}^*_{\max}$.

Proof: Substitution of $\hat{Y}^*$ on the left of Eq. (22), with $\hat{a}_1 = \hat{a}_4$, yields two equations, since $\hat{Y}$ is a complex variable. The solution to these equations satisfies Eqs. (67) and (68).

10. Fabry–Perot Type Filters

The concept of the Fabry–Perot (FP) interferometer is useful in the design of single-cavity filters of the type MDM' where M and M' are metal layers and D is the dielectric spacer of metric thickness h_t and refractive index n_t. The radiant transmittance of such a filter is derived in standard texts, such as Stone (*17*). It is helpful to refer to a specific example in writing down the equations. Figure 6 shows an FP type filter that is asymmetrical for two reasons: first, the thicknesses of the metal layers M_1 and M_2 are different. Even if they were the same, the filter is asymmetrical because the metal film M_2 is bounded by the glass substrate and the other film is contiguous to the incident medium of air. The transmittance of the FP is

$$T = T_{\max}(1 + F \sin^2 \eta)^{-1} \tag{71}$$

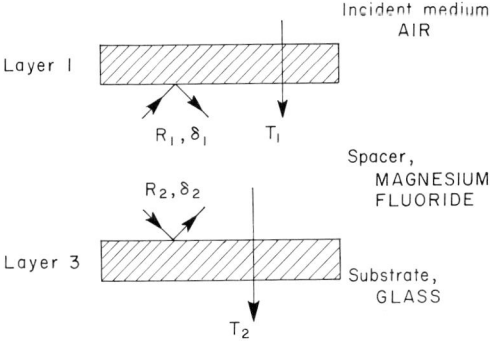

FIG. 6. The nomenclature that is used to describe the internal reflections (R_1 and R_2), phase shifts upon reflection (δ_1 and δ_2), and transmittances (T_1 and T_2) of a Fabry–Perot type filter.

where

$$T_{\max} = T_1 T_2 (1 - R_m)^{-2} \tag{72}$$

$$F = 4R_m(1 - R_m)^{-2} \tag{73}$$

$$\eta = 2\pi\sigma n_t h_t + \delta_m \tag{74}$$

and where R_m is the geometric mean of the reflectances

$$R_m = (R_1 R_2)^{1/2} \tag{75}$$

and δ_m is the arithmetic average of the phase shift upon reflection

$$\delta_m = \tfrac{1}{2}(\delta_1 + \delta_2) \tag{76}$$

Figure 6 illustrates how the reflectance, phase shift, etc., are measured. In this example, we assume that the spacer is magnesium fluoride, $n_t = 1.38$. T_1 is the transmittance from air through the film M_1 into the spacer. In other words, it is the same transmittance that would be measured if the film M_1 were deposited on a substrate of index 1.38. T_2 is the transmittance of M_2 from the spacer into the substrate, that is, from an incident medium of index 1.38 through the metal into the substrate of index 1.517. R_1 and δ_1 are, respectively, the radiant reflectance and phase shift upon reflection that are measured from the spacer as an incident medium and air as the emergent medium. R_2 and δ_2 are, respectively, the radiant reflectance and phase shift upon reflection when the film M_2 is viewed from an incident medium of index 1.38.

The usual practice in bandpass filter design is to choose the film thicknesses M_1 and M_2 as well as n_t and then to determine h_t such that T is a maximum at a particular wavelength. Equation (71) is a maximum when

$$\eta = m\pi \tag{77}$$

where the integer m is the order of interference. The equation that is obtained when Eq. (73) is solved for h_t can take various forms depending on the sign convention that is used for the phase shift upon reflection, which is defined in Eq. (11). The convention used here is that δ lies in the first or second quadrants in the ρ plane. This means that $\delta < 180°$ and therefore the optical thickness of the first-order spacer is less than if quarter-wave stacks of dielectric films were used in lieu of the metal films, in which case $\delta_m = 180°$. The wavenumbers σ_m of the transmission bands of the filters are given by

$$2\pi\sigma_m n_t h_t + [\pi - \delta_m(\sigma)] = m\pi \tag{78}$$

Writing $\delta_m(\sigma)$ emphasizes the fact that the phase shift is not constant but varies with frequency. This variation is principally due to the dispersion of the optical constants of the metal in the MDM-type filter. This is discussed in more detail in Section IV,1.

Standard texts (*17*) show that the resolution is

$$Q = \tfrac{1}{2}m\pi F^{1/2} \tag{79}$$

where F is defined in Eq. (73) and m is the order. The resolution can be increased by either using mirrors with a higher reflectance or by incorporating a higher order spacer. This is discussed further in Section IV,3.

11. The Transmittance of a Symmetrical Multilayer

The concepts of the FP filter and ψ developed in the foregoing section are design tools that are adequate to concoct most multiple-cavity metal–dielectric interference filters. There is one simplification that arises when the filter is symmetrical. It supplies some useful insight into how its spectral transmittance is influenced by the admittance of the emergent medium.

As shown in Fig. 7, we consider the special case where a stack of absorbing layers is surrounded by identical dielectric multilayers. The former is called the absorbing stack and is symmetrical, as for example the designs, a, aba, or abcba, and so on. The properties of the dielectric multilayers are repre-

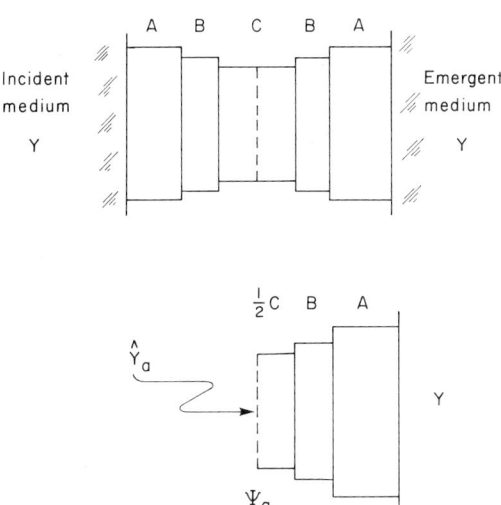

FIG. 7. The center layer c of the filter abcba (upper) is bisected to form the half-filter (lower).

sented by an admittance Y. Thus we can envisage the system as the absorbing stack surrounded on both sides by media of admittance Y, as shown in Fig. 7.

The next step (8) is to bisect the absorbing stack so that the system consists of half-filters of the design $\frac{1}{2}$a, $\frac{1}{2}$ba, $\frac{1}{2}$cba, etc., that are adjacent to an emergent medium of admittance Y, as depicted in Fig. 7. This system can be regarded as a spacerless FP filter. If the admittance at the face of the half-filter stack is $\hat{Y}_a = x_a + jz_a$ and its net flux ratio is ψ_a, then the transmittance of the symmetrical multilayer (8) is

$$T = \psi_a{}^2(1 + z_a{}^2 x_a{}^{-2})^{-1} \qquad (80)$$

This shows that for a given ψ_a, the maximum transmittance occurs when the admittance of the half-filter is pure real. Both Y_a and ψ_a depend on the matrix elements of the half-filter as well as the admittance Y of the emergent medium. The absorbing stack a, aba, etc. is symmetrical and hence the diagonal elements of its characteristic matrix are equal. In order to avoid confusion, we use a special set of symbols to describe the matrix of this stack:

$$\begin{bmatrix} g_1 + jh_1 & g_3 + jh_3 \\ g_2 + jh_2 & g_1 + jh_1 \end{bmatrix} \qquad (81)$$

It is possible to express Eq. (77) in terms of the Y. After Y is transformed into the ρ plane via Eq. (8), the resulting equation for T is a rational polynomial that is a quartic in both u and w:

$$T = [1 - (u^2 + w^2)]^2 P^{-1} \tag{82}$$

where

$$P = Q_0 + Q_{10}u + Q_{01}w + Q_{11}uw + Q_{20}u^2 + Q_{02}w^2 + Q_{12}uw^2$$
$$+ Q_{21}u^2w + Q_{22}u^2w^2 + Q_{30}u^3 + Q_{03}w^3 + Q_{40}u^4 + Q_{04}w^4 \tag{83}$$

The coefficients Q_0, etc. are expressed in terms of the matrix elements in Table III.

TABLE III

THE COEFFICIENTS OF THE POLYNOMIAL IN Eq. (83) IN TERMS OF THE ELEMENTS OF THE MATRIX IN Eq. (81)

$Q_0 = g_1(g_1 + g_2 + g_3) + \tfrac{1}{2}g_2g_3 + \tfrac{1}{4}(g_2{}^2 + g_3{}^2) + h_1(h_1 + h_2 + h_3) + \tfrac{1}{2}h_2h_3 + \tfrac{1}{4}(h_2{}^2 + h_3{}^2)$

$Q_{01} = 2[g_1(-h_2 + h_3) + g_2(h_1 + h_3) - g_3(+h_1 + h_2)]$

$Q_{02} = 2g_1{}^2 - 3g_2g_3 + \tfrac{1}{2}(g_2{}^2 + g_3{}^2) + 2h_1{}^2 - 3h_2h_3 + \tfrac{1}{2}(h_2{}^2 + h_3{}^2)$

$Q_{03} = 2[g_1(-h_2 + h_3) + g_2(h_1 - h_3) + g_3(-h_1 + h_2)]$

$Q_{04} = g_1(g_1 - g_2 - g_3) + \tfrac{1}{2}g_2g_3 + \tfrac{1}{4}(g_2{}^2 + g_3{}^2) + h_1(h_1 - h_2 - h_3) + \tfrac{1}{2}h_2h_3 + \tfrac{1}{4}(h_2{}^2 + h_3{}^2)$

$Q_{10} = 2g_1(g_2 - g_3) + g_2{}^2 - g_3{}^2 + 2h_1(h_2 - h_3) + h_2{}^2 - h_3{}^2$

$Q_{11} = 4[-g_1(h_2 + h_3) + h_1(g_2 + g_3)]$

$Q_{12} = 2g_1(-g_2 + g_3) + g_2{}^2 - g_3{}^2 + 2h_1(-h_2 + h_3) + h_2{}^2 - h_3{}^2$

$Q_{20} = -2g_1{}^2 + g_2(1.5g_2 - g_3) + 1.5g_3{}^2 - 2h_1{}^2 + h_2(1.5h_2 - h_3) + 1.5h_3{}^2$

$Q_{21} = 2[g_1(-h_2 + h_3) + g_2(h_1 - h_3) + g_3(-h_1 + h_2)]$

$Q_{22} = 2g_1(-g_2 - g_3 + g_1) + g_2(g_3 + \tfrac{1}{2}g_2) + \tfrac{1}{2}g_3{}^2 + 2h_1(-h_2 - h_3 + h_1)$
$\quad + h_2(h_3 + \tfrac{1}{2}h_2) + \tfrac{1}{2}h_3{}^2$

$Q_{30} = g_2(-2g_1 + g_2) + g_3(2g_1 - g_3) + h_2(-2h_1 + h_2) + h_3(2h_1 - h_3)$

$Q_{40} = g_1(-g_2 - g_3 + g_1) + g_2(\tfrac{1}{2}g_3 + \tfrac{1}{4}g_2) + \tfrac{1}{4}g_3{}^2 + h_1(-h_2 - h_3 + h_1)$
$\quad + h_2(\tfrac{1}{2}h_3 + \tfrac{1}{4}h_2) + \tfrac{1}{4}h_3{}^2$

As an example, Fig. 8 shows the isotransmittance contours for a silver film 0.05 wavelength in thickness. The point of maximum transmittance lies in the third quadrant in the ρ plane and is located at the coordinates of $\psi_{\max}$. Several examples in Sections V and VI illustrate how such data are used in filter design.

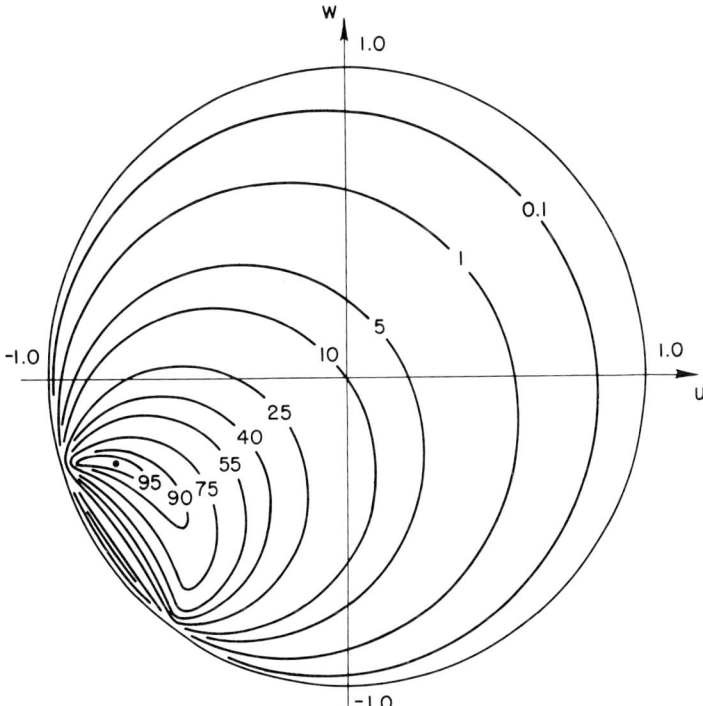

FIG. 8. Contours of isotransmittance of a silver film as a function of the amplitude reflection coefficient of the medium on its emergent side. It is 0.05 wavelength thick and has an optical constant $0.06 - j3.75$.

III. General Considerations in Bandpass Filter Design

1. Summary of the Attributes of Bandpass Filters

It is useful to establish some terminology to describe the attributes of a bandpass filter, as for example the spectral transmittance curve shown in Fig. 9. Its attributes are as follows: (1) The maximum transmittance T_{max} in its passband. (2) The wavelength λ_0 at which T_{max} occurs. (3) The spectral width of the passband, which is expressed in terms of its bandwidth $\Delta\lambda_{1/2}$ at $0.5 T_{max}$ or its deciwidth $\Delta\lambda_{1/10}$ at $0.1 T_{max}$. (4) The resolution

$$Q = \lambda_0/\Delta\lambda_{1/2} \qquad (84)$$

(5) The offband rejection, which is nebulously defined as the attenuation outside the passband. This is shown as T_0 in Fig. 9. (6) The contrast, which

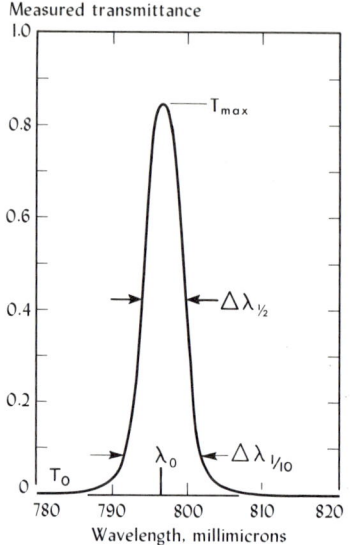

FIG. 9. The spectral transmittance of a bandpass filter, showing its maximum transmittance T_{max}, bandwidth $\Delta\lambda_{1/2}$, deciwidth $\Delta\lambda_{1/10}$, and offband transmittance T_0.

is the ratio

$$C = T_{max}/T_0 \qquad (85)$$

T_0 is defined more explicitly below. (7) The angle shift, which is the change of λ_0 as the angle of the incidence flux changes. (8) The angular field, which is related to the angle shift. The passband broadens when the filter is illuminated with convergent flux. The angular field is a measure of the maximum permissible cone angle of this flux. (9) The shift of λ_0 with temperature. (10) The order of interference m, which is the number of half-wavelengths in the resonant cavity.

In some instances the absorptance of a transmission bandpass filter is important because it causes it to overheat. We should also remember that A is usually different from the two sides of a filter.

For example, the characteristics of a filter in terms of its λ_0, T_{max}, Q, and so on, may be excellent and the filter could function well when its transmittance is measured in the low-flux environment of a spectrophotometer. However, suppose that this filter is then placed in an intense beam of flux that extends over a broad spectral region. Its absorptance can cause it to overheat and perhaps even suffer *auto-da-fé*. The user has no recourse but to add multilayer dielectric heat reflectors to mitigate the absorption.

2. The Choice of Metals for Bandpass Filters

Having developed an elaborate design theory and established the terminology, we finally pose the question, Why use metal films in the design of bandpass filters? Most designers will agree on the answer: Metal layers offer the most effective way of obtaining a substantial offband rejection over a broad spectral region.

Filters that attenuate over a wide spectral region are termed "metal blockers" in the trade. In most instances there is little point in producing this offband rejection at the expense of a low peak transmittance in the passband. This trade-off between rejection and passband transmittance is determined by the filter design and the types of metal films that are used.

The concept of ψ_{max} is invaluable in choosing both the type of metal and its thickness in designing a transmission bandpass filter because ψ_{max} is related to its T_{max} at λ_0. Once this latter parameter is chosen, the ψ_{max} of a film depends only on its thickness and optical constants. We should remember that the optical constants of a metal layer depend on its thickness and this is why the n and k values for various metals cannot be applied to films that are too thin. A rule of thumb is that layers that are thinner than 10 nm consist of aggregates of particles rather than a continuous film. The optical constants of the aggregated film are considerably different from those of the bulk material. Like all rules of thumb, the exceptions and conditions under which it is inapplicable are legion. For example, a silver film that is deposited at temperature greater than 150°C shows an aggregated structure even when its thickness is substantial. Its n and k differ considerably from those of films deposited at lower temperatures.

Although it is valuable, ψ_{max} is not the exclusive criterion for choosing a metal film for a bandpass filter. The *reductio ad absurdum* would be to use a dielectric film with a k of 0.01. Here the ψ_{max} is quite close to unity but the problem is that its offband rejection is poor. The offband rejection can be quantified by computing the transmittance T_0 of a film with a medium of unit admittance on its incident and emergent sides. In other words, T_0 is the transmittance of an unbacked film that is surrounded with air. Starting with this definition of T_0, the contrast as defined in Eq. (85) can be used as a merit function in choosing metal films for bandpass filters. As an illustration, consider the weakly absorbing dielectric film cited above. Suppose that its optical thickness is 0.25 wavelength and its optical constant is $\hat{n} = 2.0 - j0.01$. Its ψ_{max} is virtually 1.0 and its T_0 is 6.4% which produces a contrast of $(0.64)^{-1} = 1.56$. This is quite poor. As Berning and Turner (8) observed, materials with a large k/n ratio are the likely candidates for MDM filters.

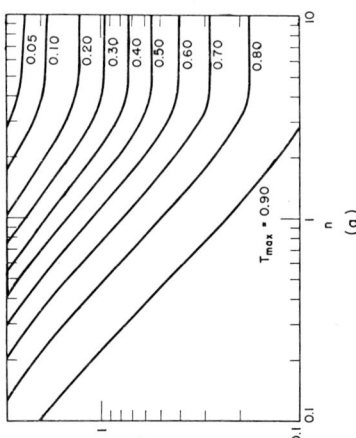
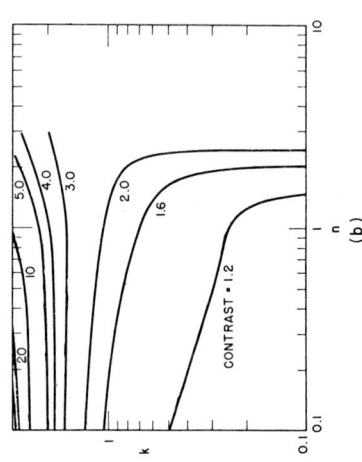

FIG. 11. (a) T_{max} and (b) contrast contours on the complex $\tilde{n}$ plane for $h = 0.10\lambda$.

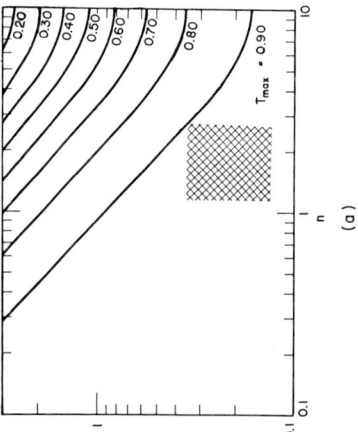
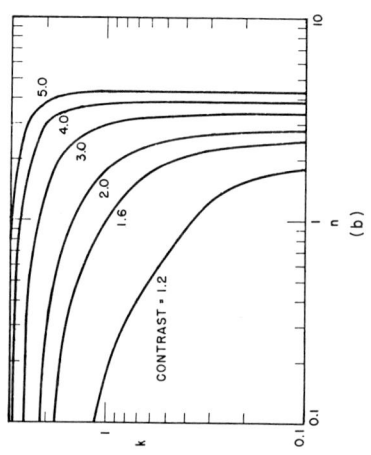

FIG. 10. (a) T_{max} and (b) contrast contours on the complex $\tilde{n}$ plane for $h = 0.05\lambda$.

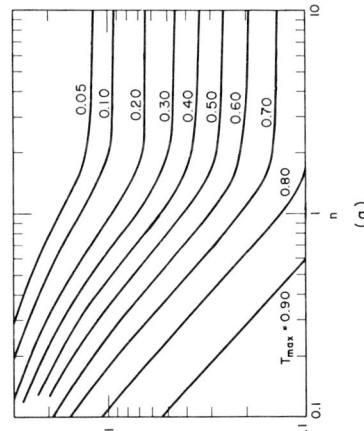
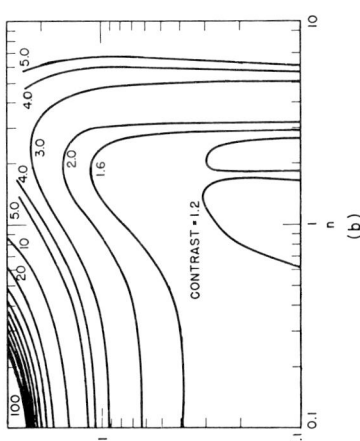

FIG. 13. (a) T_{max} and (b) contrast contours on the complex $\tilde{n}$ plane for $h = 0.20\lambda$.

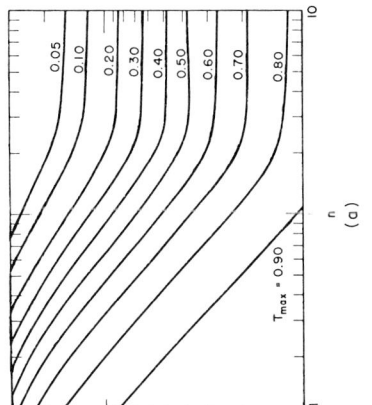
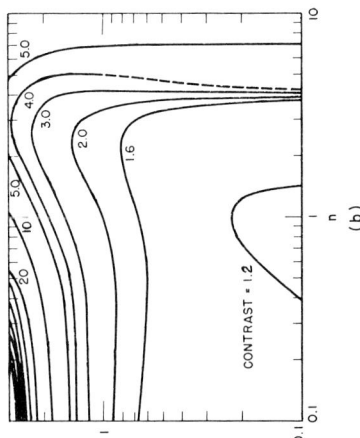

FIG. 12. (a) T_{max} and (b) contrast contours on the complex $\tilde{n}$ plane for $h = 0.15\lambda$.

There are several ways (14) in which the contrast and ψ_{max} of single-metal layers can be presented. Figures 10, 11, 12, and 13 show curves of ψ_{max} and contrast versus n and k for thickness/wavelength ratios of 0.05, 0.1, 0.15, and 0.20, respectively. Weakly absorbing films lie in the cross-hatched region in Fig. 10. As mentioned previously, they have a ψ_{max} that is close to one but has poor contrast. Figures 14, 15, 16, and 17 plot ψ_{max} and contrast versus contours of n and k for the same thickness ratios cited above.

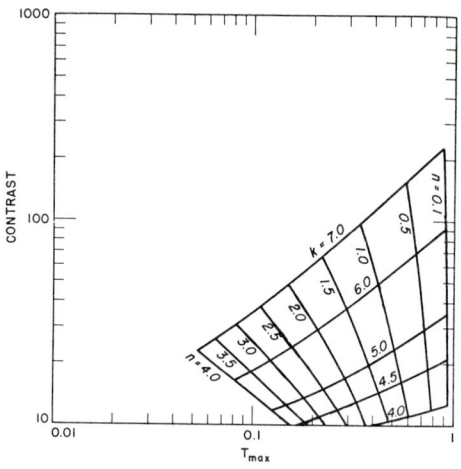

FIG. 14. Complex $\hat{n}$ mesh on contrast versus T_{max} coordinates for $h = 0.05\lambda$.

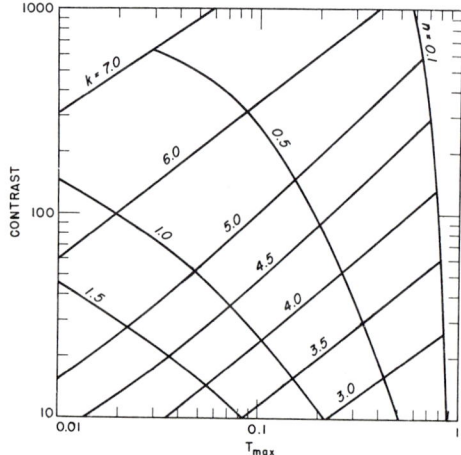

FIG. 15. Complex $\hat{n}$ mesh on contrast versus T_{max} coordinates for $h = 0.10\lambda$.

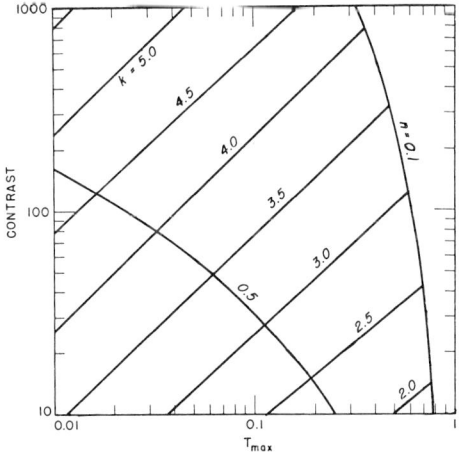

FIG. 16. Complex $\hat{n}$ mesh on contrast versus T_{max} coordinates for $h = 0.15\lambda$.

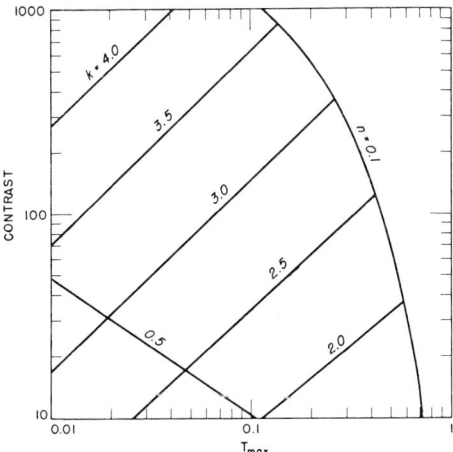

FIG. 17. Complex $\hat{n}$ mesh on contrast versus T_{max} coordinates for $h = 0.20\lambda$.

As an example, we consider whether or not a chromium film ($n \cong 3$, $k \cong 5$) would be suitable to use in a filter at a wavelength of 500 nm. We choose a metric thickness of 25 nm which means that h/λ is 0.05. Figure 14 shows that the ψ_{max} is approximately 8% and the contrast is 12%. This is not a likely candidate. Consider a silver film ($n \cong 0.1$, $k \cong 2$) and $h/\lambda = 0.2$. Figure 13 shows that its ψ_{max} is close to 60% and its contrast is more than 30.

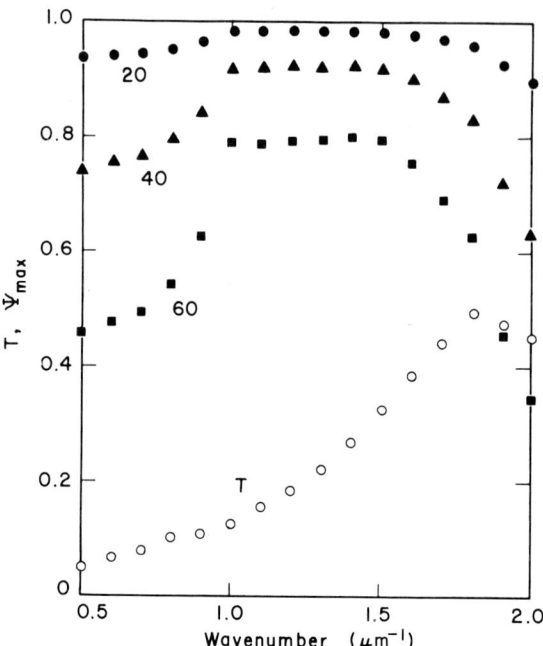

FIG. 18. The computed ψ_{max} for a gold film versus wavelength for a thickness of 20 nm, 40 nm, and 60 nm. The lower open circles are the transmittance of a film 20 nm thick surrounded on both sides by a medium of unit refractive index. The optical constants from Hass and Hadley (18) are used.

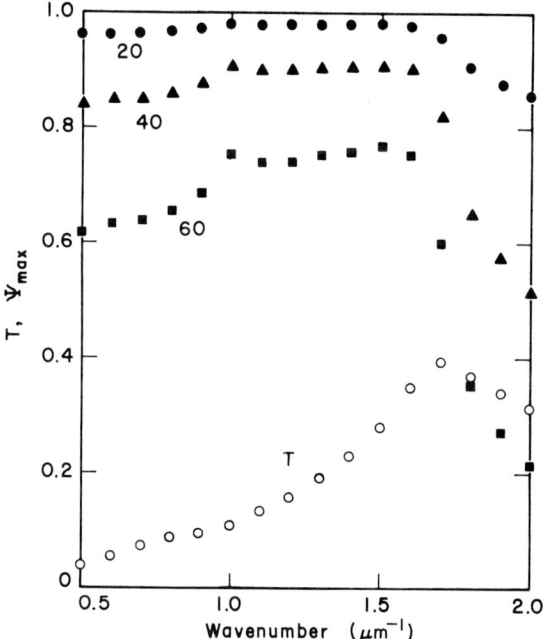

FIG. 19. Same as Fig. 18 except that the film is copper.

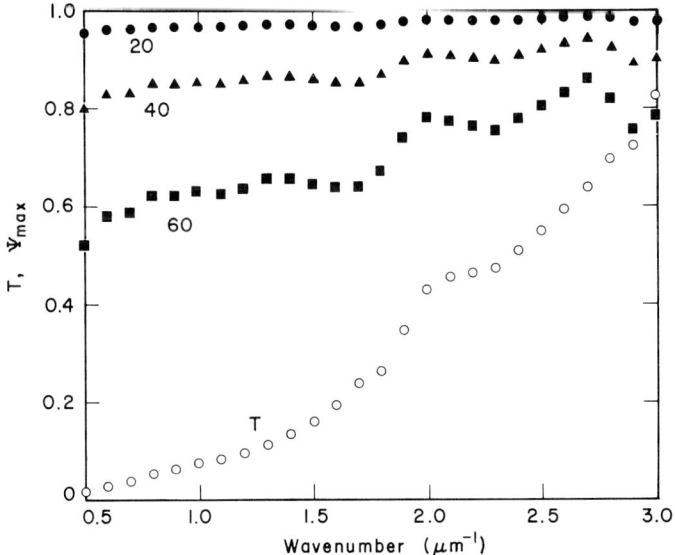

FIG. 20. Same as Fig. 18 except that the film is silver.

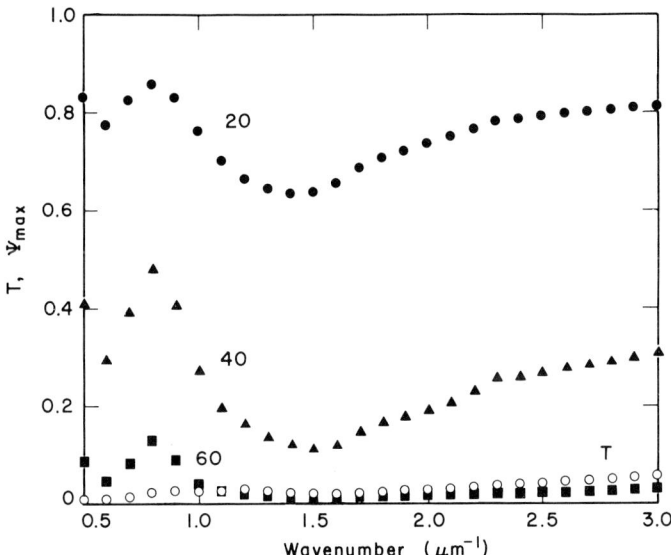

FIG. 21. Same as Fig. 18 except that the film is aluminum.

Another useful design tool (*10*) is to compute the ψ_{max} of metal films of various thicknesses using published data (*18*) for the optical constants. If the ψ_{max} of a given material is too small, then there is little motivation to investigate other attributes such as the contrast.

Figures 18 and 19 show the ψ_{max} versus wavenumber for films of gold and copper. The abscissa extends from 0.5 μm^{-1} ($\lambda = 2.0$ μm) to 2.0 um^{-1} ($\lambda = 0.5$ μm). Both of these materials show a substantial ψ_{max} in the red and near infrared and hence are suitable candidates for bandpass filters in these spectral regions. Figures 20 and 21 show the ψ_{max} of silver and aluminum in the spectral region from 0.5 μm^{-1} ($\lambda = 2.0$ μm) to 3.0 μm^{-1} ($\lambda = 333$ nm). The fact that silver has a substantial ψ_{max} and high reflectance (which is not shown in these graphs) makes it useful as a coating for Etalon plates in the red and near infrared portions of the spectrum. The reflectance of aluminum increases in the blue and near uv spectral regions, and it is the best material to use in interference filters in these spectral regions. Gold, copper, and silver are suitable in the infrared out to the 4-μm region.

IV. Single-Cavity Filter Design

The simplest type of bandpass filter is the MDM type of Fabry–Perot filter. Subsequent sections show how the D spacer layer can be replaced with a subassembly of nonabsorbing films.

1. The Design of an MDM Filter

The design procedure for the MDM is illustrated by choosing the thicknesses of the layers in bandpass filters that transmit at 400 and 700 nm, respectively. The equations are outlined in Section II,10 and the symbols R_1, T_1, and ψ_a are used here with the same meaning.

The first step is to select the metal layers and their thicknesses. Silver is commonly used for filters in the visible and near infrared portions of the spectrum. The thickness of the metal layers is influenced by the trade-off between the offband rejection and T_{max}. The thicker layer gives better offband rejection but a lower T_{max}. Thicknesses of 25 and 35 nm are chosen in this illustrative example.

The next step is to decide whether the coating is asymmetrical, as shown in Fig. 1, or symmetrical. In the latter case both silver layers are bounded by the same emergent medium, when viewed from the spacer layer. This is accomplished by cementing a cover glass on the filter after it has been deposited.

The initial design is made for order number $m = 1$. Section IV,3 discusses the consequences of increasing the order number.

The next step is to choose the refractive index of the spacer layer. A low index material is usually used for reasons that are discussed below. However, for the sake of pedagogy, we design separate filters with low index and high index spacers.

The phase shift δ_m that appears in Eq. (78) is computed and thence the optical thickness of the spacer layer $n_t h_t$ is found. The designer can verify that he has chosen the correct spacer layer thickness by computing the admittance of the half-filter $\frac{1}{2}$DM glass. According to Eq. (79), this admittance should be pure real at λ_0 if $T = T_{\max}$.

The design is now complete and Table IV shows some attributes of both the M layers and the complete filter. The δ_m, R_1, T_1 and ψ_a are computed for a single silver layer with the spacer material as an incident medium and glass as the substrate. The parameters $T_{\max}$, A, R, and ψ are computed for the entire filter. The Q is computed from Eqs. (73) and (79) using the R_1 in the table. The Q of the actual filter is also influenced by the dispersion of the optical constants of the silver, particularly in the blue part of the spectrum. Table V lists the optical constants that are used in the computations.

TABLE IV

COMPUTED ATTRIBUTES OF AN MDM FILTER THAT IS SURROUNDED ON BOTH SIDES BY AN INDEX OF 1.517[a]

Parameter	Design wavelength							
	$\lambda_0 = 400$ nm				$\lambda_0 = 700$ nm			
Metric thickness of silver film	25 nm		35 nm		25 nm		35 nm	
n_t	1.38	2.30	1.38	2.30	1.38	2.30	1.38	2.30
δ_m	108°	76°	109°	80°	142°	121°	146°	125°
$h_t n_t$ (nm)	120	84.4	121	88.4	276	234	284	243
R_1	0.43	0.42	0.62	0.61	0.80	0.73	0.89	0.85
Q	3.6	3.5	6.5	6.2	14	10	27	19
T_1	0.51	0.52	0.31	0.32	0.15	0.20	0.061	0.086
ψ_a	0.89	0.89	0.82	0.82	0.73	0.73	0.56	0.56
A	0.19	0.19	0.31	0.30	0.40	0.40	0.48	0.51
R	0.01	0.01	0.02	0.03	0.07	0.07	0.21	0.18
$T_{\max} = \psi_a^2$	0.80	0.80	0.67	0.67	0.53	0.53	0.31	0.31
ψ	0.81	0.80	0.68	0.69	0.57	0.57	0.39	0.38

[a] The optical constant of the silver is listed in Table V on p. 104.

TABLE V

THE OPTICAL CONSTANTS VERSUS WAVELENGTH THAT ARE USED IN THE COMPUTATIONS

Wavelength (μm)	n	k	Wavelength (μm)	n	k
Aluminum			Silver		
0.2000	0.110	2.20	0.6200	0.177	3.99
0.2200	0.140	2.35	0.6400	0.176	4.20
0.2400	0.160	2.60	0.6600	0.175	4.40
0.2600	0.190	2.85	0.6800	0.173	4.58
0.2800	0.220	3.13	0.7000	0.172	4.76
0.3000	0.250	3.33	0.7200	0.171	4.92
0.3200	0.280	3.56	0.7400	0.170	5.08
0.3400	0.310	3.80	0.7600	0.174	5.24
0.3600	0.340	4.01	0.7800	0.184	5.42
0.3800	0.380	4.32	0.8000	0.193	5.58
0.4000	0.400	4.45	0.8200	0.201	5.73
0.4200	0.440	4.67	0.8400	0.209	5.88
0.4400	0.481	4.88	0.8600	0.217	6.02
0.4600	0.543	5.13	0.8800	0.224	6.15
0.4800	0.605	5.37	0.9000	0.231	6.28
0.5000	0.669	5.56	0.9200	0.238	6.41
Silver			0.9400	0.244	6.52
0.5000	0.110	2.64	0.9600	0.250	6.64
0.5200	0.119	2.89	0.9800	0.256	6.74
0.5400	0.135	3.20	1.0000	0.262	6.85
0.5600	0.158	3.48	1.2000	0.335	8.14
0.5800	0.180	3.68	1.4000	0.417	9.64
0.6000	0.179	3.76	1.6000	0.495	11.06

The data in Table IV show the effect of the refractive index of the spacer. The first is that the phase shift upon reflection is smaller for the high index material and hence it is thinner, particularly in the blue. Equation (80) shows that $T_{\max} = \psi_a^2$. Since ψ_a is independent of the spacer index, $T_{\max}$ is also independent. The resolution of the filter with the high index spacer is less than that of the low index spacer, which in turn can be attributed to the influence of the spacer index on R_1. For example, Table IV shows that at $\lambda_0 = 700$ nm the resolutions of the filters with high index and low index spacer layers are 10 and 14, respectively. Aside from the degradation of the bandwidth, another disadvantage of the high index spacer is that it is absorbing in the blue part of the spectrum. However, a redeeming feature of the filter with high index spacer is that its angle shift is less than the one that contains the low index material.

2. The Design of Wedged Filters

Figure 22 shows an MDM filter in which the thickness of the spacer layer varies across its surface and the wavelength of its passband shifts accordingly. The spacer thickness is usually chosen to produce second-order interference in the red. This results in an overlap of the third-order blue (3 × 400 nm) and the second-order red (2 × 600 nm), but the unwanted blue can be eliminated easily by auxiliary absorption filters. The wedge is not precisely linear because of the influence of the phase shift upon reflection, which in turn depends on the dispersion of the optical constants of the silver.

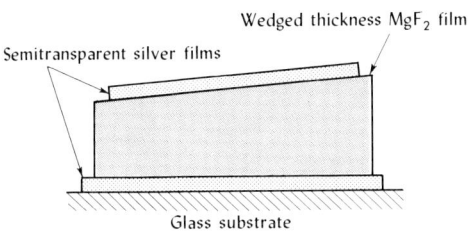

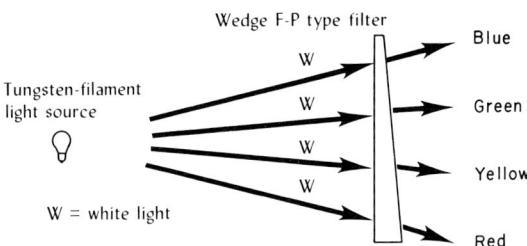

FIG. 22. The construction of a wedged MDM filter (upper).

It is also possible to deposit the wedge on the circumference of a circle rather than in a linear fashion. Several techniques can be used. One method is to open a "Japanese fan"-type shutter while the spacer is being evaporated. Another method uses rotating shutters (19, 20). Figure 23 depicts the measured T_{max} and $\Delta\lambda_{1/2}$ versus wavelength of a circular wedge-type filter. The bandwidth varies between 9 nm in the red to nearly double that value in the blue. The transmittance also increases markedly in the blue. Both effects are due to the dispersion of the optical constants of the silver, which decreases the reflectance R_1 in the blue.

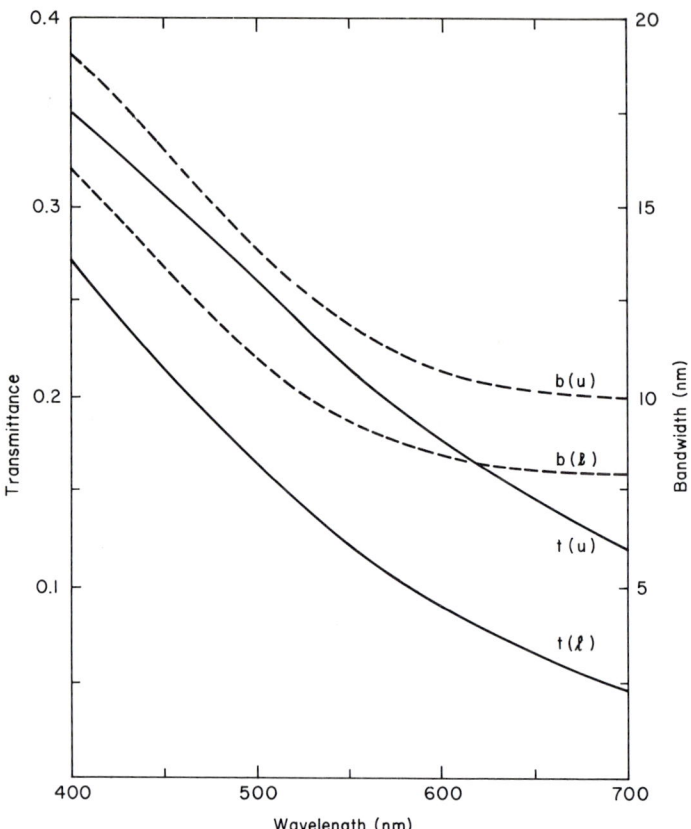

FIG. 23. The measured upper limit (u) and lower limit (l) of the transmittance t and bandwidth b of a circular MDM filter. From the data of the Israeli Electro-Optical Co.

Suppose that the dispersion were such that R_1 were independent of wavelength. Then the Q would also remain constant and $\Delta\lambda_{1/2}$ would decrease by a factor of two between $\lambda_0 = 800$ nm and $\lambda_0 = 400$ nm, which is contrary to the behavior shown in Fig. 23.

3. Filters with Augmented Spacers

The augmented spacer refers to a nonabsorbing stack that either replaces D or augments it (21). Its function is usually to decrease the spectral width of the passband. The T_{max} of the filter is unaltered provided the admittance at

λ_0 of
stack M substrate
is the same as
D M substrate

where D has been chosen using the procedure in Section IV,1. Section IV,1 discusses a specific example in which the D layer is replaced with an equivalent layer. Several examples of augmented designs are given below.

The simplest case to consider is where a layer of optical thickness $\lambda_0/2$, λ_0, $3\lambda_0/2$, and so on is added to the spacer. Such a layer does not modify the admittance at λ_0. An example of such a design is

glass M HH D M glass

A special case is where the HH layer has the same index as the D layer. The result is that the spacer is thickened and the order of interference m [in Eq. (77)] is now 2. Figure 24 depicts the transmittance of the first-order and second-order filters. As expected, the second-order filter has half the bandwidth and hence double the resolution of the filter with the first-order spacer layer.

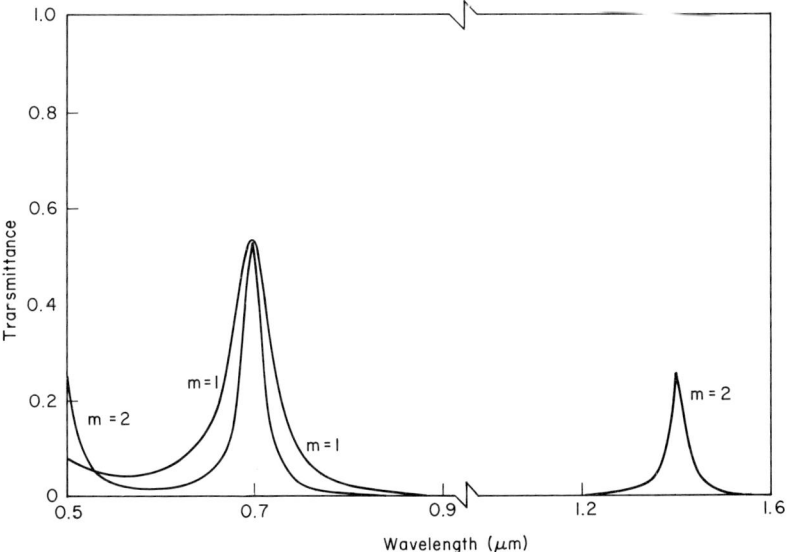

FIG. 24. The computed spectral transmittance of MDM filters, using dispersive optical constants (Table V) for the 25-nm thick silver layer M and 1.38 for the spacer layer D index. The optical thickness of the spacer of the first-order filter ($m = 1$) is 276 nm and 626 nm for the second-order ($m = 2$) filter. The substrate and incident medium index are both 1.517.

A still further reduction of the bandwidth can be achieved by adding layers of quarter-wave optical thickness to the interior of the cavity. This is accomplished as follows. First, we start with an MDM filter that we have already designed that has a passband at $\lambda_0 = 700$ nm. As shown in Table IV, the optical thickness of the magnesium fluoride spacer layer is 276 nm when the silver thickness is 25 nm. This design is symmetrical because the silver layers are identical, and the index of the incident and emergent media have the same index because the filter is cemented. Since the transmittance is a maximum at λ_0, according to Eq. (80) the admittance of the half filter

$$\tfrac{1}{2}\text{D M substrate}$$

is pure real at λ_0. The addition of a layer of optical thickness $\lambda_0/4$ adds some multiple of 180° to the phase shift upon reflection and hence the admittance is still pure real. Such filters can be called *augmented cavity filters*.

The simplest design is

$$\text{cement M } \tfrac{1}{2}\text{D HH } \tfrac{1}{2}\text{D M substrate}$$

where the optical thickness of the H is $\lambda_0/4$ and its index is 2.30. The H layers can be separated and a low index spacer can be inserted in the center

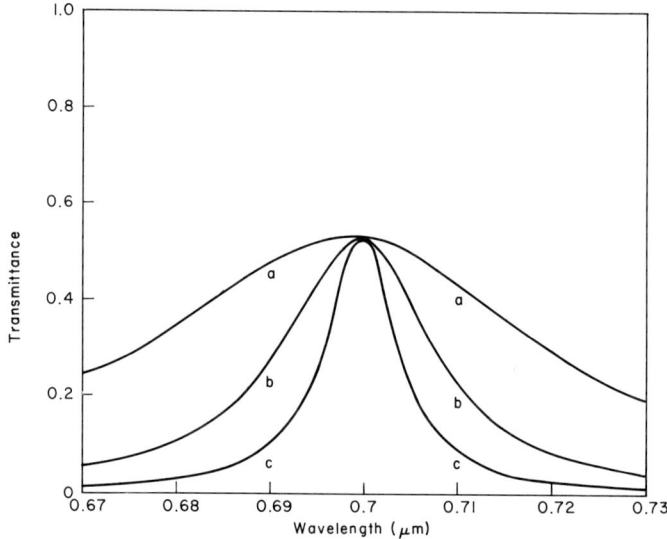

FIG. 25. The computed spectral transmittance of the designs: (a) MDM (b) M$\tfrac{1}{2}$DHH$\tfrac{1}{2}$DM (c) M$\tfrac{1}{2}$DHLLH$\tfrac{1}{2}$DM. The substrate, incident medium, M, and D (first-order) are defined in the caption to Fig. 24. H and L represent layers of index 2.30 and 1.38, respectively, of optical thickness 175 nm.

to produce the design

cement M $\tfrac{1}{2}$D H LL H $\tfrac{1}{2}$D M substrate

where the L layer of index has the same optical thickness as the H. As shown in Fig. 25, the bandwidth becomes increasingly narrow as more quarter-wave layers are added to the cavity.

We observe that the maximum transmittance of these augmented filters is the same as the simple MDM, provided that the additional layers are all nonabsorbing. The reason for this is that the ψ of each of the metal layers is unaltered by the addition of these dielectric films. According to Theorem III, the ψ of the film adjacent to the substrate is unaffected because the admittance on its emergent side is still that of glass. As mentioned earlier, the addition of the quarter-wave layers does not alter the admittance on the emergent side of the other metal film, and hence its ψ is the same regardless of whether it is in an MDM filter or in one of the augmented designs. Consequently, Eq. (80) obtains and $T_{\max} = \psi_a^2$ for each of these designs.

The disadvantage of increasing the order of interference of the cavity, whether by using a second-order spacer or by addition of the quarter-wave layers, is that spurious passbands appear at longer wavelengths. For example, Fig. 26 shows the transmittance (in absorbance units) versus wavelength for a first-order and higher-order filter. The transmittance of the first-order

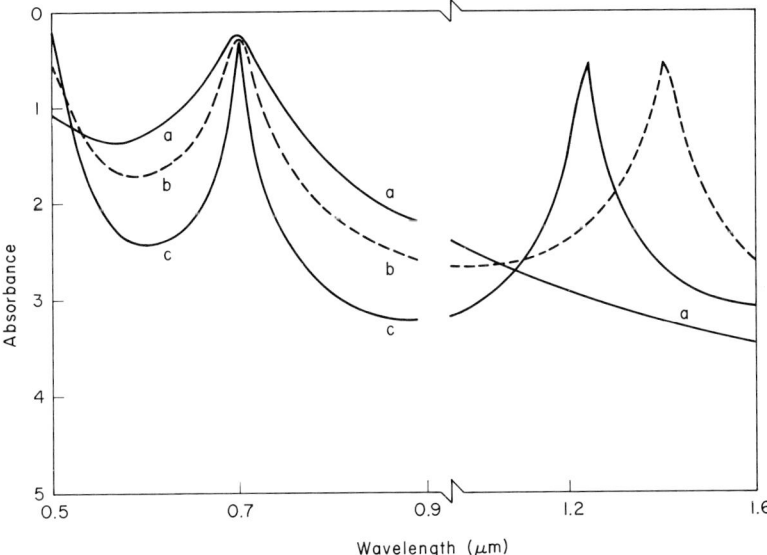

FIG. 26. The computed absorbance of (a) first-order and (b, dashed) second-order filters captioned in Fig. 24. Curve c is the same design as (c) cited in Fig. 25.

filter decreases monotonically at longer wavelengths. In the case of the filter that has a second-order spacer at 700 nm, a passband appears at double this wavelength. The filter with the quarter-wave layers in its cavity also has spurious transmission bands at longer wavelengths, although not at some simple multiple of 700 nm.

Whether these long-wavelength transmission "leaks" are significant depends upon the application. For example, if the filter were used in conjunction with a detector, then the spurious passbands would be intolerable if the detector has an appreciable responsivity at those longer wavelengths.

4. Filters with Equivalent Layer Spacers

The foregoing MDM filters were designed by choosing a D layer with the proper phase thickness so that the transmittance was optimized at λ_0. If this layer is replaced by an equivalent layer that has the same phase thickness, then the same transmittance is attained at λ_0.

An equivalent layer is a symmetrical combination of three dielectric layers in which both the refractive indices and phase thicknesses of the outer layers are identical. It has the form ABA. Space does not permit enumeration of the properties of such layers. These are adequately summarized by Thelen (22), who also lists previous publications on this subject. Recently, Ufford and Baumeister (23) developed graphical methods which aid in designing equivalent index combination for multilayer filters.

The combination of layers ABA is determined by four parameters, namely, the phase thicknesses β_a, β_b and refractive indices n_a, n_b of the outer and inner layers, respectively. Given these parameters, standard equations (22) can be used to compute the phase thickness and refractive index of the equivalent layer. Once we specify the phase thickness of the equivalent layer, then only three of the four parameters β_a, β_b, n_a, and n_b are independent (24). Even with this restriction, this leaves us with a vast number of possible combinations of thicknesses and indices from which to choose.

A judicious choice of thickness and index will improve the performance of a bandpass filter at nonnormal incidence. Specifically, it is possible to minimize the polarization splitting of the passbands (24). This is illustrated with an MDM filter designed for the spectral region near 290 nm. As shown in Fig. 27, aluminum is used for the M layers and magnesium fluoride is used for the spacer in this conventional filter. The spectral transmittance is shown at normal incidence (curve O) and at 45° for the TM polarization (curve P) and the TE polarization (curve S). The peak transmittance of the s polarization is at a shorter wavelength than the p polarization. The objective of the design is to minimize this polarization splitting of the passbands.

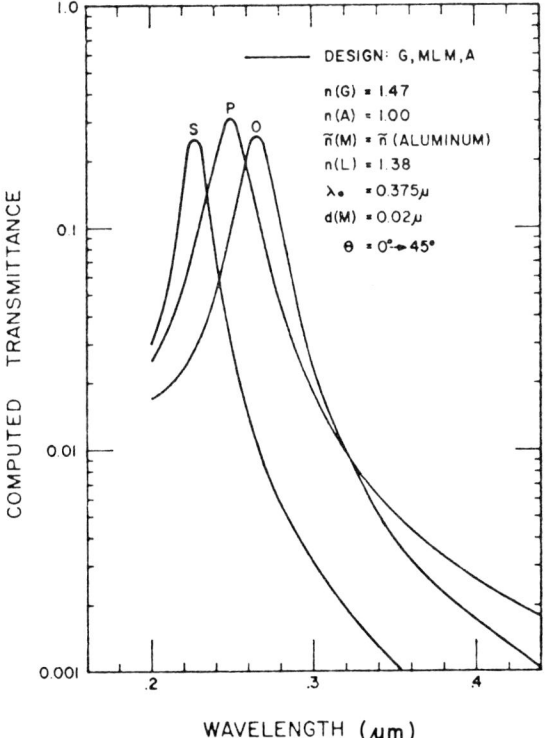

FIG. 27. The computed spectral transmittance of an ultraviolet bandpass filter. Design G, MLM, A.

The optical thickness of the spacer is 94 nm; the equivalent layer should approximate this thickness. It is shown (24) that the polarization splitting is minimized when the relationship

$$n_a = n_b\sqrt{3} \qquad (86)$$

is satisfied and the phase thickness of A is one-half that of B.

If magnesium fluoride ($n = 1.38$) is used for the A layer, then Eq. (86) requires that the index of the B layer should be 2.39. Any thin film materials with such a high index are liable to be quite absorbing in this spectral region. We compromise and use 2.0 for the n_b.

Figure 28 shows both the design and spectral transmittance of the MDM in which an equivalent layer is used as the spacer. Additional layers (24)

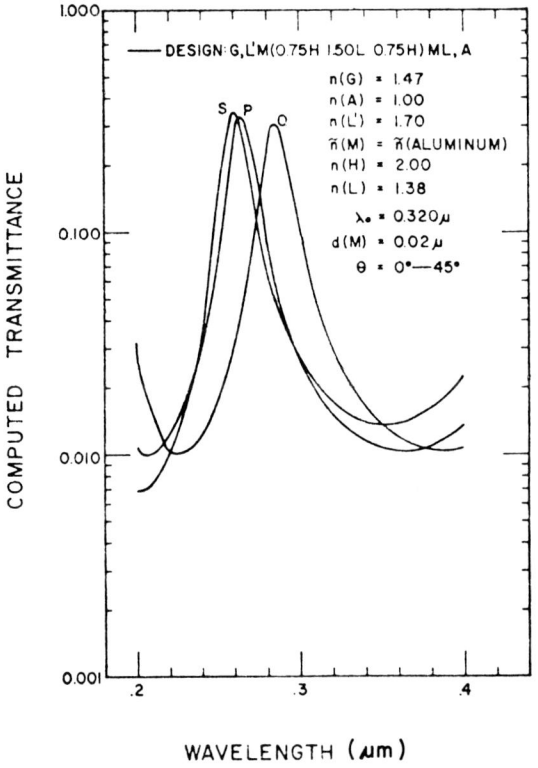

FIG. 28. The computed spectral transmittance of an ultraviolet bandpass filter. Design G, L'M (0.75H 1.50L 0.75H) ML, A.

have also been added to minimize the polarization effects of the incident medium and the substrate. The result is that the passbands for each linear polarization now virtually overlap, even at angles of incidence as large as 45°.

V. One-M Filter Design

1. THREE-LAYER SYMMETRICAL SYSTEMS

The term one-M is used generically to describe filters that contain only one metal layer. For reasons that are explained below, they are also called induced transmission filters. This term can be misleading, however, because the principle of induced transmission can be applied equally well to an MDM or an MDMDM filter, as illustrated in Section VI.

The simplest type of filter is a heat reflector that contains a single layer of gold. As shown in Fig. 29, the dispersion of the optical constants of gold is such that even a single layer has considerable transmittance in the visible part of the spectrum. The addition of single dielectric layers to both sides of the gold increases this visible transmittance still further.

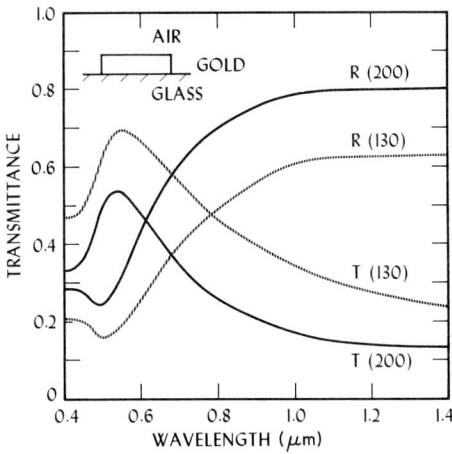

FIG. 29. The computed reflectance R and transmittance T of gold layers of thickness 130 Å and 200 Å. The dispersive optical constants are from Hass and Hadley (*18*).

It is most convenient to choose a symmetrical design, in which the incident medium and substrate have the same refractive index. This means that the coating should be cemented between sheets of glass with optical cement of refractive index identical to that of the glass. If films are added to both the incident and emergent sides of the gold, they should also be identical to maintain the symmetry.

The virtue in using a symmetrical system is that the isotransmittance contours described in Section II,11 can be used. Figure 30 depicts such data at a wavelength of 550 nm for a gold film 20 nm in thickness. The isotransmittance contour of 95% means that it is possible to attain a transmittance that is as high as 95% provided dielectric stacks—called *matching stacks*—are added to both the incident and emergent sides of the gold.

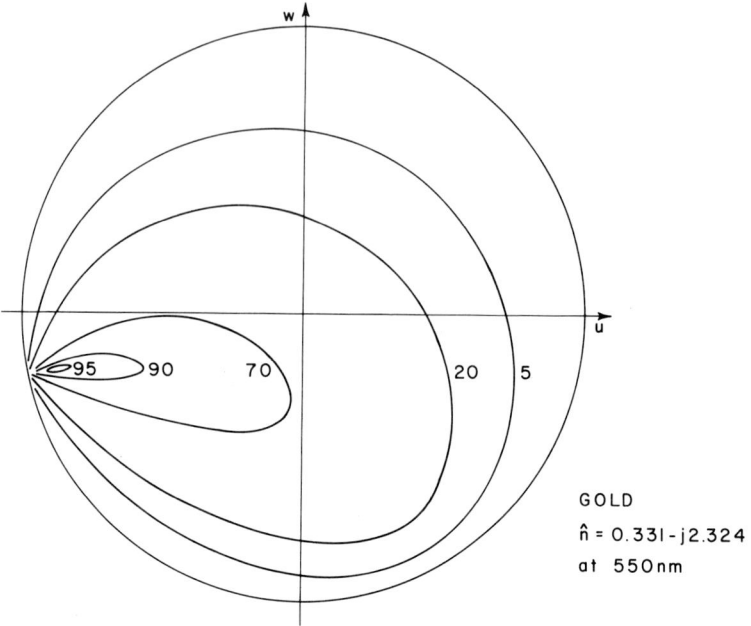

Fig. 30. Isotransmittance contours of a gold film 20 nm thick as a function of the amplitude reflection coefficient of the medium on its emergent side.

The problem is that such a matching stack must contain at least five or six layers and hence its admittance would vary quite rapidly with wavelength. Although the admittance would be properly matched at 550 nm, it would be badly matched at other wavelengths and the net transmittance over the entire visible part of the spectrum would be decreased considerably.

A compromise is to use a single high index layer to match the admittance and thus to enhance the transmittance over a broad region. Holland (25) describes coatings in which bismuth oxide is used for the dielectric layer. A transmittance of 75% was attained in the visible part of the spectrum, when the oxide layer had an optical thickness of 103 nm, which is $\frac{3}{16}$ of a wavelength at 550 nm. This has an amplitude reflection coefficient at 550 nm that is well within the 70% contour in the third quadrant of Fig. 30. This is why this optical thickness enhances the transmittance.

It is difficult to use these contours to find any optimum thickness for the dielectric layer because the isotransmittance contours shown in Fig. 30 change rapidly with wavelength. This is due to the dispersion of the optical constants of gold and, to a lesser extent, to the change of h/λ. Hence the optimum thickness for the dielectric layer is somewhat arbitrary, depending on whether the transmittance is to be optimized in either the blue or red part of the spectrum. Figure 31 shows the computed transmittance for the design

cement D gold D glass

for two thicknesses of the D layers.

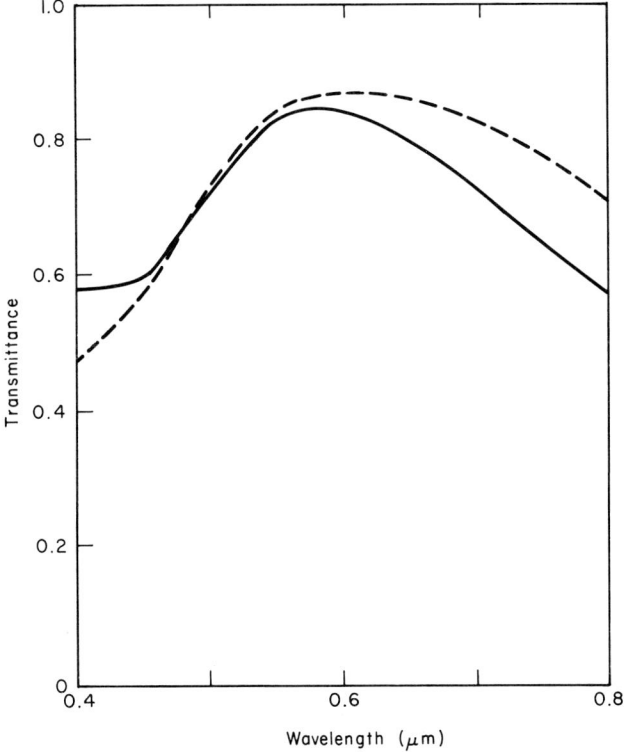

FIG. 31. The computed transmittance of design DMD surrounded by media of index 1.517. The M is 20-nm-thick gold; the optical constants are from Hass and Hadley (*18*). The optical thickness of the D is 63 nm (solid line) and 100 nm (dashed line); its index is 2.30.

These coatings have essentially the same reflectance in infrared as the naked film, as shown in Fig. 29. Thus the dielectric layers do not inhibit the infrared reflectance of the gold and these coatings function as heat reflectors.

A major deterrent to the use of gold is its cost, which at a commercial price of $175 per ounce is $0.20 per square foot for a 20-nm thick layer. As pointed out in Section III,2, gold and silver have properties that are similar to gold in the infrared. It is not surprising that both silver (26) and copper (27) have also been used for infrared heat reflectors.

2. Procedure for One-M Filter Design

The configuration of a one-M filter is

incident, (AR stack), metal, (MA stack), substrate

where the single metal layer is imbedded in nonabsorbing dielectric stacks AR and MA. *Incident* refers to the incident medium, which is air in an asymmetrical design and optical cement in a symmetrical configuration. The considerations listed in Section III,2 help the designer to choose the proper metal. The remaining problem is to design the dielectric stacks.

Macdonald (28) suggests a cut-and-try method. Berning and Turner (8) analyzed the problem by bisecting the metal layer and computing the admittance and ψ of the half-filter. The matching stack was altered until the transmittance, as computed via Eq. (80), was optimized. The authors prefer to use the computation of the combination of isotransmittance contours and the admittance of the matching stack to design such filters. The procedure is described below.

(1) The first step is to design the matching stack MA that produces the admittance that is needed to attain ψ_{max} at λ_0. Once the optical constant of the metal is determined and its thickness is chosen, the matrix elements in Eq. (20) are computed and Y_{max} is found from Eqs. (67) and (68). The ψ of the ensemble

metal (M stack) substrate

is ψ_{max} at λ_0 or at least close to it, depending on how precisely the MA stack is designed.

(2) The dielectric AR stack is added to its emergent side which reduces its reflectance to zero at λ_0. According to Theorem IV in Section II,9, the transmittance of the completed filter is ψ_{max}.

Due to the restriction on the refractive index of the layers in the MA and AR stacks, as well as thickness errors in the coatings, the admittance of these stacks never exactly achieves that of Eqs. (67) and (68). The advantage of using isotransmittance contours is that the designer can plot the admittance and immediately visualize how changes in this admittance will alter the transmittance of the entire filter.

Finally, it should be pointed out that the filter can be considered symmetrical, even though the incident medium and substrate are not of the same index, provided that the combination of the substrate MA stack is optically equivalent to the incident medium and AR stack. An example illustrates the concept of optical equivalence. Suppose a one-M filter is designed reverse order so that the emergent medium is air, rather than glass. Let us assume that the three-layer combination

$$A\ B\ C\ \text{air}$$

produces an admittance Y at λ_0.

Now suppose this three-layer stack is deposited on a glass of index 1.82 that is coated with a layer L of index $1.35 = (1.82)^{1/2}$ and optical thickness $\lambda_0/4$. The admittance of

$$A\ B\ C\ L\ \text{glass}$$

is then the same as the foregoing design. Because the admittances are the same, we say that they are optically equivalent at λ_0. This means that when these are incorporated in a one-M filter

$$\text{air}\ C\ B\ A\ M\ A\ B\ C\ L\ \text{glass}$$

the isotransmittance contours can be used to compute the transmittance.

Since the AR and MA stacks described earlier are optically equivalent, we refer to them generically as matching stacks.

3. Examples of One-M Filters

Filters containing aluminum films are particularly attractive for the ultraviolet region because the aluminum attenuates more at longer wavelengths. Thus in the parlance of filter terminology, such a filter is self-blocked. The aluminum-dielectric-aluminum ultraviolet filter with 20-nm thick aluminum films and a first-order spacer typically has a bandwidth of 15 nm. The bandwidth can be decreased by using a thicker spacer layer and higher order of interference (6). As pointed out in Section IV,3, this narrower bandwidth is achieved at the expense of the long-wave blocking of the filter because low-order interference transmission peaks are now present.

A bandpass filter with a single metal layer is attractive because it can have a narrower bandwidth than the conventional MDM. Furthermore, it can achieve this narrow bandwidth with first-order interference and thus it is self-blocked at longer wavelengths.

As an example, let us design three different bandpass filters in which the peak transmittance in the passband is at $\lambda_0 = 254$ nm but which have different spectral passband widths and offband rejection properties. In each of these filters, the metal layer is aluminum of 25 nm metric thickness. Its dispersive optical constants are listed in Table V.

The first step is to compute the matrix elements for the aluminum film. The isotransmittance contours are computed via Eqs. (82) and (83) and are shown in Fig. 32. Strictly speaking, they have this appearance only at λ_0. They change shape at $\lambda \neq \lambda_0$ for two reasons: first, the matrix elements change with wavelength, and second, the optical constants of the aluminum are dispersive. However, in the case of a metal film, the shapes of the isotransmittance contours change relatively little for a 10% change in wavelength. This means that to a good approximation, the transmittance of the one-M filter can be found by superimposing a parametric plot of the admittance (versus wavelength as a parameter) of the matching stack on the isotransmittance contours. If this parametric admittance intersects the point at which $T = T_{max}$, then the designer is assured that the transmittance of the one-M filter is optimized. Otherwise, from the admittance plot one can

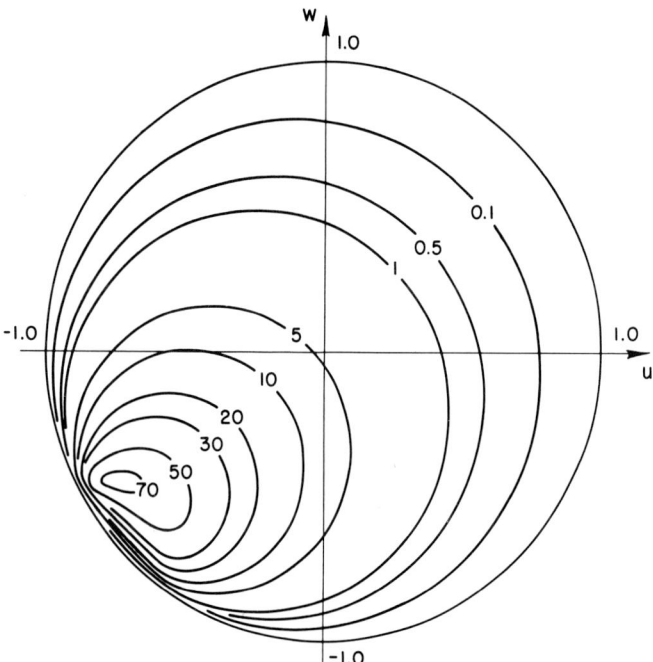

FIG. 32. Same as Fig. 30 except that the aluminum film is 25 nm thick, the wavelength is 254 nm, and its optical constant is $0.20 - j2.80$.

predict the transmittance of the one-M filter and also the change in transmittance that would be introduced by errors in the thickness or refractive index of the layers in the matching stacks. The parametric admittance plot also furnishes information about the spectral width of the passband of the filter and also its rejection properties.

This is illustrated by the design of the three different one-M filters. The isotransmittance curves (e.g., Fig. 32) furnish information on the maximum attainable transmittance at λ_0, which is 73.9%. This is obtained when the admittance of the matching stack is $\hat{Y}_{max} = 0.96 + j3.35$. Three different matching stacks are designed which have substantially this $\hat{Y}_{max}$ at λ_0. Consequently, when these stacks are combined with the aluminum layer, each of the resulting one-M filters has the same peak transmittance in its passband. However, due to the different design of the matching stacks, the "dispersion" of the admittance near λ_0 is considerably different and hence their spectral bandwidths differ. The matching stacks are referred to as # 1, # 2, and # 3 and their designs are listed in Table VI. Design # 1 consists of a quarter-wave stack of cryolite (index 1.35) and thorium oxyfluoride (index 1.55). Most of the high index layers in the quarter-wave stack for design # 2 are MgO, which has an index of 1.80 near $\lambda = 250$ nm (29). Lead fluoride could also be used (30). The design for stack # 3 consists of a quarter-wave stack of MgO and cryolite in conjunction with an additional group of quarter-wave and half-wave layers that function like an interference filter. Due to the position of the half-wave layer in design # 3, the admittances of design # 2 and design # 3 are exactly the same at λ_0.

TABLE VI

THE DESIGNS OF MATCHING STACKS AND ONE-M FILTERS

Designation	Design[a]
Stack # 1	0.74 H(LH)^{8}LQ
Stack # 2	1.77 LH(LB)^{4}Q
Stack # 3	1.77 LH(LB)^4(LH)^{5}Q
Bandpass # 1	air (HL)^{7}H 1.77 L M stack # 1
Bandpass # 2	air (LB)^5 1.74 H M stack # 2
Bandpass # 3	air (HL)^5(LH)^5(LB)^5 1.74 H M stack # 3

[a] H, L, and B represent films of optical thickness $\lambda_0/4$ and of index 1.55, 1.36, and 1.80, respectively. M is an aluminum layer 25 nm in thickness. At $\lambda_0 = 254$ nm its optical constant is $\hat{n} = 0.2 - j2.80$. Its optical constants at other wavelengths are listed in Table V. Q is the massive fused quartz substrate, index 1.47.

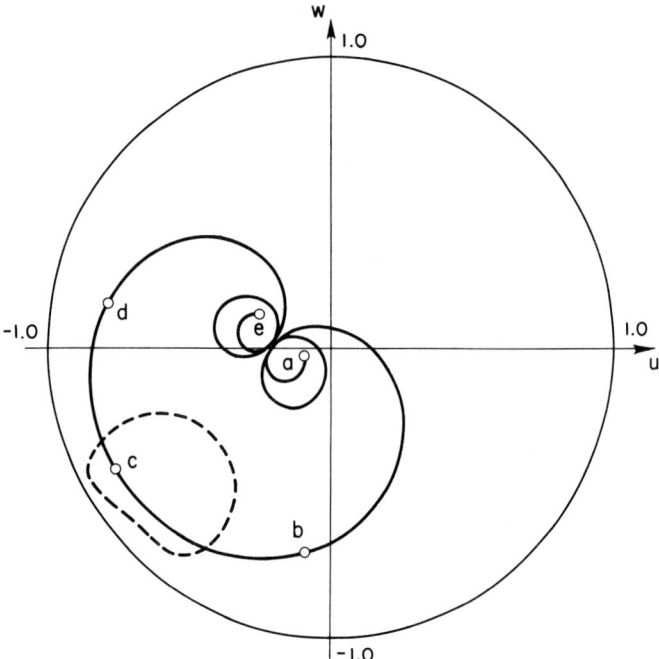

FIG. 33. A parametric plot of the amplitude reflection coefficient of design #1 (listed in Table VI) at $\lambda_0/\lambda = 0.9$ (a), 0.95 (b), 1.0 (c), 1.05 (d), and 1.1 (e).

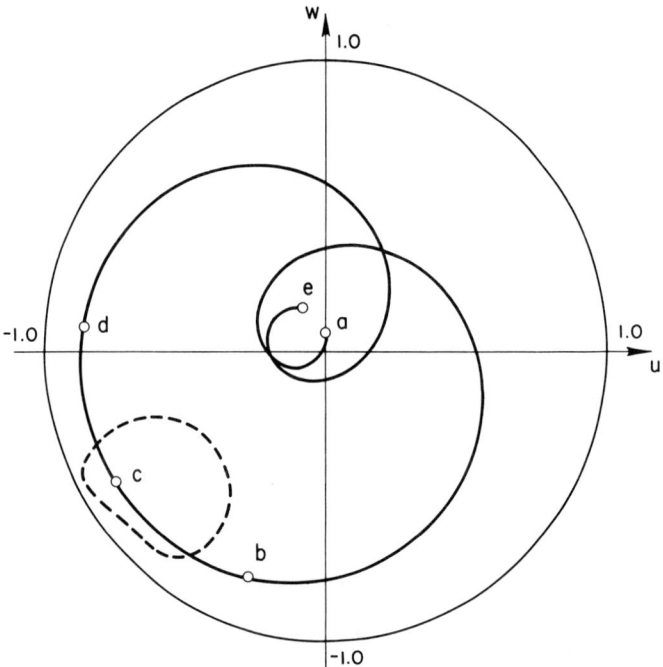

FIG. 34. Same as Fig. 33, for design #2.

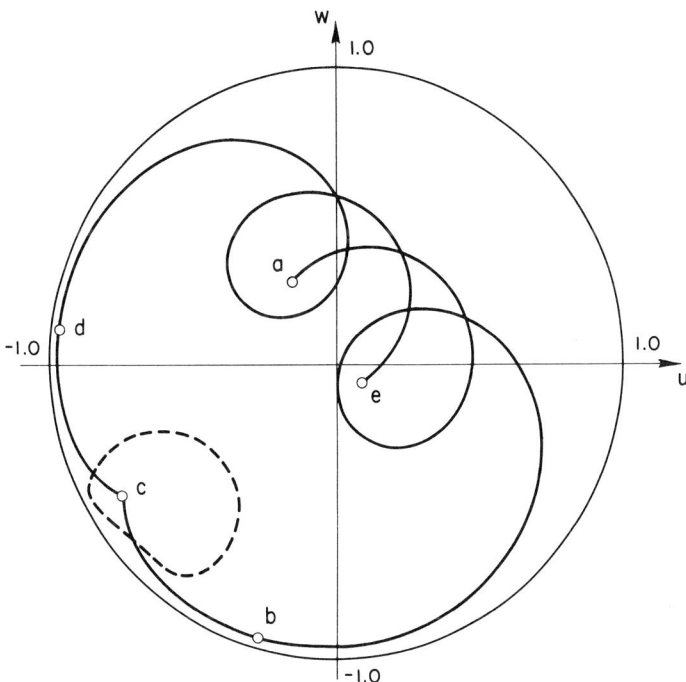

FIG. 35. Same as Fig. 33, for design #3.

The admittance versus wavelength (as a parameter) of each of these stacks plotted on the amplitude reflection plane is shown in Figs. 33, 34, and 35. In other words, a vector drawn from the origin to a point on a curve represents the (complex) amplitude (in air) reflection coefficient of the stack at that wavelength. In each case, the admittance at λ_0, which corresponds to the point c on these curves, is centered at $\hat{Y}_{max}$. When the aluminum layer is embedded in these stacks, the transmittance versus λ can be estimated by superimposing the contours in Fig. 32 with the admittance in Figs. 33–35. The $T = 30\%$ contour is shown as a dashed line in Figs. 33–35.

The parametric admittance plots of #1 and #2 are quite similar, due to the fact that they are both quarter-wave stacks. However, the dispersion of the phase shift upon reflection (and hence the dispersion of the admittance) is greater for a quarter-wave stack with a small index mismatch (e.g., stack #1) than for a stack with a larger index mismatch (e.g., stack #2).

Inspection of these admittance plots makes it evident that, to a good approximation, the spectral passband width of the one-M filter is inversely proportional to the phase dispersion of the admittance. Hence it is expected

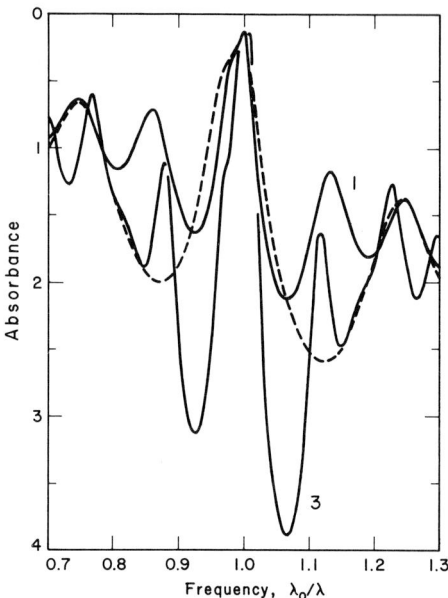

Fig. 36. The computed spectral absorbance of bandpass filters #1, #2, and #3, as defined in Table VI.

that the passband width of the filter that incorporates matching stack # 1 would be narrower than the filter that utilizes matching stack # 2. This is evident in Fig. 36 which depicts the spectral transmittance of the resulting filters.

The dispersion of the phase shift upon reflection for stack # 3 is about the same as that of the other two stacks. One significant difference, however, is that the amplitude reflectance has an inflection point at λ_0, which is shown as point c in Fig. 35. The admittance is very close to the outer perimeter of the chart from point c to point d. At the outer perimeter of the chart, the transmittance drops off extremely rapidly and hence in this region the transmittance is quite low. This effect is illustrated in Fig. 36. The minimum transmittance (i.e., maximum rejection) of #3 is at least one absorbance unit greater than that of other filter designs. Another manifestation of the inflection of the parametric admittance at λ_0 is that the spectral width of the passband of the one-M filter is slightly narrower. Figure 37 depicts the spectral transmittance, reflectance, and Ψ of filter design #2. In each case the reflectance is zero at λ_0, which is the case if the admittance matching has been properly accomplished.

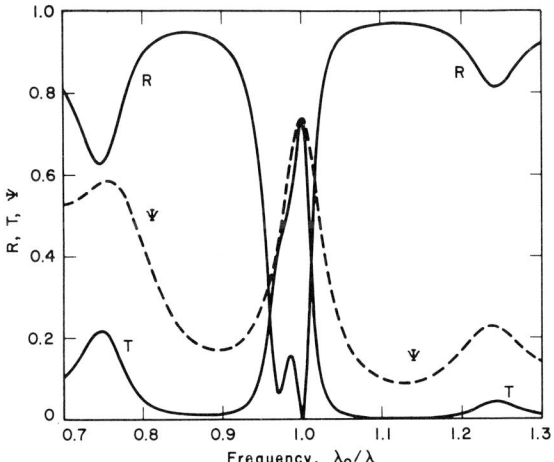

FIG. 37. The computed transmittance T, reflectance R, and net flux ratio ψ of bandpass filter #2 (see Table VI).

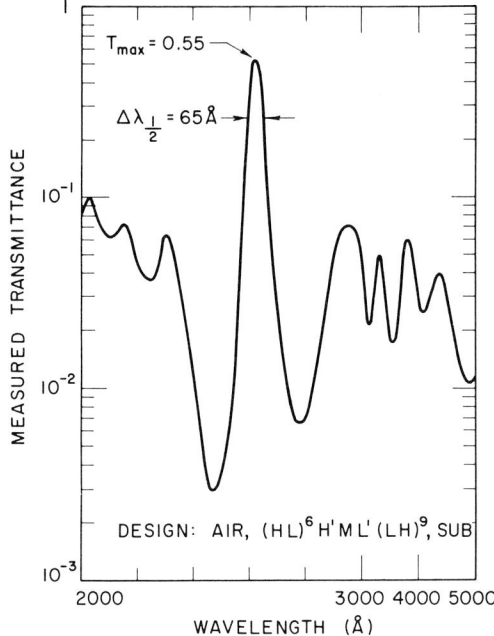

FIG. 38. The measured spectral transmittance of a one-M filter composed of aluminum M (30 nm thick), cryolite L, and thorium oxyfluoride H (9, 14).

The measured spectral transmittance of a one-M filter (9, 14) is shown in Fig. 38. Its construction is similar to design # 2 cited in Table VI, with the exception that the layers adjacent to the aluminum are cryolite, rather than high index materials. Its maximum transmittance of 55% compares favorably with the computed transmittance of 63% for this design.

VI. Multiple Cavity and Other Designs

1. An Example of Multiple-Cavity Filter Design

Multiple-cavity filters such as the designs MDMDM or MDMDMDM contain more metal than a single-cavity MDM filter and consequently have a considerably lower transmittance in the offband part of the spectrum. The problem is that the T_{max} of the filter is also reduced by the additional

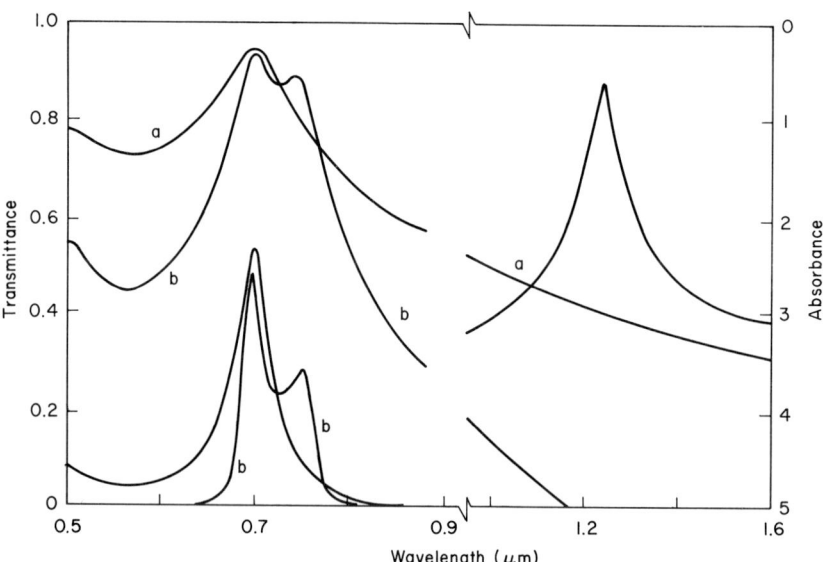

Fig. 39. The computed transmittance (two lower curves) and absorbance (upper curves) of (a) MDM and (b) MD'M'D'M, where M and D are defined for the first-order filter in the caption to Fig. 24. M' is silver, 35 nm thick, and D' is magnesium fluoride, 291 nm in optical thickness.

metal layers. As an example of a double-cavity filter, Fig. 39 shows the spectral transmittance of an MDM'DM filter. Its long-wavelength attenuation is superior to the MDM, which is shown for comparison in the same figure. Such multiple-cavity filters are rarely designed to produce a narrow bandwidth. This is why the spurious transmission "hump" that appears at 745 nm is not of serious concern in such a filter.

However, any improvement in the maximum transmittance of 47% would be welcomed. Section VI,3 discusses how this is done. Before launching into this section, we should describe briefly how this filter is designed.

The filter $M_1DM_2DM_1$ is symmetrical because it is cemented; the first and last layers have the same thickness and composition. A consequence is that the D layers are identical. This would not be the case if the filter were asymmetrical. The optimum thickness for the D layer is found by computing the ψ of the quasi-half-filter

$$M_2 \ D \ M_1 \text{ substrate}$$

and adjusting the optical thickness of the D until it is a maximum. The other two layers are then simply added, using the same thickness for D.

2. Augmented MDM Filters

As discussed in Section IV,3, the addition of nonabsorbing layers of the proper thickness inside of the cavity reduced the bandwidth but did not affect the maximum transmittance because they did not affect the ψ. The addition of only a few layers to the exterior of the cavity can provide a modest improvement in the $T_{\max}$ via the phenomenon of induced transmission.

For reasons that were mentioned previously, it is convenient to analyze symmetrical designs, such as

$$\text{cement } H' \ M \ D \ M \ H' \text{ glass}$$

The thickness of M layer is the same as in Section IV,3, namely, 25 nm. The passband should be centered at 700 nm.

The first step is to adjust the optical thickness of the H' layer so that the ψ of the half-filter

$$\tfrac{1}{2}D \ M \ H' \text{ glass}$$

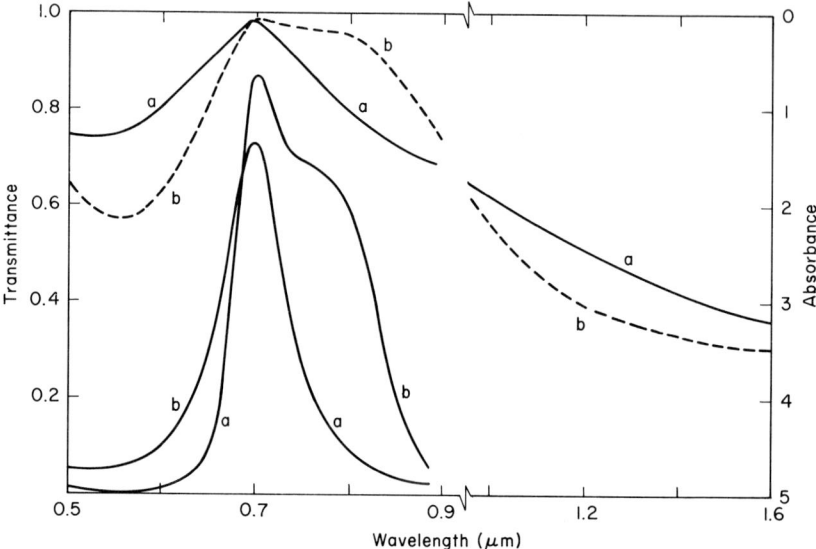

FIG. 40. The computed transmittance (lower curves) and absorbance of the designs: (a) H'MD'MH'; and (b) HLH'MDMH'LH, where M and the substrate are cited in the caption to Fig. 24. D', D, and L are of index 1.38 and optical thickness 284 nm, 305 nm, and 175 nm, respectively. The index of H' and H is 2.30, and their optical thicknesses are 149 and 175 nm, respectively.

is optimized. This occurs when the optical thickness is 149 nm. The thickness of the layer $\frac{1}{2}$D is adjusted until the half-filter's admittance is pure real, which occurs when the optical thickness of D is 284 nm. The half-filters are combined and the spectral transmittance is shown in Fig. 40. The addition of the external H' layers to the simple MDM increases its T_{max} from 53% to 72%. Another consequence is a broadening of the bandwidth from 51 to 80 nm.

How many more layers should be added to this extra-cavity stack to enhance the ψ and hence the T_{max}? In order to answer this, we must first find the admittance at λ_0 that is needed to produce ψ_{max}. From Eqs. (67) and (68), we compute that $Y_{max} = 3.32 - j7.69$. The magnitude of the amplitude reflection is computed via Eq. (9) (with $n_0 = 1.0$), and from Eq. (13) we find that the standing wave ratio V of the matching stack should be approximately 21. The V of the stack HLH (on a glass substrate) is

$$\frac{n_H^4}{n_L^2 n_s} = 9.7 \tag{87}$$

at λ_0 and a similar equation for the five-layer stack HLHLH shows that its V is 27. Consequently, a three-layer stack would undermatch the silver and the five-layer stack would overmatch it. We opt for the three-layer matching stack, which has the virtue that it is easier to produce. Figure 40 shows the spectral transmittance of the entire nine-layer filter. Its bandwidth is broader than that of the five-layer filter mentioned above. The $T_{\max}$ is 87%, which is only 3% less than the optimum of $\psi_{\max}^2 = 0.95^2 = 90\%$.

With some patience and also access to refractive indices that are intermediate to those used here—2.30 and 1.38—we could produce a matching stack that has precisely the admittance $Y_{\max}$ and thus this optimum would be obtained. Several authors (*10, 31, 32*) describe MDM filters with extra-cavity matching stacks that attain a $T_{\max}$ that is very close to the optimum.

3. Augmented Double-Cavity Filter

The final example of bandpass filter design is to enhance the maximum transmittance of the double-cavity type filter described in Section VI,1. Silver layers of the same thicknesses are used and the passband is at the same wavelength.

The $\psi_{\max}$ for a 25-nm thick silver layer is 0.96 and is 0.897 when the thickness is 35 nm. Therefore, the $T_{\max}$ of the aforementioned double-cavity filter is $0.96^2 \times 0.897 = 0.83$. In other words, the $T_{\max}$ of the simple MDMDM filter could be increased to 83% by the addition of the appropriate extra-cavity matching stacks.

The following steps are used in its design: First, the stack H′ L H is added to the emergent side of the 25-nm thick silver film. This stack is identical to the one described in Section VI,2 and, although it does not optimize ψ, it still increases it substantially. The next step is to compute the ψ of the quasi-half-filter

$$M_2 \ D \ M_1 \ H' \ L \ H \ \text{substrate}$$

The optical thickness of the D is adjusted until the ψ is a maximum. The additional layers H L H′ M D are added to its emergent side and the filter design is complete. The spectral transmittance of the completed eleven-layer double cavity is shown in Fig. 41. At longer wavelengths, the interference effects of the extracavity stacks are minimal and the transmittance approaches that of the simple MDMDM filter, which is shown for comparison.

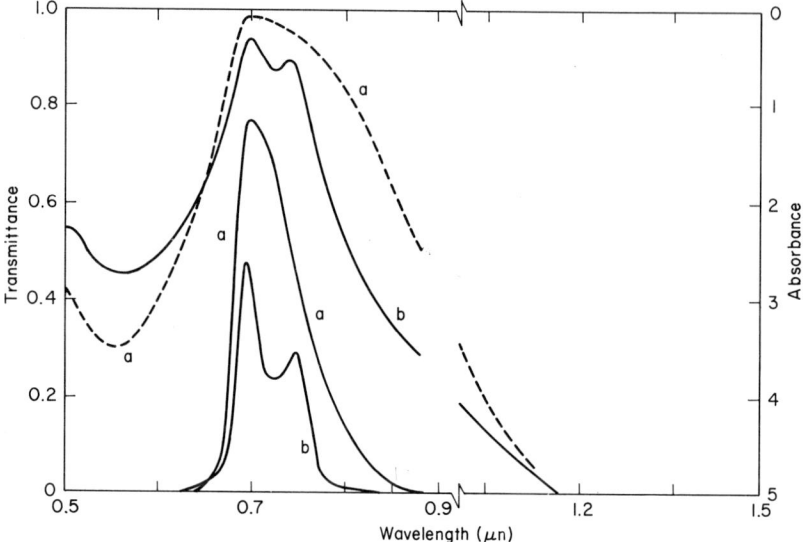

FIG. 41. The computed transmittance (lower curves) and absorbance of the design (a) HLH′MD″M″D″MH′LH and design (b) MD′M′D′M. H, L, H′, and M have the same meaning as in the caption to Fig. 40. M″ is silver, 35 nm in thickness. D″ is index 1.38, optical thickness 301 nm.

VII. Reflection Filters

All of the filters studied in the previous sections functioned by selective transmission. The untransmitted flux was either absorbed or reflected. It is possible also to construct filters that selectively reflect in certain parts of the spectrum. The nonreflected flux is absorbed by the opaque metal substrate.

One type of filter makes use of the fact that intrinsic semiconductor films are transparent at longer wavelengths and absorb strongly at wavelengths less than

$$\lambda_0 = 1.24/W_g \tag{88}$$

where W_g is its bandgap in electron volts. As an example, suppose we require a coating to absorb the solar energy in the visible part of the spectrum but simultaneously to reflect strongly (and hence have a low emissivity) in the infrared. This low emissivity reduces its radiation losses. As shown in Fig. 42, the coating consists of a semiconductor film deposited on an opaque

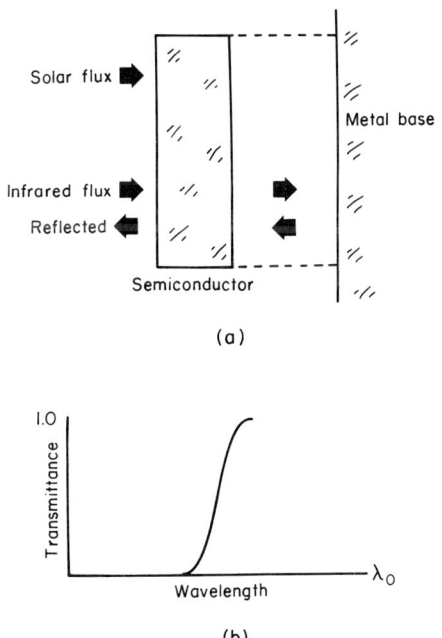

FIG. 42 (a) A solar absorber consists of a base metal overcoated with a semiconductor, whose spectral transmittance is shown in (b).

metal base layer. For the sake of illustration, the film is separated from the metal base. The graph in Fig. 42 shows the spectral transmittance of the semiconductor, which is relatively transparent at wavelengths greater than λ_0. Hence, these wavelengths are reflected by the metal base with little attenuation. The solar flux, on the other hand, is strongly absorbed by the semiconductor and hence this flux does not "see" the metal.

Figure 43 shows the spectral reflectance of silicon on an aluminum base layer (7). Details on the preparation of such coatings are given in a review article by Hass (33).

There are many possible embellishments on such coatings. For example, the reflectance of the semiconductor can be reduced by the addition of one or more layers to its incident side to antireflect it, as depicted in Fig. 44. Figure 45 shows the reflectance of a copper base layer that is coated with a germanium film, which in turn is antireflected with a SiO layer (7). Designs of this type, both with and without the antireflection coatings, have been suggested as solar thermal heat absorbers (34, 35).

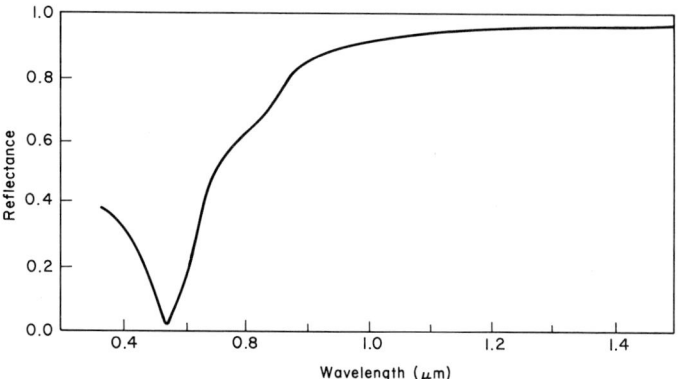

FIG. 43. The spectral reflectance of a reflection filter that consists of an opaque layer of aluminum overcoated with silicon. [From Fig. 4 in Hass et al. (7).]

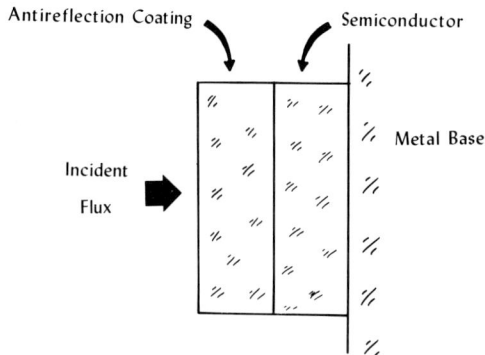

FIG. 44. A reflection filter consists of an antireflected semiconductor film deposited on a metal base.

Interference effects in MDM-type structures can be utilized in reflection filters. Figure 46 shows the spectral reflectance of a filter that appeared in a U.S. patent that was filed in 1946 (36). The top layer is semitransparent aluminum and the spacer is magnesium fluoride. Four- and five-layer structures have been developed for solar thermal energy absorbers (37). In particular, a coating consisting of alternate layers of aluminum oxide and molybdenum (38) is stable at temperatures as high as 930°C. Its reflectance is shown in Fig. 47.

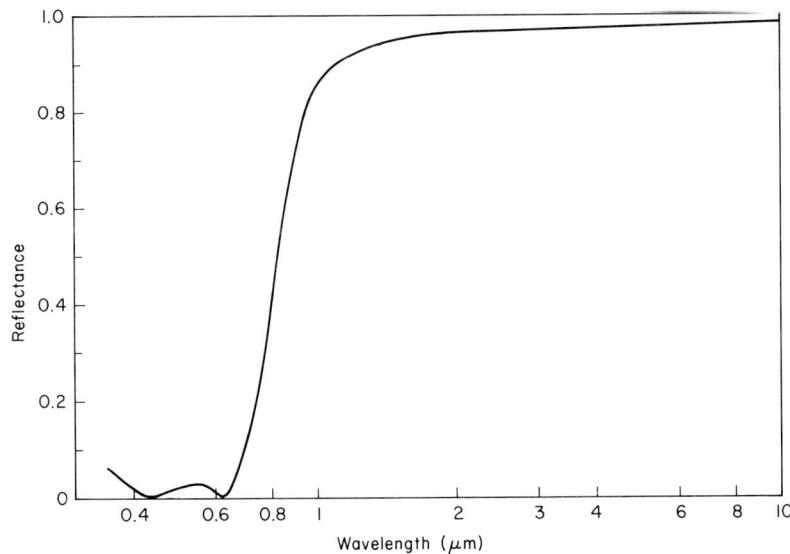

FIG. 45. The spectral transmittance of the design shown in Fig. 44, where the outer layer is silicon monoxide and the inner layer is germanium. [From Fig. 6 in Hass et al. (7).]

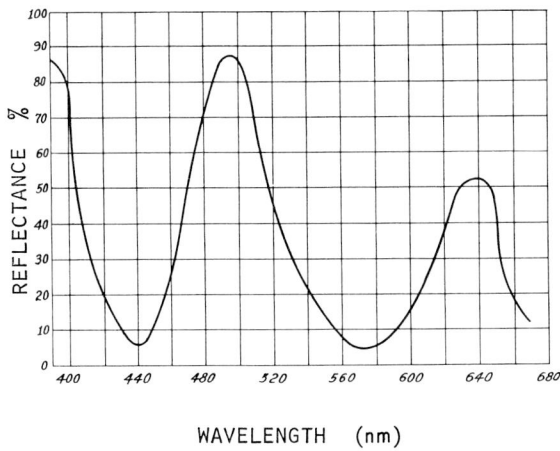

FIG. 46. The measured spectral reflectance of a filter consisting of a semitransparent aluminum film, a magnesium fluoride spacer layer, and an opaque aluminum film (36).

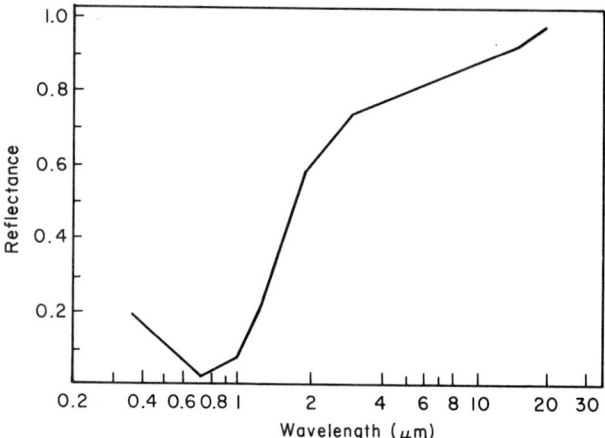

FIG. 47. A solar absorber consisting of five alternate layers of aluminum oxide and molybdenum. The exterior layer is aluminum oxide and the substrate is molybdenum sheet. The thicknesses are listed in Example II in Schmidt (37).

VIII. The Production of Filters

There are several obstacles to making any sweeping generalizations about the procedures that should be followed. The first is that the production methods are quite varied. Although most of the filters that have been described are produced by evaporation in a vacuum, there is no reason why they could not be sputtered, provided the proper instrumentation was developed to control the thicknesses of the layers. Vacuum evaporation systems vary in their sophistication—from simple glass bell jars with 6-inch oil diffusion pumps, to much larger stainless steel tanks with 20-inch diffusion pumps. The other obstacle is that there is a wide variety of materials used in coatings, ranging from zinc sulfide to titanium dioxide. Each requires a different evaporation technique.

1. DEPOSITION TECHNIQUES FOR METALS

Although filters have been developed that contain metals such as nickel (39), the majority of the bandpass filters use silver in the visible and near infrared (1, 2, 5), and aluminum in the ultraviolet (6, 40) to wavelengths as short as 120 nm (41, 42). These films should be deposited as rapidly as possible onto a relatively cold substrate. This prevents large grain growth, which is particularly noticeable in silver. These smoother films scatter much less flux and produce better filters.

If the substrates are positioned more than 50 cm from the source, electron beam evaporation is a practical way to deposit aluminum rapidly. A multiple turret E-gun usually provides for a rotary stage; each crucible may be moved in turn into the electron beam. The crucibles that contain the aluminum are fabricated from HDA (43), which is a composite of titanium diboride and boron nitride. Using 4 kW of power, we have deposited a 15-nm thick layer in less than 2 sec on a substrate that is 60 cm from the crucible.

There are two means of controlling the thickness: (1) a quartz crystal and (2) optical transmission. In the first system, a quartz crystal is driven by an electronic oscillator. The crystal is "naked" inside of the vacuum system so that the evaporants are deposited directly on its surface. The mass on its surface alters its resonant frequency. This shift in frequency is proportional to the thickness of the evaporant, provided the physical density does not vary with time. Most of the commonly used commercial crystal monitors are similar to the one described by Glang (44). Their oscillation frequencies are in the neighborhood of 5 MHz.

The frequency shift of the crystal Δf is

$$\Delta f \cong 0.3\rho h \qquad (89)$$

where ρ is the mass density (in gm/cm^3) and h is the thickness (in Å). By substituting the density of aluminum (2.7 gm/cm^3) into this equation, we get a rule of thumb that each angstrom of thickness produces one hertz of frequency change.

The frequency shift can be determined with either analog or digital circuitry. Modern instruments use digital circuitry to measure the period over a 1-sec averaging time. These are unusable for rapid evaporations, since the thickness could be sampled only five times during a 5-sec deposition. This problem can be circumvented by constructing faster circuitry than is available in these commercial instruments. In one custom unit (45), a 48-MHz local oscillator is used to measure the period. The instrument is capable of measuring the thickness to a precision of 1 Å over an averaging time of 0.1 sec. Not the least of the problems in using a quartz crystal with a thermal evaporation source is that the heat can cause the count to indicate a "negative" thickness as large as 2 or 3 nm.

The thickness can also be monitored during deposition by measuring the change in the radiant transmittance as a slab of glass or fused quartz is coated with the metal layer. The wavelength that is used for monitoring need not be the wavelength at which filter is prepared. For example, a convenient wavelength to monitor the thickness of aluminum is the resonance emission line at 254 nm from a mercury discharge lamp. One virtue of this wavelength is that the aluminum is more transparent at shorter wavelengths than at longer wavelengths. This means that the aluminum layer will attenuate the longer wavelengths from the lamp, as for example the emission lines near 360 nm. At 254 nm, a 30-nm thick layer of aluminum reduces the transmittance of fused quartz to approximately 2.5%, i.e., one-fortieth of its initial value.

One method of avoiding problems associated with the nonlinearity of the electronics is to feed the signal from the photodetector into a potentiometer. The voltage from the "tap" on the potentiometer is then measured by an amplifier. This is used to deposit the 30-nm thick aluminum layer as follows. After the light source for the transmission monitoring has stabilized, the potentiometer is set at one-fortieth of its initial scale setting. The meter reading on the amplifier output is noted. Then, when the aluminum is evaporating rapidly, the shutter is opened and the potentiometer tap is switched to full scale. The amplifier meter initially reads above full scale. The aluminum film then attenuates the beam until the meter reading reaches the previously noted value, at which time the film is the proper thickness and the evaporation source is shuttered. In practice, although the meter needle is moving rapidly, the transmittance after completion of the aluminum layer is usually within 10% of the target value.

2. Reflection or Transmission Monitoring

An effective method of monitoring the *optical thickness* of dielectric layers is to measure the reflectance *in situ* as the layer is being deposited. The italics are used to emphasize a virtue of this method—namely, that it measures the phase thickness of a layer—the same parameter that appears in Eq. (18). The advantage is that even if the refractive index of the layer changes, reflection monitoring adjusts the metric thickness so that at least the correct β is obtained for that layer. The same is not true of other systems, such as the quartz crystal monitor.

A rather specialized reflection monitoring system is shown in Fig. 48. A light source is imaged on the monitoring place via a spherical mirror. The reflected flux is collected by another mirror and transferred to a photomultiplier. A stack of interference filters selects the wavelength. The reflection optics makes the system easy to align in the ultraviolet part of the spectrum. Regardless of whether a filter stack or a monochromator is used

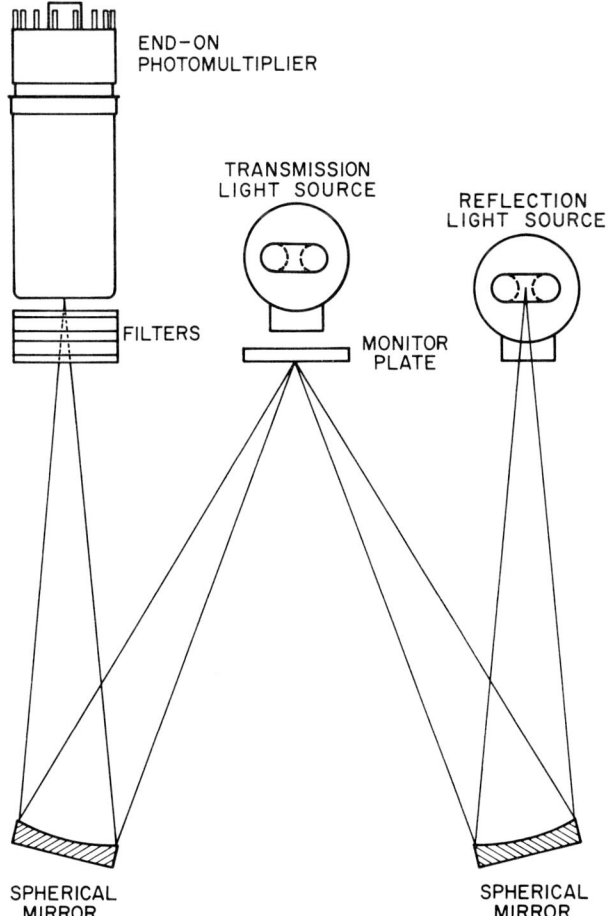

FIG. 48. A reflection monitoring system for the deposition of ultraviolet filters (*14*).

to select the wavelength, the system should be free of scattered flux and also produce a relatively narrow bandwidth.

In many monitors, it is arranged so that the flux can be reflected from either the filter itself or from separate monitor plates. There are several advantages to monitoring on the filter itself. The first is that the rate of formation of a film is usually different, depending on whether it is deposited on a fresh glass (or quartz) substrate or deposited on the surface of other films. This is caused by the different nucleation rates of the film during its early stages of formation. It can also be produced by a temperature differential between the monitor plate and filter. Regardless of the cause, this error can be eliminated by monitoring on the filter itself, rather than on separate monitor plates.

Another advantage of monitoring the filter itself is that the system has a "memory" of what was evaporated previously, thus making it possible to correct for past mistakes. This is particularly true in the deposition of quarter-wave stacks (46), as for example the quarter-wave layers that are used in the filters cited in the caption to Fig. 25.

One final comment should be made about the reflection monitor shown in Fig. 48. The angle of incidence is not zero and therefore the layers are deposited slightly thicker than they should be. Low-index layers of index 1.38 would be 1.8% too thick and the high-index layers (index 2.30) would be only 0.6% too thick. Such small corrections are important when high-precision filters are fabricated. Macleod (47) described a monitoring system that can function in either the transmission or reflection mode, with the flux normally incident upon the filter.

3. Examples of Optical Monitoring

Examples illustrate how optical monitoring is accomplished by displaying the reflectance R and transmittance T of several filters as they are deposited. Several *caveats* should be mentioned about these computed curves. The first is that the quoted R and T values do *not* take into account second-surface reflection losses. That is, the transmittance of a *single* air–glass interface is 95.7%, whereas the second-surface reflection loss reduces this to approximately $(0.957)^2 = 91.6\%$. Thus to iterate, all of the R and T values are quoted for a *single surface*.

The second *caveat* is that optical constants of all films change with thickness and do not stabilize until they attain at least 10–30 nm depending on the material. This is why the curves in Figs. 49 and 50 are fictional when the layers are thin.

Another *caveat* relates to the monitoring wavelength. Just because the operator monitors the filter's transmission at $\lambda_0 = 700$ nm *in vacuo* does not assure that it will be centered at this wavelength after the bell jar is opened. The coating's adsorption of water vapor and other contaminants shifts the passband—usually to longer wavelengths. The cementing of the cover glass to the filter permits the cement to permeate the relatively porous layers, thereby shifting the passband (48–50). Regardless of the wavelength at which the passband is eventually located, we monitor these two filters at a nominal $\lambda_0 = 700$ nm in this example.

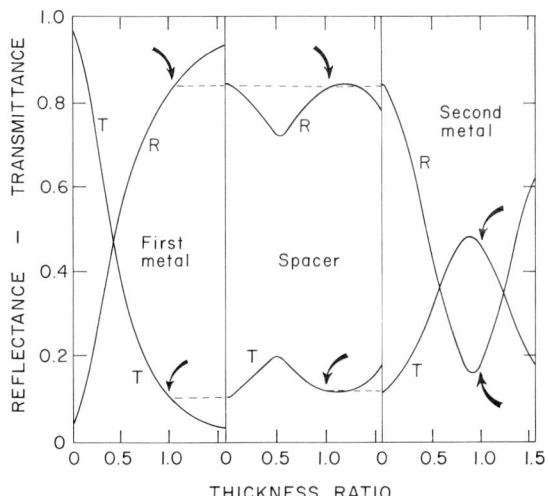

FIG. 49. The computed reflectance R and transmittance T of an MDM filter, as each layer is deposited. The abscissa is the ratio of the thickness to the correct value. The optical thickness of the spacer of the first-order filter ($m = 1$) is 276 mm.

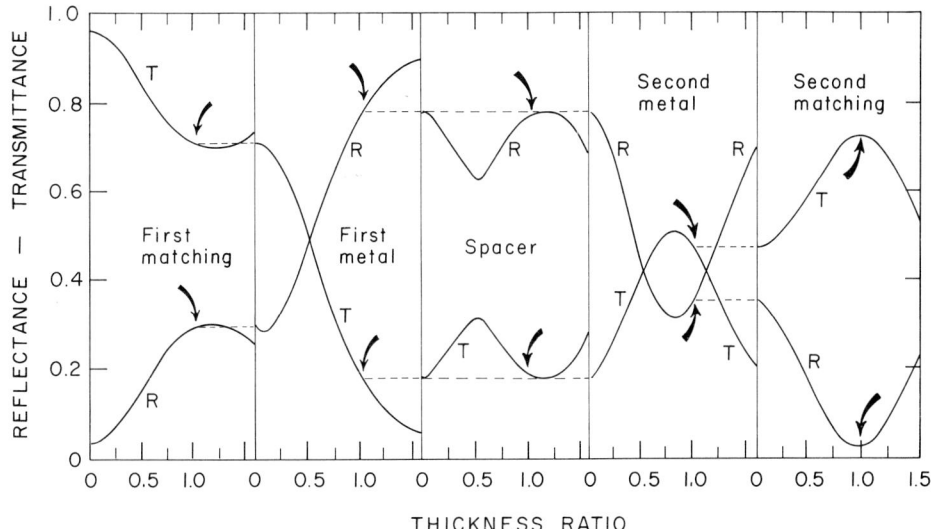

FIG. 50. Same as Fig. 49, except that the design is the five-layer MD'M'D'M cited in the caption to Fig. 39.

The first example is the monitoring of the first-order MDM filter cited in the caption to Fig. 24. As shown in Fig. 49, the initial transmittance of the bare glass substrate is 95.8% and the reflectance is 4.2%. The more reliable way to monitor the first metal layer of silver is via transmission; the deposition is terminated when T reaches 12%, as indicated by the arrow in the figure. The abscissa on each part of the figure is the fraction of the correct thickness. The scale extends from 0 to 1.5 and thus we can see what happens when an "overshoot" occurs. In the case of the metal, the T just keeps decreasing monotonically.

The next step is to deposit the magnesium fluoride spacer layer. The prescription is to deposit this film until either the transmittance or the reflectance attains the value it had initially. This is useful because it implies that the scale reading on the photometer need not be calibrated in absolute units. This would mean that if, before the start of the deposition of the magnesium fluoride, the photometer scale reading is, say, 56 units, then the deposition should be terminated when the same reading of 56 recurs. The reason for this is given later. This method also has the advantage of producing a usable filter, even though its $T_{\max}$ might not be exactly what is required. For example, suppose the silver was deposited too rapidly and, due to a time-lag in the shuttering, was deposited too thickly. Thus its thickness might be 27 nm or even 30 nm, rather than the desideratum of 25 nm. Regardless of its thickness, the spacer optical thickness will be deposited correctly, provided that the prescription mentioned earlier in this paragraph is followed.

The second metal layer of silver is then deposited. The reflectance decreases to a minimum of 15% and the deposition is terminated when it increases slightly to 18%. The transmittance increases from its initial value of 12% to a maximum of 49% and then decreases to 46%. We might ask, Why did we not halt the evaporation when the maximum transmittance was attained? We would have, indeed, if the filter were *not* to be cemented and, thus, air were the incident medium for the final filter. But since the filter is cemented, we must coast past this maximum.

If the filter were uncemented, then the second metal layer should be a bit thinner than the first metal film. The optical monitoring process makes these compensations automatically.

The second example is the monitoring of five-layer design H'MD'MH' cited in the caption to Fig. 40. It is preferable to monitor the first layer in reflection, because the change in the photometer scale reading is the greatest. It is also important that some calibration of the photometer be made before the deposition is started. Let us assume that this will be monitored in reflectance and that the second-surface reflection from the back side of the monitor plate has been eliminated. This is accomplished by either rough-grinding the reverse side of the monitor plate or coating it with a black laquer.

The single surface reflectance of 4.2% (for glass of index 1.517) furnishes a calibration for the photometer. The reflectance at which the evaporation is to be terminated is 29.7, i.e., a ratio of 7.1:1. The procedure is to set the gain of the photometer until it reads some arbitrary value and then terminate the deposition when it increases by a factor of 7.1. For example, the initial meter reading could be set at 10 units, and the deposition halted when it attains 71 units. Although a similar calibration could be made for the transmission monitor, the reflection method is more reliable.

Using the procedure outlined in the previous paragraph, the 2.30 index layer of zinc sulfide is then deposited. The deposition of the first metal layer of silver also involves a calibration of the transmission photometer. The decrease in transmittance is from 70.3 to 18.1%—a ratio of 3.7:1. Thus the photometer could be set at 100 units before at the start of the silver deposition and the source is shuttered when it reads 27 units.

The deposition of the magnesium fluoride D spacer layer follows precisely the same procedure as cited earlier in Section VIII,3.

The deposition of the second metal layer is easier because there is a "check point" when the transmittance attains a minimum and starts to increase. The T starts at 19.2%, increases to 51.2% (a ratio of 2.7:1) and decreases to 47.1% (a ratio of 2.47:1 of the initial reading).

Suppose the transmission photometer is set at 100 units at the start of the deposition. Translating ratios into scale readings, the minimum should reach 37 units and then increase to 41 units. Now the operator can note the minimum reading that is actually attained during the deposition and exercise his prerogative of adjusting the point at which the deposition is terminated. Suppose the indicated minimum was not the desideratum of 37 but rather 39. Then a good compromise would be to terminate the deposition at two units higher, namely, at $41 + 2 = 43$.

The final layer H' is simply deposited until either a maximum transmittance or a minimum reflectance is attained.

Finally, we justify the statement made earlier in this section about the optimum thickness of the spacer layer—namely, that it should have the same radiant reflectance as the uncoated metal layers on its emergent side. Suppose that we are fabricating an MDM filter and have just deposited the first metal layer. The admittance of this layer is Y. In the case of the 25-nm silver layer, this is $\hat{Y} = 0.73 - j3.9$. We convert this admittance into an amplitude reflection coefficient via Eq. (8). The ρ_1 for this silver layer is plotted in Fig. 51. In Section IV,3, we noted that when one-half of the

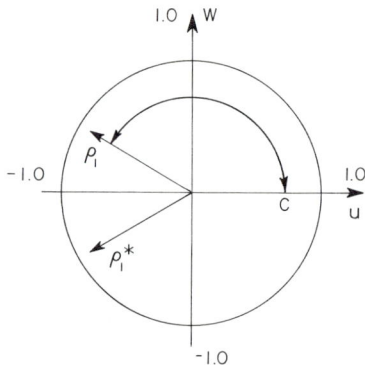

FIG. 51. The amplitude reflection coefficient ρ_1 after the deposition of a metal layer and the coefficient ρ_1^* after the spacer has been correctly applied in the construction of an MDM filter.

spacer layer is deposited on its surface, then the admittance is pure real and lies on the positive real axis, as shown as point c in Fig. 51. Due to symmetry, it follows that complete spacer D transforms the amplitude reflection coefficient ρ_1 into its complex conjugate ρ_1^*. The same is true of the admittance; the D layer transforms $\hat{Y}$ at the metal surface into a value of $\hat{Y}^* = 0.73 + j3.9$. Since the radiant reflectance is independent of the phase of ρ, it achieves the same value as the uncoated metal when the thickness of the spacer is correctly adjusted.

4. Cementing and the Addition of Blocking Filters

The cementing of the cover glass on the completed filter has several purposes. First, this seals it and helps to stabilize the wavelength of its transmission band. It is also necessary to protect "soft" coating materials such as cryolite, zinc sulfide, lead fluoride, and silver from being abraded or scratched. The sealing of the edges of the filter with the cement also prevents it from being damaged by water, whether in the form of high humidity or a liquid.

Auxiliary filters usually must be added to attenuate the flux in the regions where the filter has transmission "leaks." Consider, for example, the filters shown in Fig. 24. The transmittance is increasing rapidly below 500 nm due to the onset of second- or third-order transmission bands. Absorption filters are an effective way of reducing the transmission in the short-wavelength region. It is possible to mix a chemical dye in the optical cement. However, sometimes these dyes are not chemically stable or are unavailable in certain spectral regions. The only recourse is to use absorbing glass for the blocking. It would also have to be used if the filters were of interferometric optical quality.

Another technique, which was pioneered by Geffcken (*51*), is to cement two MDM filters together. Figure 52 shows the transmittance of tandem array of such filters. The scale of the ordinate is double logarithmic. It is tempting to imagine that the transmittance T_a of the tandem array of filters is simply

$$T_a = T_1 T_2 \tag{90}$$

i.e., the product of the transmittances of the two components. This does not hold true because interreflections between the surfaces, as shown in Fig. 52, are neglected. If we assume that the multiply reflected beams from the surfaces add incoherently and that the detector collects all of them, then the transmittance is

$$T_a = \frac{T_1 T_2}{1 - R_1 R_2} \tag{91}$$

where T_1, T_2, R_1, and R_2 are given in Fig. 6. Equation (91) reduces to Eq. (90) in the case where either of the coatings is nonreflecting and hence the denominator of Eq. (91) is unity.

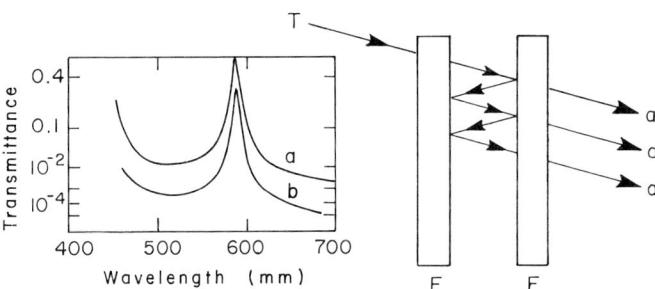

FIG. 52. (*Left*) The measured spectral transmittance of a single MDM filter (a) and two filters cemented together (b) (*51*). (*Right*) The beams of flux that multiply reflect between two such filters.

As an example, the upper curves in Fig. 53 depict the transmittance (in absorbance units) of a pair of one-M filters (*52, 53*). If Eq. (90) were to hold true, then the absorbance of the tandem array of the two filters would be the sum of the two components. Such is usually not the case, due to their reflectance. Figure 37 shows that the reflectance of the one-M filters is quite substantial in the offband part of the spectrum. The lower curve in this figure

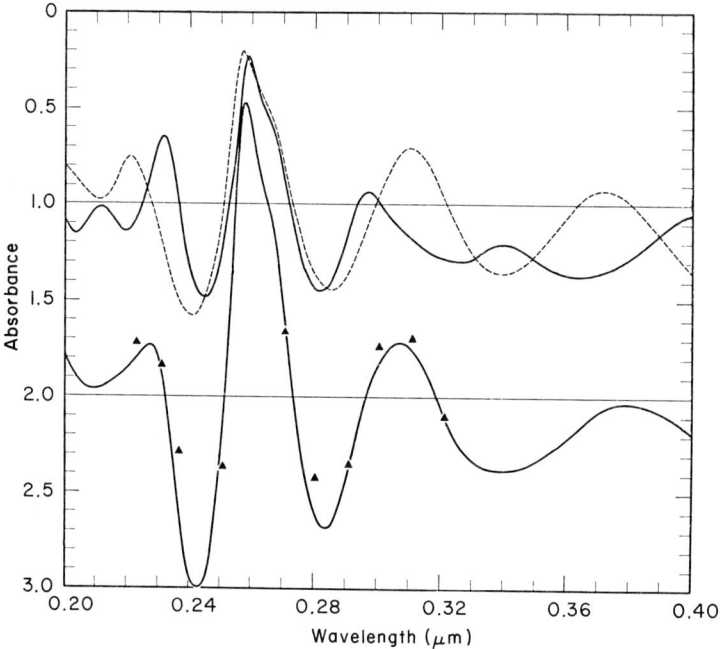

Fig. 53. The measured transmittance (in absorbance units) of one-M filters (upper curves) and the two filters placed in tandem (lower curve). The triangles are the computed transmittance.

shows the measured transmittance and the triangles are the values computed from Eq. (91), using measured values for T_1, T_2, and R_1, R_2.

There are no simple equations that can be applied if three or more filters are combined in the tandem array. Rather, a matrix theory (53) must be used to find the T_a of these arrays. These multicomponent arrays can include filters that contain metals, all-dielectric stacks, and absorbing glasses or dyes.

IX. Future Developments

The emphasis in this article has been on filters that are practical. This is why we usually assumed that the thicknesses of the metal films were always in excess of 20 nm and hence that the optical constants had stabilized. If it were possible to deposit ultrathin layers of metals such as silver and still maintain optical constants identical to those of the bulk, it would be possible to fabricate a multiple-cavity filter of the design

M D M D M D M · · ·

which grades from very thin M layers at the exterior to thicker layers in the interior. Such a bandpass filter would exhibit an extraordinarily large T_{max} and an offband transmittance of at least 6 absorbance units (54). Perhaps future advances in epitaxial deposition will make this a reality. The curves in Figs. 10–17 present convincing evidence that the alkali metals would make superb materials for one-M or multicavity filters. Only an extraterrestrial environment is suitable for such filters because of their chemical reactivity with water and oxygen.

ACKNOWLEDGMENTS

The National Aeronautics and Space Administration supported much of the research on ultraviolet filters (9, 14, 31, 32, 41, 45, 52, 53).

Figures 10–17, 27, 28, 38, and 48 are from Verne R. Costich.

References

1. W. Geffcken, Ger. Patent 716,153 (1942).
2. L. Hadley and D. Dennison, *J. Opt. Soc. Am.* **37**, 451 (1948).
3. A. Hermansen, *K. Dan. Vidensk. Selsk.* **29**, No. 13 (1955).
4. H. Wolter, in "Handbuch der Physik" (S. Flügge, ed.), Vol. 24, p. 500. Springer-Verlag, Berlin and New York, 1956.
5. C. Dufour, *J. Phys. Radium* **11**, 413 (1950).
6. D. J. Schroeder, *J. Opt. Soc. Am.* **52**, 1380 (1962).
7. G. Hass, H. Schroeder, and A. F. Turner, *J. Opt. Soc. Am.* **46**, 31 (1956).
8. P. H. Berning and A. F. Turner, *J. Opt. Soc. Am.* **47**, 230 (1957).
9. P. Baumeister, V. Costich, and S. Pieper, *Appl. Opt.* **4**, 911 (1965).
10. P. Baumeister, *Appl. Opt.* **8**, 423 (1969).
11. P. Berning, in "Physics of Thin Films" (G. Hass, ed.), Vol. 1, pp. 69–121. Academic Press, New York, 1963.
12. F. Abelès, in "Advanced Optical Techniques" (A. C. S. Van Heel, ed.), p. 178. North-Holland Publ., Amsterdam, 1967.
13. P. Smith, "Electronic Applications of the Smith Chart." McGraw-Hill, New York, 1969.
14. V. Costich, Ph.D. Thesis, Inst. of Optics, Univ. of Rochester, Rochester, 1965.
15. P. Berning, *J. Opt. Soc. Am.* **46**, 779 (1956).
16. P. M. Morse and H. Feshbach, "Methods of Theoretical Physics," p. 360. McGraw-Hill, New York, 1953.
17. J. Stone, "Radiation and Optics," p. 405, McGraw-Hill, New York, 1963.
18. G. Hass and L. Hadley, in "American Institute of Physics Handbook" (D. Gray, ed.) 3rd Ed., Sect. 6g. McGraw-Hill, New York, 1972.
19. A. Thelen, *Appl. Opt.* **4**, 977 (1965).

20. J. Apfel, *Appl. Opt.* **4**, 983 (1965).
21. F. A. Turner, *J. Phys. Radium* **11**, 444 (1950).
22. A. Thelen, in "Physics of Thin Films" (G. Hass and R. E. Thun, eds.), Vol. 5, p. 47. Academic Press, New York, 1969.
23. C. Ufford and P. Baumeister, *J. Opt. Soc. Am.* **64**, 329 (1974).
24. V. R. Costich, *Appl. Opt.* **9**, 866 (1970).
25. L. Holland, "Vacuum Deposition of Thin Films," p. 507. Chapman & Hall, London, 1956.
26. J. C. C. Fan, F. J. Bachner, G. H. Foley, and P. M. Zavracky, *Appl. Phys. Let.* **25**, 693 (1974).
27. R. M. Gelber, U.S. Patent 3,758,185 (1973).
28. J. Macdonald, "Metal-Dielectric Multilayers." Amer. Elsevier, New York, 1971.
29. J. Apfel, *J. Opt. Soc. Am.* **56**, 553A (1966).
30. G. Honcia and K. Krebs, *Z. Phys.* **156**, 117 (1959).
31. R. L. Maier, *Thin Solid Films* **1**, 31 (1967).
32. R. L. Maier, M.S. Thesis, Inst. of Optics, Univ. of Rochester, Rochester, 1966.
33. G. Hass, in "Applied Optics and Optical Engineering" (R. Kingslake, ed.), Vol. 3. Academic Press, New York, 1965.
34. H. Tabor, *Bull. Res. Counc. Isr. Sect. A* **5**, 119, 1956.
35. M. J. E. Golay, U.S. Patent 3, 000, 375 (1961).
36. K. F. Tripp, U.S. Patent 2, 590, 906 (1952).
37. R. N. Schmidt, U.S. Patent 3, 272, 986 (1966).
38. R. E. Peterson and J. R. Ramsey, *J. Vac. Sci. Technol.* **12**, 174 (1975).
39. J. H. Apfel and R. M. Gelber, U.S. Patent 3, 679, 291 (1972).
40. B. Bates and D. J. Bradley, *Appl. Opt.* **5**, 971 (1966).
41. D. H. Harrison, *Appl. Opt.* **7**, 210 (1968).
42. A. Halherbe, *Appl. Opt.* **13**, 1275 (1974).
43. Trade name of the National Carbon Company.
44. R. Glang, in "Handbook of Thin Film Technology" (L. I. Maissel and R. Glang, eds.), pp. 1–107. McGraw-Hill, New York, 1970.
45. J. Boles, M.S. Thesis, Inst. of Optics, Univ. of Rochester, Rochester, 1975.
46. V. R. Costich, *J. Opt. Soc. Am.* **62**, 1354A (1972).
47. H. A. Macleod, "Thin-Film Optical Filters," p. 236. Amer. Elsevier, New York, 1969.
48. J. Meaburn, *Appl. Opt.* **5**, 1757 (1966).
49. S. A. Furman, *Opt. Spectrosc.* **28**, 412 (1969).
50. A. M. Title, T. P. Pope, and J. P. Andelin, Jr., *Appl. Opt.* **13**, 2675 (1974).
51. W. Geffcken, *Z. Angew. Phys.* **6**, 249 (1954).
52. R. Hahn, M.S. Thesis, Inst. of Optics, Univ. of Rochester, Rochester, 1966.
53. P. Baumeister, R. Hahn, and D. Harrison, *Opt. Acta* **19**, 853 (1972).
54. J. H. Apfel, *Appl. Opt.* **11**, 1303 (1972).

Surface Plasma Oscillations and Their Applications

H. RAETHER

Institute of Applied Physics
University of Hamburg
Hamburg, Germany

I. General Considerations on Surface Plasmons 145
   1. Introductory Remarks 145
   2. Nonradiative Surface Plasmons on a Thin Film 149
   3. Radiative Surface Plasmons on a Thin Film 159
   4. Interaction of Nonradiative and Radiative Plasmons via Roughness 164
   5. Surface Plasmons on Spheres and on Voids 166
   6. Comparison with Guided Light Modes 168
II. Excitation of Radiative Surface Plasmons 171
   1. General Remarks 171
   2. Excitation by Electrons 173
   3. Excitation by Light 180
III. Excitation of Nonradiative Surface Plasmons 199
   1. Excitation by Electrons: Electron Energy Losses 199
   2. Excitation by Electrons: Light Emission via Roughness 219
   3. Excitation by Light 223
   4. Excitation by Evanescent Light Waves (ATR Method) 230
   5. Closing Remarks 255
   References 255

I. General Considerations on Surface Plasmons

1. INTRODUCTORY REMARKS

Surface plasmons exist in the boundary of a solid (metal or semiconductor) whose electrons behave like those of a quasi-free electron gas. These plasmons represent the quanta of the oscillations of surface charges, which are produced by exterior electric fields in the boundary. For large wave vectors (K_x) these plasma waves behave like real surface waves, their electromagnetic field being concentrated around the boundary at a region of the order of $1/K_x$ or $\lesssim 10$ Å. For small wave vectors these fields extend far into space and resemble more and more those of a photon propagating along the boundary. This type of electromagnetic excitation of a solid has an energy of about 10 eV for large K_x and an energy width of about 1 eV, which corresponds to a relatively short lifetime of about some ten oscillations or 10^{-14}–10^{-15} sec (*1–4a*).

Oscillations of the electron density, for example, of the density of the valence electrons, also occur in the interior of the solid. These are called volume plasma oscillations and can be excited by electrons or light. They are not discussed here (*5*).

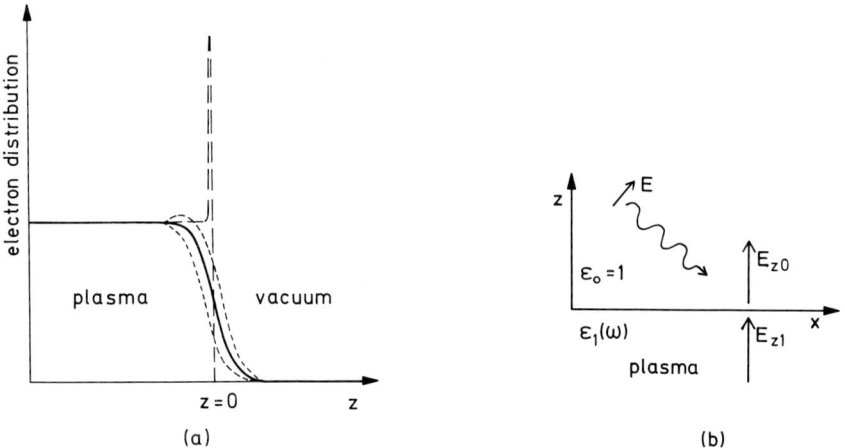

FIG. 1. Electron distribution in the plasma–vacuum boundary ($z = 0$). The peak in (a) represents a surface charge at a given moment produced by the difference of the normal components of an incoming electric field inside and outside the plasma, see (b). The oscillations of this surface charge produce the field of the surface waves. In a more realistic representation the electron density is not a δ-function, but is smoothed out.

If an exterior electric field acts on a plasma boundary, it produces surface charges due to the discontinuity of the normal component of the electric field at the boundary between the two media with the dielectric functions ε_0 and ε_1 (see Fig. 1)

$$\sigma = E_{z1} - E_{z0} \tag{1}$$

Their oscillations move as surface plasma waves along the surface

$$\sigma(x, t) = \sigma \exp i(K_x x - \omega t) \tag{2}$$

as shown in Fig. 2. K_x is the wave vector along the boundary. To a first approximation the electron density decays at the metal surface like a step function, and the charges oscillate within the screening length (Thomas–Fermi length) of a few angstroms [see the broken line (long dashes) in Fig. 1a]. In a more detailed picture the electron density in the metal edge has to be represented as a smooth curve $n(z)$ (6) (see solid line in Fig. 1a). This is valid in the static case. The charge oscillations or surface plasmons are schematically displayed in Fig. 1a as broken lines (short dashes). A theory of the oscillating density profile does not yet exist (see Section I,2,c and p. 202).

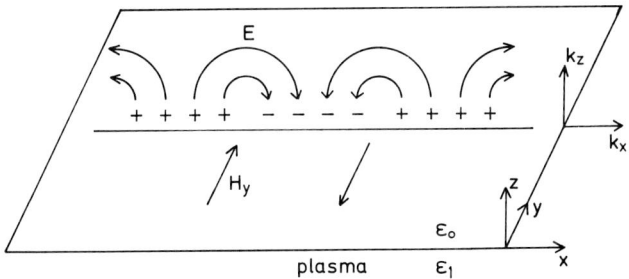

FIG. 2. Scheme of a p-polarized surface wave on a boundary of a semi-infinite plasma. It is called a p-polarized wave if the E field oscillates in the (x, z) plane. The charge oscillations in the boundary are coupled with electric (E_x, E_z) and magnetic fields (H_y). Exchange of E and H leads to an s-polarized wave.

The charge oscillations are coupled with high frequency electromagnetic fields extending into space, as Fig. 2 demonstrates. The dependence of the normal component E_z on z is given by

$$E_z = \text{const} \exp i(K_x x - \omega t) \cdot \exp iK_z \cdot z \quad (3)$$

which characterizes two types of surface waves: nonradiative and radiative ones. If K_z outside the plasma (K_{z0}) is imaginary, the fields decrease exponentially as $\exp(-|K_{z0}|z)$, so that these waves transport no energy away from the boundary, whereas E_z has oscillatory or radiative character if K_{z0} is real. Since the wave vectors K_x and K_{zv} are coupled in accordance with

$$K_x^2 + K_{zv}^2 = \varepsilon_v(\omega/c)^2 \quad (4)$$

where ε_v is the dielectric function of the medium ($v = 0$ outside the plasma and $v = 1$ inside the plasma), we obtain the important relation

$$K_{zv} = [\varepsilon_v(\omega/c)^2 - K_x^2]^{1/2} \quad (5)$$

Now we can characterize the different oscillations as follows, where ε_0 (medium outside the plasma) shall be real.

If
$$\sqrt{\varepsilon_0}(\omega/c) < K_x, \tag{6a}$$

then K_{z0} is imaginary which results in nonradiative plasmons.

If
$$\sqrt{\varepsilon_0}(\omega/c) > K_x, \tag{6b}$$

then K_{z0} is real which results in radiative plasmons.

Both types of waves travel along the boundary of the plasma. They are guided waves, in contrast to the free electromagnetic waves or photons.

The eigenfrequency ω of these waves and the wave vector (K_x) in Eq. (2) are related by a dispersion relation $\omega(K_x)$, which is different for the two types of surface waves, since K_z has a different character. This is also true for either a boundary of a semi-infinite plasma or a plasma film. Most of the experiments with plasmons were performed with thin films, such as silver, aluminum, and magnesium. Since many phenomena are more pronounced in thin films than on a boundary of a semi-infinite plasma, we will consider mainly plasmons on a thin plane-parallel plasma slab.

The relation $\omega(K_x)$ for surface waves can be derived by applying Maxwell's equations together with the continuity conditions for **D** (electrical displacement) and **H** (magnetic field) in various ways. For example, we regard electric field waves inside a plasma film of thickness d, traveling in both z directions ($\pm$) and traveling away from the film surface outside the film. The continuity conditions yield a homogeneous system of equations which, if the following condition is fulfilled, has solutions for p-polarized fields:

$$D = (\varepsilon_1 K_{z0} + \varepsilon_0 K_{z1})(\varepsilon_2 K_{z1} + \varepsilon_1 K_{z2})$$
$$+ (\varepsilon_1 K_{z0} - \varepsilon_0 K_{z1})(\varepsilon_2 K_{z1} - \varepsilon_1 K_{z2})e^{2iK_{z1}d} = 0 \tag{7}$$

and for s-polarized fields:

$$(K_{z0} + K_{z1})(K_{z1} + K_{z2}) + (K_{z0} - K_{z1})(K_{z1} - K_{z2})e^{2iK_{z1}d} = 0 \tag{8}$$

We have assumed that the plasma slab is bounded on both sides by media of different dielectric functions ε_2 and ε_0. In the following sections we will derive the dispersion relations for different layer systems. If the dispersion relation is fulfilled the amplitude of the density oscillations increases and produces strong electromagnetic fields at the surface. This resonance behavior leads to interesting experiments which are described in the following sections.

2. Nonradiative Surface Plasmons on a Thin Film

a. Symmetric Layer System. A plasma slab with the dielectric function $\varepsilon_1(\omega)$ can be covered on both sides with media of equal dielectric constants ε_0 (e.g., $\varepsilon_0 = 1$: air or vacuum). This system is symmetric (see Fig. 3). If the two media are different, as in Fig. 56 (inset) on p. 230, we have an asymmetric layer system. In the following we shall treat the symmetric system.

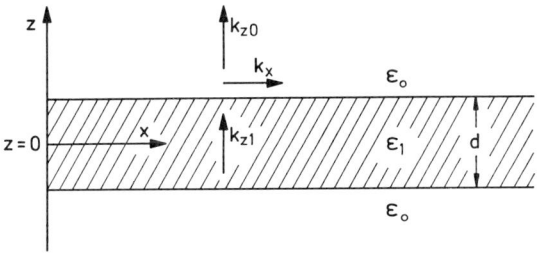

FIG. 3. A plane parallel plasma film with dielectric function $\varepsilon_1(\omega)$ bounded on both sides by a medium with the dielectric function $\varepsilon_0 (\geq 1)$.

The nonradiative plasma modes on this slab have been described by Ritchie (*1*). They are characterized by the exponential decrease of their electromagnetic fields both outside and inside the plasma film, or K_{z0} and K_{z1} are both imaginary.

Equation (7) can be written for *p*-polarized fields:

$$L^+ = \varepsilon_1 K_{z0} + \varepsilon_0 K_{z1} \tanh \frac{1}{2i} K_{z1} d = 0 \quad (9)$$

$$L^- = \varepsilon_1 K_{z0} + \varepsilon_0 K_{z1} \coth \frac{1}{2i} K_{z1} d = 0 \quad (10)$$

There are no solutions for *s*-polarized plasma modes since the electric field component E_y is continuous and thus no surface charges are produced.

Equations (9) and (10) lead to two solutions for ω: a high frequency mode ω_+ [Eq. (9)] and a low frequency mode ω_- [Eq. (10)]. As we shall see later they have electric field components $E(z)$ asymmetric and symmetric about $z = 0$. The existence of these two solutions can easily be derived from Eq. (7) if we neglect retardation. This means we replace K_{z1} and K_{z0} in Eq. (7) by iK_x using Eq. (5) with $\varepsilon_2 = \varepsilon_0$ and $c = \infty$ to obtain

$$\frac{\varepsilon_0 + \varepsilon_1}{\varepsilon_0 - \varepsilon_1} = \mp e^{-K_x d} \tag{11}$$

In the following we derive the equations for a free electron gas. In this case we use as the dielectric function in the film

$$\varepsilon_1 = \varepsilon_{r1} + i\varepsilon_{i1} = 1 - \omega_p^2/[\omega^2(1 + 1/i\omega\tau)] \tag{12}$$

or if $\omega\tau \gg 1$

$$\varepsilon_{r1} = 1 - \omega_p^2/\omega^2 \tag{13}$$

$$\varepsilon_{i1} = (1/\omega\tau)(\omega_p^2/\omega^2) \tag{14}$$

with the plasma frequency

$$\omega_p^2 = (ne^2)/(m\,\Delta) \tag{15}$$

where n is the electron density, $\Delta = (4\pi 9 \times 10^{11})^{-1}$ C/V·cm, and τ is the relaxation time.

If we introduce this function into Eq. (11), the splitting of the frequency, as a function of $K_x d$, becomes (1)

$$\omega_\pm = [\omega_p/(1 + \varepsilon_0)^{1/2}](1 \pm e^{-K_x d})^{1/2} \tag{16}$$

ω_+ coming from the tanh relation represents the high frequency mode. With large $K_x d$ both surface modes approach the well-known value

$$\omega_s = \omega_p/(1 + \varepsilon_0)^{1/2}$$

or

$$\omega_s = \omega_p/\sqrt{2} \quad \text{if} \quad \varepsilon_0 = 1 \tag{17}$$

This splitting of the p-mode into two frequencies $\omega_\pm$ is due to the fact that the plasma oscillation fields of equal frequency on the two sides of the film interact with each other and split up; the thinner the film compared with the plasma wavelength ($\lambda = 2\pi/K_x$), or the smaller $K_x d$, the more pronounced the splitting.

At lower K_x values, retardation becomes important (2), so that Eq. (16) has to be replaced in the case of a free electron gas (7) by

$$\omega_\pm(K_x) = \frac{\omega_p}{[1 + \varepsilon_0(K_{z1}/K_{z0})]^{1/2}} \left(\frac{1 \pm e^{-iK_{z1}d}}{1 \mp \gamma e^{-iK_{z1}d}}\right)^{1/2} \tag{18}$$

where

$$\gamma = \frac{\varepsilon_0 K_{z1} - K_{z0}}{\varepsilon_0 K_{z1} + K_{z0}} \tag{19}$$

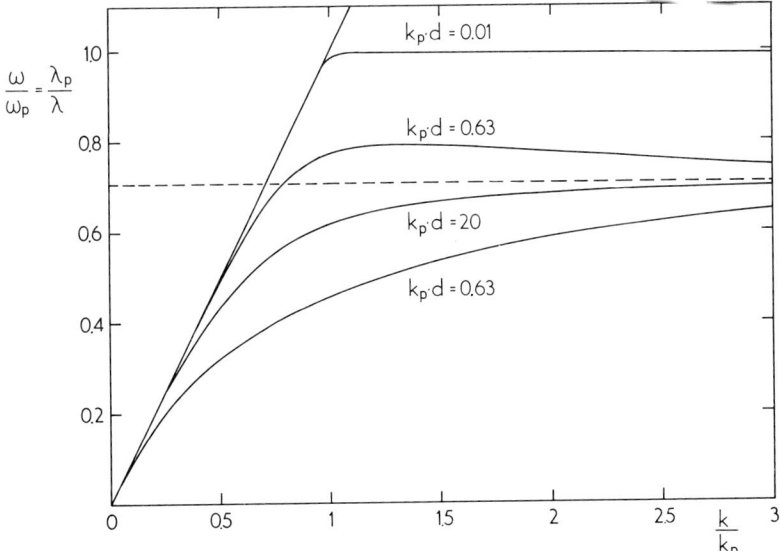

FIG. 4. Dispersion relation of nonradiative plasmons in a free electron gas for a thin plane-parallel slab of thickness d. $K_p = 2\pi/\lambda_p$, the wave vector at the plasma wavelength. The product $K_p d$ is varied.

Figure 4 shows the characteristic dependence for different values of $K_p d$ ($K_p = \omega_p/c$) calculated using Eq. (18). The nonretarded relation for $K_p d \to \infty$ is shown by a dotted line in Fig. 4. At very low K_x values the two frequencies $\omega_\pm$ are related to K_x by

$$K_x = (\omega/c)[\varepsilon_0 \varepsilon_1/(\varepsilon_0 + \varepsilon_1)]^{1/2} \tag{20}$$

and approach the light line

$$K_x = (\omega/c)\varepsilon_0^{1/2} = K_l \tag{21}$$

However, K_x of the plasmons always remains larger than that of light of the same frequency by an amount

$$K_x - K_l = (\omega/c)[1/(2|\varepsilon_r|)] \tag{22}$$

Further we see that $K_x \to \infty$ is equivalent to

$$\varepsilon_1 = -\varepsilon_0$$

or

$$\varepsilon_1 = -1 \quad \text{if} \quad \varepsilon_0 = 1 \quad \text{(vacuum)} \tag{23}$$

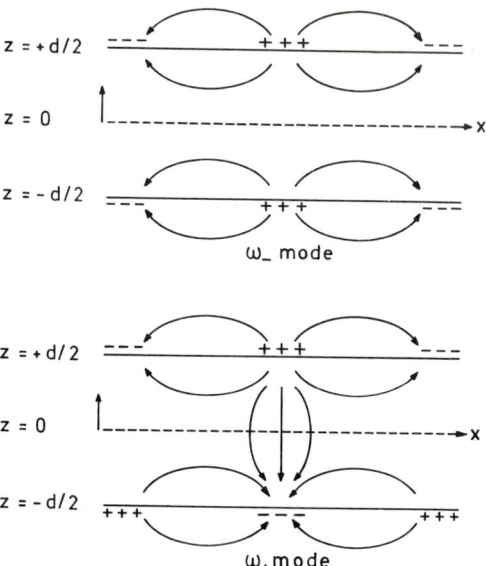

FIG. 5. The distribution of charges and electric fields of the ω_+ and ω_- mode for one wavelength in a thin plasma slab.

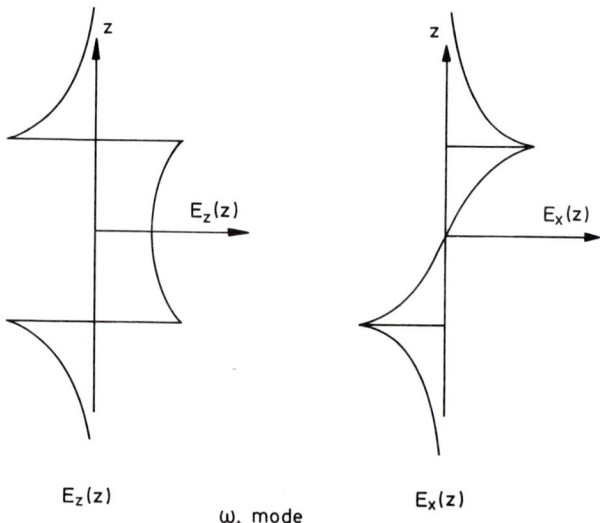

FIG. 6. The strength and direction of the E_z and E_x components as functions of z for the ω_+ mode. Outside the slab, along the z direction, these components decay exponentially. The discontinuity of E_z at $z = \pm d/2$ is due to the surface charges as indicated in Fig. 5. Inside the slab hyperbolic functions determine the decay of the field which is characteristic for the nonradiative plasmon.

whereas small K_x values are equivalent to

$$\varepsilon_1 \to -\infty \qquad (24)$$

The character of the modes just described becomes more evident by looking at the fields connected with them. In Figs. 5 and 6 the charges and the field with its x and z components are shown. This field system propagates in the K_x direction as $\exp i(K_x x - \omega t)$.

The dependence of $E_z(z)$ and $E_x(z)$ is shown in Fig. 6 for the ω_+ mode. It has the higher frequency, since the charges of one sign face those of the other sign leading to a higher restoring force than in the case of the ω_- mode. The dependence of the field of the ω_- mode on z is obtained by exchanging E_z and E_x inside the slab, retaining the continuity conditions at the boundaries.

If the thickness of the film d becomes large so that $|K_{z1}|d \gg 1$, we have the case of a semi-infinite boundary of a metal, and the dispersion relation for Eq. (9) or (7) becomes

$$\varepsilon_0 K_{z1} + \varepsilon_1 K_{z0} = 0 \qquad (25)$$

for p-polarized oscillations ($E_y = 0, H_x = H_z = 0$). Introducing Eq. (5) into Eq. (25) one obtains (8)

$$K_x = (\omega/c)[\varepsilon_0 \varepsilon_1/(\varepsilon_0 + \varepsilon_1)]^{1/2} \qquad (26)$$

which represents the dispersion relation of nonradiative plasmons on a semi-infinite metal surface for all values of K_x. It is identical with Eq. (20). In Fig. 4 this case is indicated by the large value of $K_p d = 20$.

A similar calculation for s-polarized fields ($H_y = 0, E_x = E_z = 0$) gives

$$K_{z1} + K_{z0} = 0 \qquad (27)$$

a condition which cannot be fulfilled for $\varepsilon_0 \neq \varepsilon_1$. This is due to the absence of E_z so that no surface charges exist.

Because $\varepsilon(\omega)$ is complex, the term K_x in Eq. (26), consists of a real part describing the spatial propagation and an imaginary part correlated with the spatial damping. If the damping is small, i.e., $\varepsilon_i \ll |\varepsilon_r|$, we obtain from Eq. (26) with $\varepsilon_0 = 1$ (9, 10)

$$\text{Re}(K_x): \quad K_{rx} = K_x = \frac{\omega}{c}\left(\frac{\varepsilon_r}{\varepsilon_r + 1}\right)^{1/2} \qquad (28a)$$

$$\text{Im}(K_x): \quad K_{ix} = \frac{\omega}{c}\frac{\varepsilon_i}{2\varepsilon_r^2}\left(\frac{\varepsilon_r}{\varepsilon_r + 1}\right)^{3/2} \qquad (28b)$$

The value of K_{ix} determines the decrease of the intensity of the oscillation along the boundary according to

$$\exp(-2|K_{ix}|x) \tag{29}$$

Thus we can obtain the length L at which the intensity has decreased to e^{-1} as

$$L = 0.5/|K_{ix}| \tag{30a}$$

In the case of a free electron gas

$$L = (\omega_p^2/2c\tau)(1/\omega^2) \tag{30b}$$

In the infrared region for $\lambda \sim 10 \ \mu m$, the value of L reaches 1.5 cm for silver, whereas in the visible region it reduces to $\sim 10^2 \ \mu m$ (*11, 12*). The intensity loss is due to internal damping.

The fields produced by the surface charges decrease inside and outside the plasma as $\exp i(K_{zv}z)$. For large values of K_x this means the field decays as

$$\exp(-|K_x|z) \tag{31}$$

so that the fields are highly concentrated at the boundary, thus representing real surface waves.

For small K_x values the field vanishes in the interior of the film as

$$\exp[-(\omega/c)(|\varepsilon_{r1}|)^{1/2}z] \tag{32}$$

since $|\varepsilon_{r1}| \gg 1$, the field is concentrated in the interior near the boundary of the plasma, whereas outside it extends far into the vacuum, or ε_0-medium, as

$$\exp[-(\omega/c)\varepsilon_0^{1/2}z] \tag{33}$$

In the language of radio waves, this is equivalent to the Sommerfeld wave.

b. Asymmetric Layer System. An important system for optical applications and electron energy loss experiments is represented by the plasma film covered on both sides with media of different dielectric constants, $\varepsilon_0 \neq \varepsilon_2$. The dispersion relation is given by Eq. (7). For a large thickness d (K_{z1} imaginary!), as well as for large K_x, Eq. (7) splits up into the two dispersion relations for the separated boundaries 0/1 and 1/2

$$K_{xv} = (\omega/c)[\varepsilon_v\varepsilon_1/(\varepsilon_v + \varepsilon_1)]^{1/2} \tag{34}$$

where $v = 0$ or 2. This gives two separate branches (see Fig. 7). In the case of a free electron gas the asymptotic values of the frequencies for $K_x \to \infty$ become

$$\omega_{sv} = \omega_p/(1 + \varepsilon_v)^{1/2} \tag{35}$$

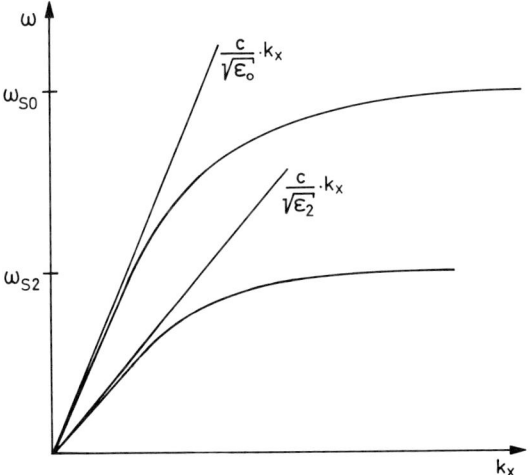

FIG. 7. The dispersion relation of nonradiative modes on a plasma film covered on both sides with substances of different dielectric functions ε_0 and ε_2; it splits into two branches especially for $2|K_{z1}|d \gg 1$.

For optical experiments, the case wherein d values are small but

$$2|K_{z1}|d > 1 \tag{36}$$

is important and shall be discussed (13). The dispersion relation of the boundary 1/2 alone is given by Eq. (34), and its solution will be called $K_x^{\infty m}$. We assume $\varepsilon_2 = 1$ (air) (see Fig. 56 inset on p. 230.) Now we approach the boundary 0/1 (e.g., glass/plasma) to the 1/2 surface (plasma/air) retaining the condition (36), and obtain the slightly changed solution:

$$K_x^m = K_x^{\infty m} + \Delta K_x^m \tag{37}$$

Here ΔK_x^m is a complex quantity whose real part $\text{Re}(\Delta K_x^m)$ displaces the resonance value of $K_x^{\infty m}$ and its imaginary part $\text{Im}(\Delta K_x^m)$ changes the damping of the system. The calculation yields

$$\Delta K_x^m = \frac{\omega}{c} \frac{2}{1 + |\varepsilon_r|} \left(\frac{|\varepsilon_r|}{|\varepsilon_r| - 1} \right)^{3/2} \exp(-2K_x^{\infty m} d) \cdot r_{01}^p(K_x^{\infty m}) \tag{38}$$

$r_{01}^p(K_x^{\infty m})$ is the complex Fresnel coefficient of the p-polarized field

$$r_{01}^p(K_x^{\infty m}) = \frac{K_{z0}/\varepsilon_0 - K_{z1}/\varepsilon_1}{K_{z0}/\varepsilon_0 + K_{z1}/\varepsilon_1} = \frac{\varepsilon_0^2 - a^2}{\varepsilon_0^2 + a^2} + i\frac{2a\varepsilon_0}{\varepsilon_0^2 + a^2} \tag{39}$$

where

$$a^2 = |\varepsilon_r|(\varepsilon_0 - 1) - \varepsilon_0 \tag{40}$$

and where ε_0 is the dielectric function of, for example, glass as in Fig. 58b on p. 233. The displacement $\text{Re}(\Delta K_x^m)$ is positive as long as

$$\varepsilon_0^2 - a^2 > 0 \tag{41}$$

This is valid for $|\varepsilon_1| \leq \varepsilon_0[(\varepsilon_0 + 1)/(\varepsilon_0 - 1)]$, a relation which holds in general except in the low energy region. In the case of silver, Fig. 8 demonstrates that ΔK_x^m is positive for $\lambda > 4500$ Å and negative for smaller wavelengths as long as the thicknesses $d > 200$ Å. At very small thicknesses the curves approach the light line LL.

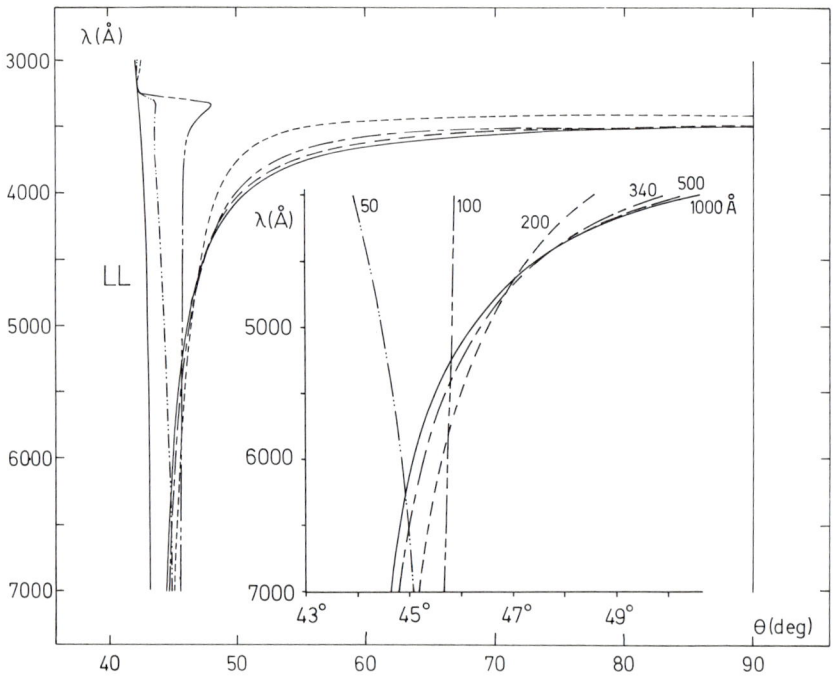

FIG. 8. Dispersion relation of silver films of different thicknesses (50–1000 Å), calculated for the substrate quartz ($\varepsilon_0^{1/2} = 1.46$ at $\lambda = 5000$ Å). The dependence on λ (Θ) is plotted. The displacement of the curves of thin films can be positive or negative compared to the thick film (1000 Å). The sign depends on whether $\lambda \gtreqless 4500$ Å. The inset shows more details.

The imaginary part of ΔK_x^m signifies that this system has in addition to the interior damping given by Eq. (28b), an additional damping term described by $\text{Im}(\Delta K_x^m)$. This is obtained by replacing in Eq. (38) the term $r_{01}^p(K_x^{\infty m})$ by its imaginary part [see Eq. (39)]. As we shall see later this additional damping is attributed to radiation damping of the asymmetric system or, in other words, the nonradiative plasmons in the boundary 1/2 become coupled with oscillatory solutions (light) in the medium 0. The reverse is also true; photons coming from the medium ε_0 can excite nonradiative plasmons in the boundary 1/2 (see Fig. 56 on p. 230). This way of exciting nonradiative plasmons is applied in the attenuated total reflection (ATR) method (see Section III,4). If we replace medium 0 by medium 2, i.e., if we return to a symmetrical system, ΔK_x^m becomes zero.

A system very similar in its physical properties to that just described is obtained by regarding ε_2 as the plasma, ε_1 as a vacuum or air gap of thickness d, and ε_0 as a medium with a dielectric function, such as, quartz with $\varepsilon_0 > 1$ (see Fig. 58a on p. 233). Both these layer systems are used in the experiments of exciting nonradiative surface plasmons by light or in experiments that study the radiation of these plasmons.

For the experiments with energy losses produced by excitation of surface plasmons, we have further an important system—a plasma film assumed for simplicity very thick (medium 2), covered with a thin contaminating film, such as, an oxide coating adjacent to the medium ε_0 (vacuum) (see Fig. 9).

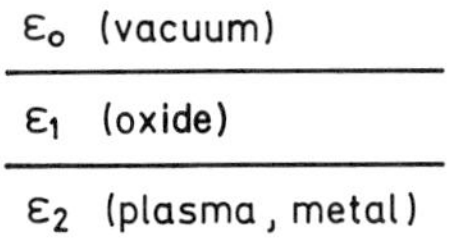

ε_0 (vacuum)

ε_1 (oxide)

ε_2 (plasma, metal)

FIG. 9. Three layer system: a semi-infinite plasma (ε_2) covered with an oxide film of finite thickness.

If the plasma or the metal (ε_2) also has a finite thickness, we have a three-boundary system (vacuum/plasma/coating/vacuum) whose dispersion relations become rather cumbersome and difficult to discuss in a general way. In these cases the relations have to be calculated with a computer for different parameters.

c. The Dispersion at Large K_x Values. In the above considerations it has been tacitly assumed that the electron density at the boundary of the metal goes to zero as a step function. Such an abrupt profile does not seem realistic, since the electron density decreases gradually to zero. Furthermore, the thermal vibration and to a certain extent surface roughness will contribute to this continuous transition of the electron density to zero.

The surface plasma oscillations for large K_x values should be a sensitive probe of surface profile. At large K_x values, or at small wavelengths (comparable with a, see below), the plasmon frequency should become dependent on the diffuseness of the metal boundary. One expects that the frequency at large K_x values becomes K_x-dependent for $K_x > K_s = \omega_s/c$ instead of approaching asymptotically the classical value ω_s for clean surfaces [Eq. (17)]. An expression

$$\omega^2 = \omega_p^2(1 + AK_x + BK_x^2 + \cdots) \tag{42}$$

can result, where the coefficients depend strongly on the form of the surface profile. Thus this question becomes an important aspect of surface physics. It has been studied with energy loss experiments of electrons.

A first theoretical approach to this problem was made by Ritchie (14) who obtained an expression

$$\hbar\omega(K_x) = \hbar\omega_s(0) + [\tfrac{3}{5}E_F(\hbar^2/2m)]^{1/2}K_x \tag{43}$$

where E_F is the Fermi energy. Here the thermal fluctuation has been taken into account but the density profile was still a steplike function. It results in a strong dependence on K_x (see Fig. 37 on p. 203) [see also (14a–c)].

Further calculations regarding the diffuseness of the boundary showed that the dispersion relation depends on the form of the electron density. To illustrate the situation, Fig. 10 shows the calculated $\omega(K_x)$-dependence on the parameter a, which measures the decay of the electron density from

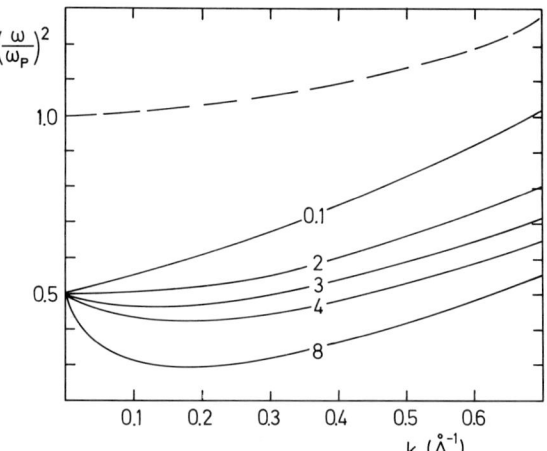

FIG. 10. Dispersion relations of the nonradiative surface plasmons on a magnesium surface at larger values of K_x. The numbers indicate the length in angstroms in which the electron density decays from the bulk value to zero. The broken line shows the dispersion relation of volume plasmons [from Bennet (15)].

the metal interior to zero outside the metal (15). It shows that larger a values are connected with negative slopes. Furthermore the theoretical considerations yield several surface frequencies if the density profile deviates from the sharp cutoff (15, 16). These theories are based on the hydrodynamic model and not a microscopic one, so corrections are expected, but it is difficult to see to what extent they influence the result. A series of articles dealing with these questions have been written [see (17)].

Recently the dispersion of surface plasmons has been discussed in connection with its sensitivity to adsorbed atoms (18). Experimental results will be discussed in Section III,1.

3. Radiative Surface Plasmons on a Thin Film

The dispersion relation of radiative surface plasmons is obtained by introducing a real K_{z0} in Eq. (7). The real value means that the field outside the film becomes oscillatory so that radiation damping is present. We consider a symmetric system and obtain the following relations for p-polarized fields with $\varepsilon_0 = 1$

$$\varepsilon_1 K_{z0} - iK_{z1} \tan K_{z1} d/2 = 0 \tag{44a}$$

$$\varepsilon_1 K_{z0} + iK_{z1} \cot K_{z1} d/2 = 0 \tag{44b}$$

These equations are identical with those given by Kliewer and Fuchs (19) and Pafomov (20). The interesting solutions are those for the p-modes which we will discuss in the following. Those for the s-modes are of no interest, being too strongly damped.

The important p-mode or Ferrell mode is described by the relations of Eq. (44a) characterized by $\varepsilon_1 \sim 0$ if we neglect interior damping. A number of interesting experiments have been made with either electrons or light exciting this radiative surface mode. From Eq. (44a), if we look for solutions with $K_x \sim 0$, we obtain

$$\tan(\varepsilon_1)^{1/2}(\omega/c)/(d/2) + i(\varepsilon_1)^{1/2} = 0 \tag{45}$$

Neglecting the interior damping, the relation is fulfilled by

$$\varepsilon_1 = 0 \tag{46}$$

or

$$\omega = \omega_p \tag{47}$$

for a free electron gas.

For small K_x values the function $\omega(K_x)$ depends on the thickness d of the film and for thick films, or $K_p d/2 \gg 1$, it becomes

$$\omega = \omega_p(1 + \tfrac{1}{2}\sin^2\Theta) = \omega_p\left[1 + \frac{1}{2}\left(\frac{K_x}{K_p}\right)^2\right] \qquad (48)$$

where

$$K_x = K_p \sin\Theta \qquad (\omega \sim \omega_p)$$

For thin films, such that $K_p d/2 \ll 1$,

$$\omega \simeq \omega_p\left[1 + \frac{1}{2}\left(\frac{K_p d}{2}\right)^2 \tan^2\Theta\right] \simeq \omega_p\left[1 + \tfrac{1}{8}(K_p d)^2\left(\frac{K_x}{K_p}\right)^2\right] \qquad (49)$$

The shape of the dispersion curve of the radiative mode is seen in Fig. 23 (p. 181) and Fig. 19 (p. 172). The consequence of a real K_z is that the modes show a more or less strong radiation damping. Introducing the lifetime τ which determines the decay of the field amplitude

$$\exp(-t/\tau) \qquad (50)$$

the value $\omega\tau$ becomes

$$\frac{1}{\omega_p \tau} = \frac{1}{K_p d}\sin\Theta\tan\Theta = \frac{1}{K_p d}\left(\frac{K_x}{K_p}\right)^2 \qquad \text{for } K_p d/2 \gg 1 \qquad (51)$$

and

$$\frac{1}{\omega_p \tau} = \frac{K_p d}{4}\sin\Theta\tan\Theta = \frac{K_p d}{4}\left(\frac{K_x}{K_p}\right)^2 \qquad \text{for } K_p d/2 \ll 1 \qquad (52)$$

Both terms become identical at $K_p d = 2$, the maximum value of damping (see also Fig. 11). If we consider the intensity, we have to double the values of $(\omega_p \tau)^{-1}$. The value $K_p d/2 = 1$ or $d = \lambda_p/\pi$ which separates the two regimes occurs for $d \sim 300$ Å in the case of Al with $\lambda_p = 720$ Å.
We see that the radiation damping decreases to zero with smaller K_x so that the character of this mode as an eigensolution becomes more pronounced. This can be understood from the field configuration of Fig. 12.
For larger K_x values, the eigenfrequency and its radiation damping as a function of K_x is shown with $W = K_p d$ as the parameter ($K_p = \omega_p/c$, $d =$ film thickness) in Fig. 11. Here $\Omega' = \omega/\omega_p$ is plotted against $\sin\Theta$ where $K_x = (\omega/c)\sin\Theta$. The vertical line $\Theta = 90°$ signifies the light line. The frequency increases with $\sin\Theta$ for values of $W > 1.7$ whereas it remains nearly constant and decreases to zero approaching the light line for $W < 1.7$. The correlated damping $\Omega'' = 1/\omega_p\tau$ passes through a maximum value which can be seen in Fig. 11 by looking at Ω'' at constant $\sin\Theta$ for different d or

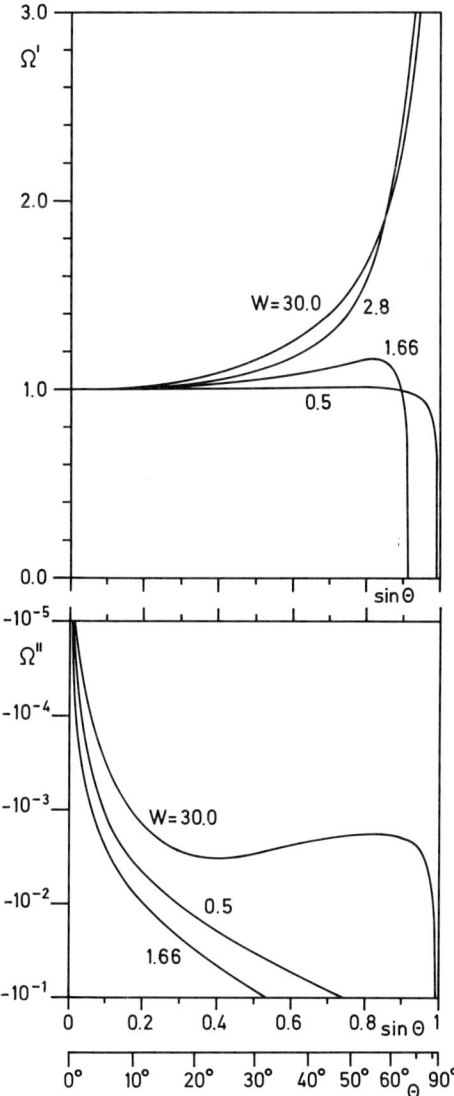

FIG. 11. The dispersion of the Ferrell mode or $\Omega^1 = \omega/\omega_p$ as a function of $\sin \theta$ (above) and the dependence of its damping on $\sin \Theta$ (below). Parameter $W = K_p d$.

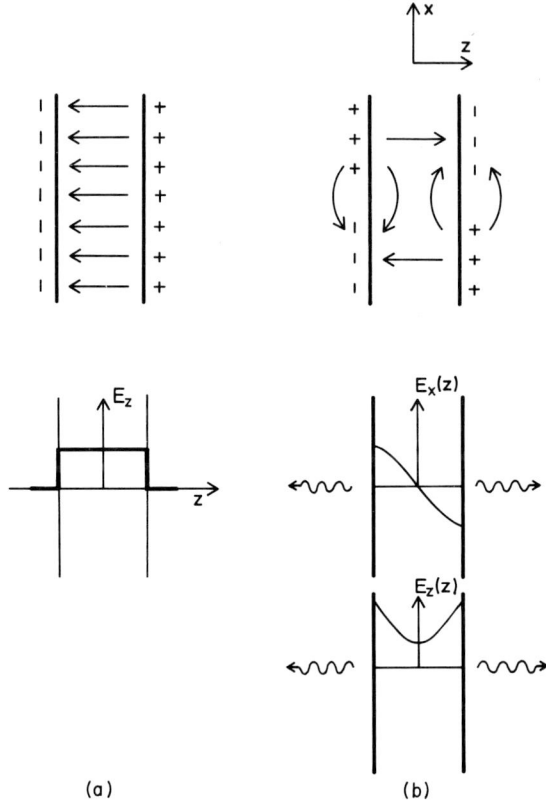

FIG. 12. (a) The tan mode $m = 0$ for $K_x = 0$. It has the field components E_{z1} = const and $E_{x0} = E_{x1} = 0$. This corresponds to the field of an infinitely extended plane condenser. (b) The field components of the tan mode $m = 0$ for $K_x > 0$ and small thicknesses. The mode is coupled to exterior oscillatory solutions.

W values. At higher K_x values the damping becomes so strong that the term eigenmodes loses its meaning. A value of $\Omega'' = 0.1$ means that after a time $t \sim 1.5\, T_p$ where $T_p = 2\pi/\omega_p$ is the time of one period of the mode, the amplitude has decreased to $1/e$ ($\Omega''\omega_p t = 1$). For comparison, the interior damping due to ε_i amounts to $\Omega'' = 0.03$ for Al at $\omega = \omega_p$.

The solutions given above are obtained under the assumption that we observe these modes at a well-defined angle Θ. Because $K_x = (\omega/c)\sin\Theta$ is complex, it is required that the quotient of the two complex quantities K_x/ω must be real or that the two terms K_x and ω have the same phase factor (constant-angle modes) (19).

Further radiative modes. The trigonometric functions in Eq. (44) imply further solutions or higher harmonics of the Ferrell mode. For small K_x and ε_1 values or small $\varepsilon_1(K_{z0}/K_{z1})$, the solution of the tan mode can be approximated roughly by

$$\tfrac{1}{2}K_{z1}d \sim m\pi \tag{53a}$$

where m takes on integral values, whereas those of the cot modes are given by

$$\tfrac{1}{2}K_{z1}d \sim (m + \tfrac{1}{2})\pi \tag{53b}$$

This means that there are standing waves in the z direction.

These expressions allow one to estimate the eigenfrequencies. For example, the frequencies of the tan modes from Eq. (53a) can be written, for a free electron gas

$$\left(\frac{\omega}{\omega_p}\right)^2 \sim 1 + m^2 \frac{\pi}{(K_p d/2)^2} + \left(\frac{K_x}{K_p}\right)^2 \tag{54}$$

At $K_x = 0$, the eigenfrequencies become, for $K_p d = 10$,

$$\begin{aligned} m &= 1 & \omega_1 &= 1.2\omega_p \\ m &= 2 & \omega_2 &= 1.6\omega_p \\ m &= 3 & \omega_3 &= 2.1\omega_p \end{aligned} \tag{55}$$

m indicates the number of knots in the $E_x(z)$ component inside the film.

For $m > 0$ the fields of these p-modes have finite amplitudes at the boundaries $z = \pm d/2$ for $K_x > 0$ as well as for $K_x = 0$, and join oscillatory solutions exterior to the film. Therefore, for $K_x = 0$, these modes also lose energy by radiation.

The tan mode $m = 0$, shows a different behavior as shown in Fig. 12. For $K_x = 0$ the E_x-field component becomes zero and inside the film $E_z = $ const. This field configuration is that of a plane condenser of infinite extension, which does not radiate, in agreement with $1/\omega_p \tau \to 0$ for $K_x \to 0$ [see Eqs. (51) and (52).] As a consequence, light of frequency ω_p falling at normal incidence ($K_x = 0$) on a plasma slab cannot excite the tan mode $m = 0$. At $K_x > 0$ however, (see Fig. 12), the field configuration changes and radiation losses or light absorption occur, but are much smaller than at the other p-modes with $m > 0$.

This strong damping can be understood since the refractive index in the interior of the slab $0 \le \varepsilon_1^{1/2} < 1$ is smaller than the index outside the slab ($\varepsilon_0^{1/2} = 1$). Light falling from the interior on the boundary is reflected but not totally so that light power is lost. The term virtual mode (*19*) is therefore more adequate.

4. Interaction of Nonradiative and Radiative Plasmons via Roughness

The radiative plasma oscillations are coupled with free electromagnetic waves (photons), so they are automatically damped by radiation. As we shall see in Fig. 20 these guided waves can transform into light by picking up momentum perpendicular to the surface $\hbar K_z$ with a continuous number of values.

The nonradiative plasmons, however, cannot transform into light by picking up $\hbar K_z$ values, their momentum being always larger than that of light of the same frequency. This is valid for smooth surfaces. However, the possibility exists that surface waves propagating along a rough surface—statistical roughness or in the simplest case a sinusoidal profile (grating)—can change its momentum into $\hbar K'_x = \hbar(K_x + K_r)$ with $K_r = 2\pi/a$, where a is the period of the sinusoidal component of the roughness. It is thus possible to transform nonradiative surface waves into radiative ones and vice versa.

This process has important consequences:

(a) Radiative surface plasmons with a wave vector K_x and frequency ω_p run in the rough surface. The K_x value may be small, so that the resultant K'_x lies inside the light line as well as on the constant part of the dispersion curve $\omega = \omega_p$ (see Fig. 13, process 0 → 1). The wave vector of the oscillation can be regarded as the tangential component of the light wave vector $K_l = \omega/c$ which hits the boundary of the plasma at an angle Θ_0 or $K_x = (\omega/c)\sin\Theta_0$. The new K'_x can, therefore, be written as

$$K'_{x1} = (\omega/c)\sin\Theta_1 \qquad (56)$$

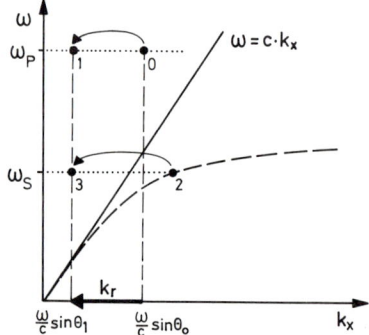

FIG. 13. Coupling of surface plasmons by roughness. The roughness wave vector K_r couples radiative waves (0 → 1) as well as nonradiative with radiative waves (2 → 3) and vice versa.

This means that the photon is emitted in the direction Θ_1 instead of the direction Θ_0 (the direction of the reflected light). It describes the fact that photons are observed not only at the reflection angle Θ_0 but also in other directions (diffuse scattering). We call this phenomenon plasma resonance emission if the light has the resonance frequency ω_p.

This is a one-dimensional model. In reality the phenomenon takes place in the two-dimensional surface, so that we have to consider, instead of a light line and a dispersion line, a light circle and a dispersion circle for $\omega = $ const and $K_\perp = (K_x^2 + K_y^2)^{1/2} = $ const, if the propagation is isotropic. Then we observe the radiation in the direction $K_\perp$ with an azimuth different from the plane of incidence (see Fig. 14). If the incoming photon is p-polarized in the plane of incidence, the radiation now has an s component, given by the projection of the component E_x on the direction perpendicular to $K_\perp$.

(b) The same can happen to nonradiative waves, as Fig. 13 indicates, by the process $2 \to 3$. This interaction leads to a coupling with photons or an

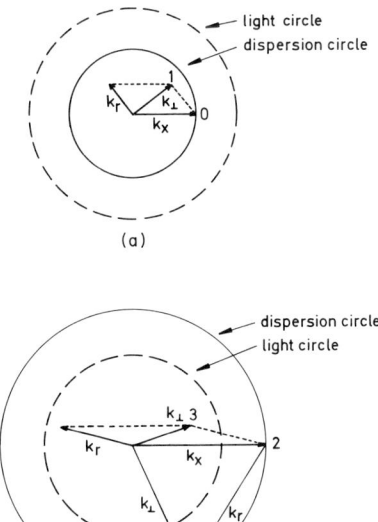

FIG. 14. (a) Coupling of the radiative oscillation K_x via K_r with the radiative oscillation $K_\perp$. For simplification the surface inside the solid line circle is regarded as a plane or the dependence $\omega(K_x)$ is neglected. (b) The same for the coupling between nonradiative and radiative modes ($2 \to 3$) (upper part) and between nonradiative modes ($2 \to 4$) (lower part).

additional damping by radiation and thus to the possibility of detecting the nonradiative plasmons by light. There are many experiments to demonstrate this coupling process. The two-dimensional representation is sketched in Fig. 14b.

The reverse process is also important. It allows one to excite nonradiative plasmons of the same frequency by light of frequency $\omega < \omega_s$ as proved by many experiments. Here the wave vector $\mathbf{K}_r$ is added to $\mathbf{K}_\perp$ so that $\mathbf{K}_\perp + \mathbf{K}_r$ reaches the dispersion relation across the light circle (process 3 → 2 in Fig. 14b).

If K_r has a value such that the tip of $K_\perp$ remains on the dispersion circle, we have the same situation as with the radiative wave of Fig. 14a.

We have seen that the intensity of plasmons moving in a given direction is reduced by the internal damping given by Im (ε). It follows from the above considerations that on rough surfaces this intensity can further be reduced by radiation as well as by scattering processes. By the radiative decay plasmons disappear and by scattering processes they leave the original direction of propagation.

5. Surface Plasmons on Spheres and on Voids

Surface plasma modes exist on boundaries other than planes. As an example we consider a plasma sphere of radius r embedded in a dielectric whose dielectric constant is ε_a. By applying the usual boundary conditions, the eigensolutions give the following relation for the modes (21)

$$\varepsilon(\omega) = -[(m + 1)/m]\varepsilon_a \tag{57}$$

where $m = 1, 2, \ldots$.

Assuming a free electron gas and neglecting retardation and putting $\varepsilon_a = 1$, one gets

$$\omega_s = \left(\frac{m}{2m + 1}\right)^{1/2} \omega_p = \frac{\omega_p}{(2 + 1/m)^{1/2}} \tag{58}$$

The corresponding wavelength is given by

$$\Lambda = 2\pi r/m \tag{59}$$

As we see there exist a high number of modes starting with the mode of lowest energy $m = 1$,

$$\omega_s = \omega_p/\sqrt{3} \quad \text{(Mie mode)} \tag{60}$$

and converging to

$$\omega_s = \omega_p/\sqrt{2} \tag{61}$$

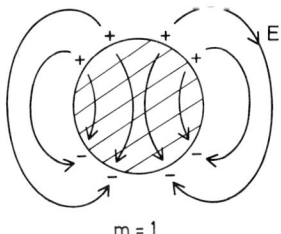

FIG. 15. Field distribution of the surface mode $m = 1$ on spheres with the dielectric function $\varepsilon(\omega)$.

at large values of m. The mode $m = 0$ gives $\omega_s = 0$. The mode $m = 1$ has a dipole field configuration, as Fig. 15 shows, and is, therefore, optically active; it absorbs and emits light. Putting $m = 2$ one gets the quadrupole mode.

One finds similar types of oscillations in voids or spherical cavities (22, 23). These voids have been observed in metals subjected to high radiation doses at elevated temperatures, as occurs in fast reactors (24). If ε_1 describes the dielectric behavior of the interior of the void, the relation for these modes is

$$\varepsilon(\omega) = -\varepsilon_1[m/(m + 1)] \tag{62}$$

In the case of a free electron gas and for $\varepsilon_1 = 1$, the frequency of these modes is given by

$$\omega_v = \omega_p[(m + 1)/(2m + 1)]^{1/2} \tag{63}$$

As with the sphere modes, those of the voids approach the plane modes for large m values. In contrast, however, void frequencies are higher than $\omega_p/\sqrt{2}$, the highest being $\omega_v = \omega_p$ for $m = 0$. This frequency has the same value as that of the bulk plasmon at $K_x = 0$, but in spite of the equality this is a surface mode; the charges oscillate in the surface of the void. The sign of the charges is the same around the interior of the void, so the mode has been called a breathing mode (monopole). An important consequence is the fact that due to this mode, $m = 0$, these voids attract each other with forces stronger than van der Waals forces (dipole forces) (23). This interaction can lead to the formation of a lattice of voids, a phenomenon which has already been observed with electron microscope methods (25).

The mode $m = 1$ has the frequency

$$\omega_v = \omega_p\sqrt{\tfrac{2}{3}} \tag{64}$$

resembling the field distribution pattern of a sphere.

Theoretical studies on small cubes (26) and cylindrical surfaces (27) have also been reported.

6. COMPARISON WITH GUIDED LIGHT MODES

As we have seen, the character of the plasma modes is determined by the z-component of the wave vector of the electromagnetic field. An imaginary K_z inside and outside the plasma slab leads to the nonradiative plasma oscillations, whereas a real K_z inside and outside the plasma film are correlated with radiative plasmons (Fig. 16a,b).

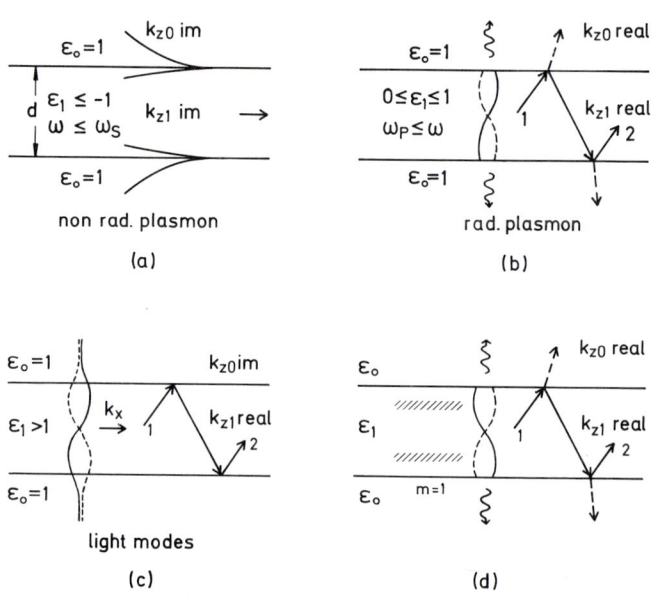

FIG. 16. Comparison of surface waves on a plasma. (a) Nonradiative plasmon, (b) radiative plasmon, (c) and (d) guided light modes.

A new kind of mode results if the film is not filled with a free electron plasma but with an insulator or semiconductor with all electrons bound, so that the refractive index $\varepsilon_1^{1/2}$ can attain values of $\varepsilon_1^{1/2} > 1$ with ε_1 assumed to be essentially real (see Fig. 16c). In the simplest case, a glass film in air fulfills this condition for visible light. If the light hits the boundary from the interior so that total reflection takes place and, in addition, the phase difference between points 1 and 2 amounts to $2\pi m$ (m whole numbers), the light intensity is confined to the film and is transported up to long distances. Under these conditions standing waves exist in the z direction according to

$$K_z d = m\pi \tag{65}$$

whereas in the x direction this field pattern moves with the phase velocity ω/K_x. This phenomenon may be called guided light modes. The number of knots in the standing wave is given by $(m - 1)$.

We see that, in contrast to the surface plasmons whose fields have their maximum in the surface, the light modes have their peak values in the interior.

By using glass fibers several kilometers long instead of thin films, this phenomenon can be applied to transport light energy over long distances. The damping of these optical transmission lines can be reduced to less than 3 dB/km. An analog to the radiative plasmons is obtained if $\varepsilon_0^{1/2} > \varepsilon_1^{1/2}$ so that no total reflection takes place (Fig. 16d). Here at every reflection a certain amount of light intensity is lost, leading to a strong radiation damping of the "mode." In the case of the radiative plasmons, the refractive index $\varepsilon_1^{1/2}$ is smaller than 1 since $\omega_p \leq \omega$.

The dispersion relations of these light modes can easily be derived from the Eq. (7) or (9) since they are eigensolutions of a thin film. However, the dielectric function $\varepsilon_1(\omega)$ is no longer given by Eq. (12) but by the refractive index of the insulator and its wavelength dependence. If we start with the symmetric case, Eqs. (9) and (10) can be written as

$$K_{z1}d = 2 \arctan \frac{\varepsilon_1}{\varepsilon_0} \frac{K_{z0}}{K_{z1}} + m\pi \tag{66}$$

If m is odd, the solutions are electric fields $E_x(z)$ symmetric to the middle plane of the film and are derived from Eq. (9). If m is even, the solutions are not symmetric and are given by Eq. (10). The s-polarized light modes whose dispersion relations do not contain the factor $\varepsilon_1/\varepsilon_0$ in Eq. (66) are as important as the p-polarized ones in contrast to the plasma modes. These equations are fundamental for the physics of wave guide modes (28, 29).

The asymmetric system is of more interest in connection with the questions treated here. If we write the dispersion relation given in Eq. (7) in analogy to Eq. (66), we obtain

$$2K_{z1}d = 2 \arctan \frac{\varepsilon_1}{\varepsilon_0} \frac{K_{z0}}{K_{z1}} + 2 \arctan \frac{\varepsilon_1}{\varepsilon_2} \frac{K_{z2}}{K_{z1}} + 2\pi m \tag{67}$$

In this form the dispersion relation can be interpreted as follows (see Fig. 16c). [The symmetric system in Fig. 16c must be replaced by an asymmetric one $(\varepsilon_0|\varepsilon_1|\varepsilon_2)$]. Constructive interferences result if the wave leaving point 1 arrives at point 2 with a phase difference $2\pi m$. The phase shifts φ_{01} and φ_{02} occurring at the reflection on the boundaries 1/0 and 1/2 are given by the two arc tan terms. They can easily be derived from the Fresnel coefficients of Eq. (39). Thus Eq. (67) can be written

$$2K_{z1}d = \varphi_{01} + \varphi_{02} + 2\pi m \tag{68}$$

The symmetric relation (66) follows immediately. This procedure represents a rather simple way of deriving the dispersion relations of Eq. (7).

The interesting asymmetric system of a wave guide film bounded on one side by vacuum and on the other side by a metal consists of a metal with ε_2, a wave guide film with ε_1, and vacuum with ε_0. Figure 17 depicts the dispersion relation of this layer system. Assuming that the ε_1 values are frequency independent, one can exchange ω with the wave guide film thickness d, the more interesting quantity, and plot the dependence of d against $\sin \Theta = [K_x/(\omega/c)]$. In this representation the lines $c/\varepsilon_0^{1/2} \cdot K_x$ and $c/\varepsilon_2^{1/2} \cdot K_x$ of the $\omega(K_x)$ diagram, shown in Fig. 7, become vertical lines with the abscissas $\varepsilon_0^{1/2}$ and $\varepsilon_1^{1/2}$. The different s- and p-modes characterized by the numbers m_s and m_p start at different values of d—in a symmetric system from the same d—and approach, asymptotically, the line $\varepsilon_1^{1/2}$. The number of light modes increases with the film thickness. The mode $m_p = 0$ alone behaves differently. It starts at $d = 0$ with

$$\sin \Theta = [\varepsilon_0 \varepsilon_2/(\varepsilon_0 + \varepsilon_2)]^{1/2} \quad (69)$$

which is identical with the dispersion relation for the free silver boundary and, for large d, approaches the vertical line [see Eq. (26)]

$$\sin \Theta = [\varepsilon_1 \varepsilon_2/(\varepsilon_1 + \varepsilon_2)]^{1/2} \quad (70)$$

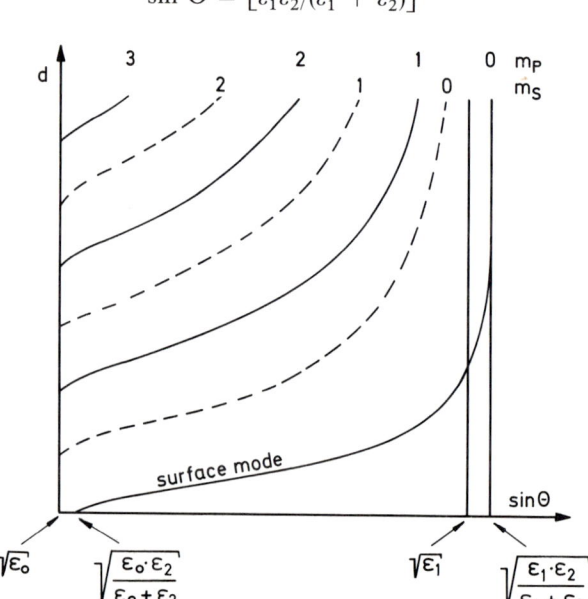

Fig. 17. Dispersion curves of guided light waves in an asymmetric system ($\varepsilon_0|\varepsilon_1|\varepsilon_2$). ε_1 is the dielectric function of the wave guiding film, ε_2 that of the metal, ε_0 that of vacuum (air).

as the limiting line which represents the dispersion curve of the metal film with a coating (ε_1) of infinite thickness. In between these limits the curve $m_p = 0$ represents the thickness dispersion. We shall come back to the experimental aspects of this question in Section III,4.

II. Excitation of Radiative Surface Plasmons

1. GENERAL REMARKS

Up to now we have discussed the properties of the different electromagnetic oscillations in a plasma film. In the following we review the different possibilities of exciting these waves and how these plasmons can be detected. The excitation is produced by the electric fields of electrons or of light coming from the exterior and transferring an energy $\hbar\omega$ and a momentum $\hbar K_x$ to the plasma system. If the energy $\hbar\omega$, together with the value of K_x, fulfills the dispersion relation $\omega(K_x)$ of the surface plasmons, the waves are excited. In general, the smaller the damping of the plasmons, the higher the amplitude of the plasma oscillations.

Another way to describe the resonance process is to say that the phase velocity of the exterior perturbing field v_{Ph} must equal the phase velocity of the plasma waves or

$$v_{Ph} = \omega/K_x \qquad (71)$$

In the case of radiative waves one has $\omega/K_x > c$, whereas for nonradiative waves $v_{Ph} < c$, with c the velocity of light.

If the perturbing fields are plane waves of light, we see immediately that radiative oscillations can be excited. Figure 18 demonstrates how the

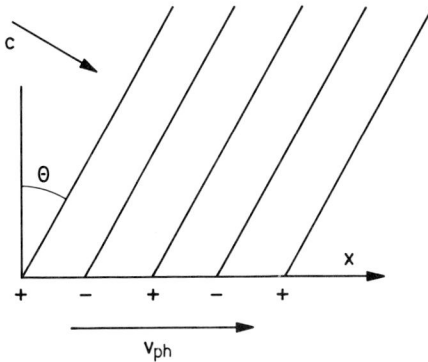

FIG. 18. Excitation of surface plasma waves by incoming plane waves. The phase velocity v_{Ph} of these waves along the boundary has a value of $\omega/K_l \sin \Theta = c/\sin \Theta > c$.

exterior electromagnetic field produces the oscillating surface charges that run along the surface with a phase velocity $(\omega/K_l \sin \Theta) > c$. This scheme demonstrates the coupling of the radiative modes with the free electromagnetic field. Nonradiative waves cannot be produced in such a way. Here the phase velocity of the incoming light has to be reduced to equal the phase velocity of the plasma waves which can be realized by different ways.

If electrons hit the surface at normal incidence ($\alpha = 0°$) they transfer both the energy $\hbar\omega$ and the momentum $\hbar K_{el} \vartheta$ to the surface, (see Fig. 19). Here K_{el} is the wave vector of the incoming electrons and ϑ the scattering angle of these electrons after crossing the boundary. Therefore, the phase velocity of the exterior perturbation $\omega/K_{el}\vartheta$ can thus have values $\gtreqless c$ depending on the choice of $K_{el}\vartheta$.

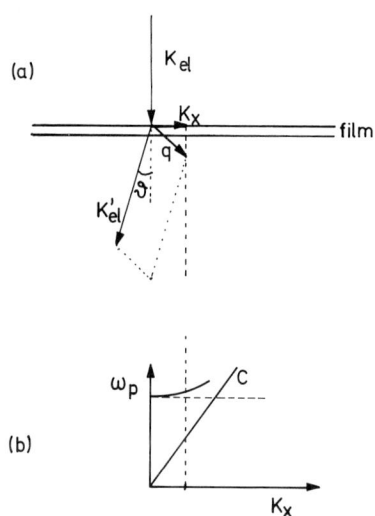

FIG. 19. (a) Schematic diagram of the excitation of radiative surface plasmons by electrons which transfer the momentum $\hbar q$ to the plasma and $\hbar K_x$ to its boundary. (b) The dispersion relation of the radiative plasmons is depicted by the curved solid line and that of the volume plasmons is indicated by the horizontal broken line.

Another problem is how to detect whether or not the excitation has taken place. This can be done by registering the energy losses of the exciting primary electrons or looking at the intensity of the light into which the radiative plasmons transform due to the inherent radiation damping, or measuring the radiation stimulated via roughness, in the case of nonradiative waves.

In all cases, we measure a resonance behavior of quantities such as the energy loss function, the emitted light intensity, the transmittance or the reflectance of light either as a function of the frequency ω or the transferred momentum $\hbar K_x$. The resonance represents the response of the system to exterior perturbation.

It should be emphasized that the plot of the frequency at which this resonance takes place as a function of the K_x values need not be identified with the dispersion relations in general. Both will approach each other the less the system is damped.

We shall describe in the following section the different experiments in which such plasmon resonances are observed by irradiating thin films containing a plasmalike electron gas (metals and semiconductor films) with electrons and/or light. First the radiative plasma oscillations are treated and then the nonradiative ones.

2. Excitation by Electrons

a. Plasma Radiation—Theoretical Remarks. Radiative plasmons can be excited by bombarding the surface of a plasma with electrons. Figure 19a shows the creation of a plasmon when electrons of energy $E_0 \gg \hbar\omega$, passing the boundary, transfer the momentum

$$\hbar q = \left[(\hbar K_{el} \vartheta)^2 + \left(\hbar \frac{\omega}{v} \right)^2 \right]^{1/2} \tag{72}$$

where K_{el} is the wave vector of an incoming electron, v its velocity, ϑ the scattering angle of the electron, and $\hbar\omega$ the energy transmitted to the plasma. Its projection $\hbar K_x$ to the boundary is given by

$$\hbar K_x = \hbar K_{el} \vartheta \tag{73}$$

for normal incidence of the electrons. The dispersion relation of the radiative plasmons correlates this K_x with an ω value which is ω_P for small K_x values (see Fig. 19b).

Lying left of the light line, the radiative surface plasmon of frequency $\sim \omega_p$ has a momentum $\hbar K_x$ smaller than light of the same frequency, $\hbar(\omega/c)$. The plasmon which can take up momentum $\hbar K_z$ of every value from the surface perpendicular to it can thus transform into light, as Fig. 20 shows. Therefore, it is possible to detect these surface oscillations by their radiation of frequency $\sim \omega_p$, as has been proposed by Ferrell (*30*). This plasma radiation emitted from the excited surface lies in general in the ultraviolet, e.g., for Ag at 3.8 eV, or in the vacuum ultraviolet region, e.g., for Al at 15 eV.

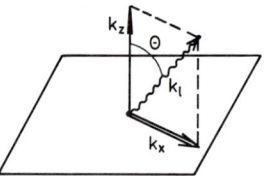

FIG. 20. Scheme of the decay of radiative plasmons of momentum $\hbar K_x$ into photons of momentum $\hbar K_l = \hbar \cdot (K_x^2 + K_z^2)^{1/2}$.

It represents a small frequency band of the so-called transition radiation. This electromagnetic transition radiation is emitted if electrons pass the boundary of a material with the dielectric constant $\varepsilon(\omega) = \varepsilon_r(\omega) + i\varepsilon_i(\omega)$ and has a broad continuous spectrum whose structure is determined by the spectral dependence of $\varepsilon_r(\omega)$. The theory of this phenomenon has been elaborated for a semi-infinite material and normal incidence of the electrons by Ginzburg and Frank (31), yielding the light intensity as a function of the frequency, angle of observation, and electron velocity. These quantities have been studied experimentally by Boersch et al. (32). Relativistic calculations have been made for thin films (20, 33) and for oblique incidence of the electrons (34, 35). These relations show the interesting phenomenon of a strong resonance peak around the plasma frequency if the imaginary part $\varepsilon_i(\omega_p) \ll 1$ and radiation damping is small. This maximum in the transition radiation is identical with the plasma radiation (36).

To get a physical understanding of the processes, we discuss briefly the nonrelativistic equation for the emitted radiation intensity produced by incoming electrons at normal incidence. If an electron of velocity v passes a thin film of thickness d of a material with the dielectric constant $\varepsilon(\omega)$ (metal or insulator) (see Fig. 21), a radiation pulse of a continuous spectrum

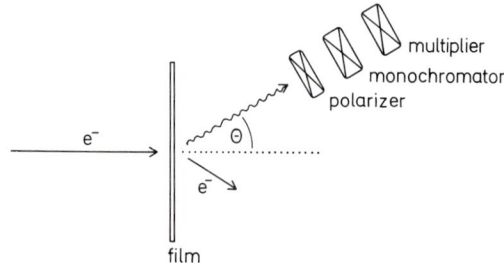

FIG. 21. Schematic diagram of an apparatus to observe the plasma radiation excited by fast electrons (25–100 keV) in a thin plasma foil (several hundred angstroms thick) at normal incidence. This radiation is analyzed by a monochromator and recorded with a multiplier at different angles Θ.

is emitted. By using Maxwell's equations, one gets for the number of photons of energy $\hbar\omega$ per energy interval $d\hbar\omega$, emitted into the solid angle $d\Omega$ in the direction Θ and per electron.

$$\frac{\partial^3 N(\hbar\omega, \Theta)}{\partial(\hbar\omega)\partial^2\Omega} = \frac{1}{\pi^2} \cdot \frac{e^2}{\hbar c} \cdot \frac{1}{\hbar\omega} \cdot |\varepsilon - 1|^2 \cdot \beta^2 \cdot \sin^2\Theta \cdot \cos^2\Theta \cdot M \quad (74)$$

$$M = \left| \frac{\sin\frac{\omega\, d}{v\,2}}{i\frac{c}{\omega}L^+} \pm \frac{i\cos\frac{\omega\, d}{v\,2}}{i\frac{c}{\omega}L^-} \right|^2 \quad (75)$$

with $\beta = v/c$, Θ the angle of observation (see Fig. 21), and $L^\pm$ the dispersion relations Eqs. (9) and (10). The plus sign between the two terms in M is valid if one observes the emission at the exit side where the electrons emerge, and the minus sign at the entrance side of the film where the electrons impinge. This differential photon number shows, roughly, the dependence on the velocity, the angle Θ, the film thickness, and the frequency given essentially by the term M.

The resonant peak in the spectral structure at $\omega \sim \omega_p$ can be derived from Eq. (74). If one assumes

$$|\varepsilon| \gg \varepsilon_i \quad \text{and} \quad \sin^2\Theta < \varepsilon_i \quad K_p d \ll 1 \quad (76)$$

then Eq. (74) becomes

$$\frac{\partial^3 N}{\partial^2\Omega\partial(\hbar\omega)} = \text{const}\frac{\gamma_t^2}{4(\omega - \omega_p)^2 + \gamma_t^2} \quad (77)$$

This equation gives a resonance behavior at $\omega = \omega_p$ if damping is not too strong. γ_t is the sum of two damping contributions, the interior damping γ_i and the radiation damping γ_r,

$$\gamma_t = \gamma_i + \gamma_r = \frac{1}{\tau_i} + \frac{1}{\tau_r} \quad (78)$$

with

$$\frac{\gamma_r}{\omega_p} = \frac{1}{\omega_p \tau_r} = \tfrac{1}{2}K_p d \cdot \sin\Theta \tan\Theta \quad (79a)$$

and

$$\frac{\gamma_i}{\omega_p} = \frac{1}{\omega_p \tau_i} = \varepsilon_i(\omega_p) \quad (79b)$$

γ_i depends only on ε_i whereas γ_r is ascribed to the radiation damping due to the coupling of these waves to light. It increases with thickness d and with the angle Θ at which the radiation is observed. In the direction $\Theta = 0$, γ_r becomes zero since there is no radiation in this direction. The damping by radiation can be so strong that the plasma peak is nearly suppressed at larger values of Θ as well as at larger thicknesses.

The restrictions given above [see Eq. (76)] are in general not fulfilled. One has, therefore, to take the exact relativistic formulas for the computer calculations.

In the following we compare the experimental and calculated results with the complete relativistic expression [see, e.g., (35)]. For the discussion we use the less cumbersome relation Eq. (74). A review of this topic has been given by Steinmann (37).

b. Experimental Results. This interesting theoretical prediction of a resonant light emission of thin plasma films has been verified in experiments by Steinmann (38) and Brown *et al.* (39) on silver films. The scheme of the experimental arrangement is shown in Fig. 21. Electrons of 50–100 keV cross a thin metal foil (thickness of ~ 100 Å). The surface plasmons excited by the electrons decay into photons and their spectral distribution, angular dependence, and polarization can be measured using the detector system shown in Fig. 21. The first experiments were performed on silver films because its plasma radiation lies in the near ultraviolet at ~ 3300 Å and its surface is rather insensitive to oxidation. Further work has verified these results on substances such as aluminum (40) magnesium (41), beryllium (42), that have a plasma of quasi-free electrons. In general these materials are easily oxidized and have their plasma frequencies in the vacuum ultraviolet region, and therefore these experiments have to be performed under ultrahigh vacuum conditions. For more details on Ag see (43, 46), on Au see (44), and on Cd, In, and Zn see (45). Table I shows a collection of data.

TABLE I

VALUES OF $\hbar\omega_p$ AND THE HALF-WIDTH $(\hbar\omega_p)_{1/2}$ OF THE PLASMA RESONANCE RADIATION

Material	$\hbar\omega_p$ (eV)	$(\hbar\omega_p)_{1/2}$ (eV)	References
Ag	3.78		(43)
Al	14.9		(40)
Be	18.1 ± 0.4	5	(42)
Cd	9.1	0.9	(45)
In	11.8		(45)
Mg	10.4 ± 0.03	0.6	(41)

Comparison of the observed with theoretical results is very sensitive to the value of $\varepsilon(\omega)$ near the plasma frequency. Since the optical constants of a thin film are often not identical with those of the bulk material, the ε values of films of similar thickness, produced under rather similar conditions in regard to substrate temperature, deposition rate, etc., have been used for the evaluation at least in Ag. The value of ε_i around the plasma frequency has higher values than in bulk material in general.

In Fig. 22 some recent measurements of the absolute value of the differential photon number for silver films are shown (46). The circles are values calculated with $\varepsilon(\omega)$ determined on films of the same thickness and produced under similar conditions (47, 48) and the observed data are represented by the dashed line. The agreement is good within an error of $\sim 10\%$.

Detailed measurements of the peak position and its half-width at different angles Θ for magnesium demonstrated two facts: (1) the peak position is displaced to higher energy values with increasing angle Θ (film thickness

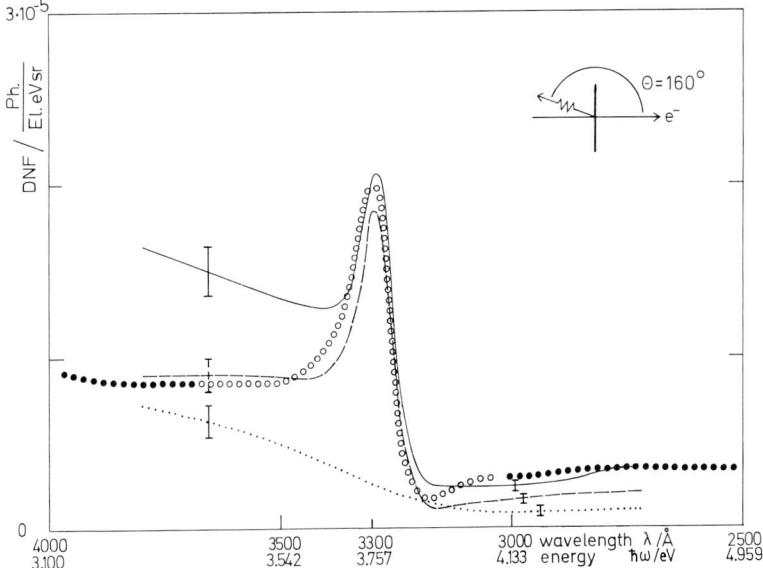

FIG. 22. Comparison of observed numbers of photons with calculated values on a silver film of 600 Å thickness bombarded with 80-keV electrons as a function of the wavelength. The open circles are calculated values using $\varepsilon(\omega)$ of thin silver films produced under similar conditions (47). The filled circles are values from Schlüter (48). The observed intensity in absolute units is given by the dashed line. The angle of observation is 20° from the film normal in the backward direction. The dotted line represents the s-polarized contribution. DNF means photons per electron-eV-steradian [from Hattendorf (46)].

200 Å), and (2) the peak broadens (41). The peak position is compared with the dependence of the frequency of the plasma oscillations for thin films on the wave vector K_x, which for $K_p d/2 \ll 1$ is given in Eq. (49). This is allowed since the electrons in Mg behave like those of a free electron gas. Plotting the peak position against $\tan^2 \Theta$ gives a thickness of 220 Å in agreement with the thickness of 220 ± 15 Å measured with a quartz oscillator. The broadening of the peak with increasing angle Θ is predicted by Eq. (52). If one evaluates the linear dependence $\Delta\lambda$ (sin Θ tan Θ), the same film thickness of 220 Å is obtained. These facts support the interpretation given above. [However, the displacement of the peak position of the photon number calculated with the relativistic equation should be compared with the observed data rather than the dispersion formula of Eq. (49), since both displacements can have different behaviors.]

By extrapolation to $\Theta = 0$, or zero radiation damping, the internal damping can be obtained. A value of 0.58 eV has been derived for Mg, and it is compared in Table II with values obtained from other plasmon experiments. The values of the radiative surface mode $\hbar\omega_p$ and the half-widths obtained with different methods agree rather well.

TABLE II

VALUES OF $\hbar\omega_p$ AND THE HALF-WIDTH $(\hbar\omega_p)_{1/2}$ OF THE PLASMA RESONANCE RADIATION OF Mg OBTAINED FROM VARIOUS EXPERIMENTS

$\hbar\omega_p$ (eV)	$(\hbar\omega_p)_{1/2}$ (eV)		References
10.4 ± 0.03	0.58	Plasma radiation (20°C)	(41)
10.21	0.6	Plasma resonance photoeffect (20°C)	(80)
10.2	0.75	Plasma resonance absorption (-196°C)	(49, 50)
10.5 ± 0.1	0.6	Light reflection experiments	(60)

The angular distribution of the intensity near the peak agrees with the theoretical values (37, 46). It corresponds to the radiation characteristic of a Hertz dipole placed on a metal surface. Along $\Theta = 90°$ as well as at $\Theta = 0$ there is no intensity for normal incidence of the electrons. The angle Θ_{max}, at which the intensity maximum of the angular distribution is observed, is wavelength dependent. Experimental work (51) has shown that Θ_{max} as a function of λ passes through a minimum at the plasma frequency, a characteristic feature of the plasma radiation.

The dependence of the differential number of photons on the thickness is a periodic function of the argument $\omega d/2v$ [see Eq. (80)]. In a thin film with $|1/L^+| > |1/L^-|$, it is directly proportional to $\sin^2(\omega d/2v)$. The argument can be written as

$$\omega d/2v = 2\pi(t/2T) \tag{80}$$

where T is the period of an oscillation and t the time of the passage of an electron through the film. At values

$$t = \tfrac{1}{2}T, \tfrac{3}{2}T, \text{ etc.} \tag{81}$$

or when the electrons remain just half of an oscillation period inside the film, a maximum of electron energy is transformed into plasma oscillation energy. This has been demonstrated at different electron velocities (52).

Whereas the nonradiative plasmons are rather sensitive to films on the boundary, especially at high K_x values, the radiative oscillations are not influenced very much by such films. For more details see (53, 37). If one measures the spectral dependence of the plasma radiation at different angles Θ, the peak flattens out more and more approaching $\Theta = 90°$ due to the increasing damping (46), [see Eq. (79)].

Theoretically the radiation is p-polarized if the planes of incidence and observation coincide. An s-polarized component is expected only if the plane of observation is different from that of incidence, but it is a small relativistic correction. The measurements of all investigators, however, report a rather important s-polarized contribution (see Fig. 22). Its origin is supposed to be bremsstrahlung, but its quantitative explanation is still being discussed (54–56). Light emitted via roughness from nonradiative plasmons which are excited by electrons can contribute to this s component (46).

In Fig. 22 the measured s-polarized intensity, which is smaller in the backward direction, has been subtracted from the measured p-polarized intensity (see the solid line in Fig. 22). The p minus s values are regarded as the observed data to be compared with the calculated data (see Fig. 22).

The absolute intensity has been measured for silver and compared with the theoretical values. Within the limits of 10%, good agreement has been found.

By neglecting the relativistic corrections in Eq. (74), a term in the denominator has been suppressed which can reach small values depending on the damping so that additional peaks have been lost. The denominator of Eq. (8) in Kröger (35) for $\alpha = 0°$ gives

$$1 - \beta^2(\varepsilon - \sin^2 \vartheta) \tag{82}$$

If this term becomes zero (ε real), it is identical with the Cerenkov condition where ϑ is the angle of the light cone. The correlated electron energy losses have been observed (57) as well as the radiation from thin Si-foils (58).

Detection by electron energy losses. Instead of observing the light emitted by the decay of the radiative plasmons, it is possible to analyze the energy losses that the primary electrons suffered after having passed through the plasma film. In order to distinguish them from those electrons which have excited the volume plasmons and lost the same energy, one makes use of the different dependence of frequency of both plasma oscillations on the wave vector. Equation 49 of Section I, as well as Fig. 19b, demonstrate this dependence on frequency. The frequency of the volume plasma oscillations is practically constant in the region between $K = 0$ and the light line (see Fig. 19b). Its dispersion relation has the form

$$\omega^2 = \omega_p^2 + \overline{v^2} K^2 \tag{83}$$

where $\overline{v^2}$ is the mean of the square of the velocity of the plasma electrons. This gives a relative change of the energy between $K = 0$ and K_p of

$$\frac{\hbar\omega(K_p) - \hbar\omega(0)}{\hbar\omega(0)} = \frac{\overline{v^2}}{2c^2} \sim 10^{-5} \tag{84}$$

if $\overline{v^2} = \tfrac{3}{5} v_F^2$ where v_F is the Fermi velocity. This small energy change means that the frequency of the volume plasmon remains practically constant in this K interval. In contrast, the surface plasmon energy of Al is displaced from 15 eV to about 19 eV at $K_p \simeq 7 \times 10^5$ cm^{-1}. This value of the wave vector K_p corresponds to a scattering angle ϑ_p of the electrons (75 keV) of of 5×10^{-5} rad ($\vartheta_p = K_p/K_{el}$). To observe this dispersion it is necessary to measure the energy losses at very small K_x values or at angles ϑ or $\lesssim 5 \times 10^{-5}$ rad. By applying special electron optical techniques, this energy change could be measured for thin Al-films (59). The quantitative evaluation, however, causes some difficulty since one does not know the thickness of the oxide layer grown on the thin Al-films under the conditions of this experiment.

3. Excitation by Light

a. Plasma Resonance Absorption. In the following we consider the excitation of radiative surface plasmons by light manifested by the phenomena of plasma resonance absorption and plasma resonance emission. Since these plasmons are coupled with photons, the reverse process—excitation of these plasmons by light—should be possible. Figure 23 shows a simple scheme to explain the excitation of the wave $\omega_p(K)$ by light hitting the plasma

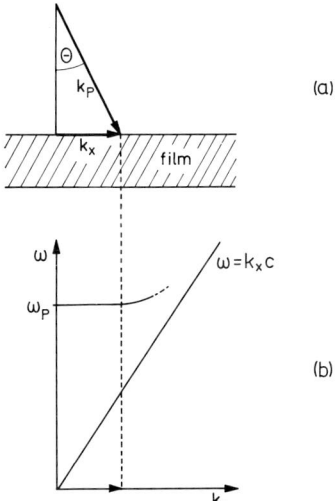

FIG. 23. Excitation of radiative plasmons by light of frequency ω_p. The component K_x of its wave vector K_l ($K_l = \omega/c = \omega_p/c$) along the boundary [see (a)] has to fulfill the dispersion relation of the radiative surface plasmons (b).

boundary at an angle Θ. The projection of $K_l = \omega_p/c$ ($\varepsilon_0 = 1$) given by $K_x = (\omega_p/c) \sin \Theta$ has to fulfill the dispersion relation of the radiative plasmons.

This excitation can be detected by measuring the transmittance of the plasma film in the frequency region around the plasma frequency. One observes a minimum if one uses p-polarized light, a phenomenon called plasma resonance absorption.

To state the situation precisely, we consider a thin film of thickness d whose electronic properties are described by the dielectric function $\varepsilon_1(\omega)$ and which is bounded to media with a dielectric constant ε_0. Light enters the slab at an angle Θ (Fig. 23). Then we calculate the Fresnel coefficients in transmittance t^p_{010} and reflectance r^p_{010} for p-polarized light [see, e.g., Kretschmann (61)]. They represent the ratio of the H_y amplitudes of the transmitted or reflected beam to that of the incident beam and have the form

$$t^p_{010} = \frac{t^p_{01} \cdot t^p_{10} \cdot \exp(iK_{z1}d)}{1 + r^p_{01} r^p_{10} \exp(i2K_{z1}d)} \quad (85)$$

$$r^p_{010} = \frac{r^p_{01} + r^p_{10} \exp(2iK_{z1}d)}{1 + r^p_{01} r^p_{10} \exp(2iK_{z1}d)} \quad (86)$$

Here the terms r_{vk}^p and t_{vk}^p are the Fresnel coefficients for one boundary given by

$$r_{vk}^p = \frac{\dfrac{K_{zv}}{\varepsilon_v} - \dfrac{K_{zk}}{\varepsilon_k}}{\dfrac{K_{zv}}{\varepsilon_v} + \dfrac{K_{zk}}{\varepsilon_k}} \tag{87}$$

and

$$t_{vk}^p = 1 - r_{kv}^p \tag{88}$$

where

$$K_{zv} = [\varepsilon_v(\omega/c)^2 - K_x^2]^{1/2} \tag{5}$$

The coefficients t_{vk}^p and r_{vk}^p are derived from the continuity conditions using the Maxwell's equations. Whereas the t_{vkl}^p and r_{vkl}^p terms can be obtained from these by adding up multiple reflections and transmissions, as indicated in Fig. 24, taking into account the phase change and damping $[\psi = \exp(iK_{zv}d)]$ on passing through the film of thickness d. For example, t_{010}^p is obtained by the sum

$$t_{010}^p = t_{01}^p e^{i\psi}(t_{10}^p + r_{10}^p e^{i\psi} r_{10}^p e^{i\psi} t_{10}^p + \cdots) \tag{89}$$

which is equal to Eq. (85).

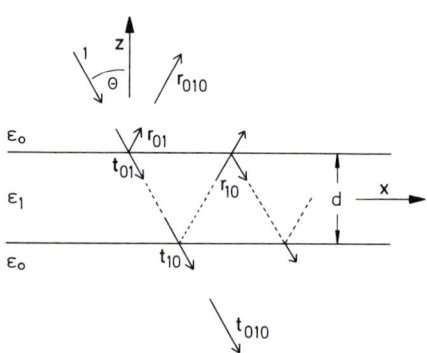

FIG. 24. A thin plasma slab is irradiated with light. The Fresnel coefficients, necessary for the calculation of the transmittance and reflectance, are shown.

A resonance situation takes place if ω approaches ω_p and the radiative plasmon is excited. This is demonstrated by developing the above expressions for a free electron gas around the frequency ω_p which is the less damped radiative mode (62, 63).

With the assumptions

$$|\varepsilon| \ll \sin^2 \Theta \qquad (90)$$

$$|K_{z1}d| \ll 1 \qquad (91)$$

which are rather restrictive, one obtains

$$t = \frac{2(\omega - \omega_p) + i\gamma_i}{2(\omega - \omega_p) + i(\gamma_i + \gamma_r)} \qquad (92)$$

with the coefficient of the interior damping

$$\gamma_i = 1/\tau_i = \omega_p \varepsilon_i(\omega_p) \qquad (93)$$

and that of the radiation damping

$$\gamma_r = 1/\tau_r = \omega_p \frac{K_p d}{2} \sin \Theta \tan \Theta \qquad (94)$$

γ_r increases with thickness and with Θ rather quickly, so that the resonance effects disappear with increasing angle and thickness as we have seen in the considerations on plasma radiation.

The transmitted intensity is obtained from Eq. (85) as

$$T^p = |t^p|^2 = 1 - \frac{(\gamma_i + \gamma_r)^2 - \gamma_i^2}{4(\omega - \omega_p)^2 + (\gamma_i + \gamma_r)^2} \qquad (95)$$

This equation has a minimum for $\omega = \omega_p$ (Fig. 25a) which can be measured conveniently. In the same way the reflected intensity can be calculated, and it becomes

$$R^p = |r^p|^2 = \frac{\gamma_r^2}{4(\omega - \omega_p)^2 + (\gamma_i + \gamma_r)^2} \qquad (96)$$

Again the same denominator as in Eq. (95) appears and has a minimum at $\omega = \omega_p$: thus the reflectance has a maximum at ω_p, which is not very pronounced except for very small ε_i values (Fig. 25b).

We see further that the absorbed energy

$$A^p = 1 - T^p - R^p = \frac{2\gamma_i \gamma_r}{4(\omega - \omega_p)^2 + (\gamma_i + \gamma_r)^2} \qquad (97)$$

reaches a maximum at $\omega = \omega_p$ (Fig. 25c). The minimum in transmission is coupled with a maximum in the absorption and reflection. This phenomenon is called the plasma resonance absorption. It does not exist for s-polarized light. The corresponding relations T^s, R^s, and A^s which do not contain the dielectric constants of the plasma show no structure around ω_p.

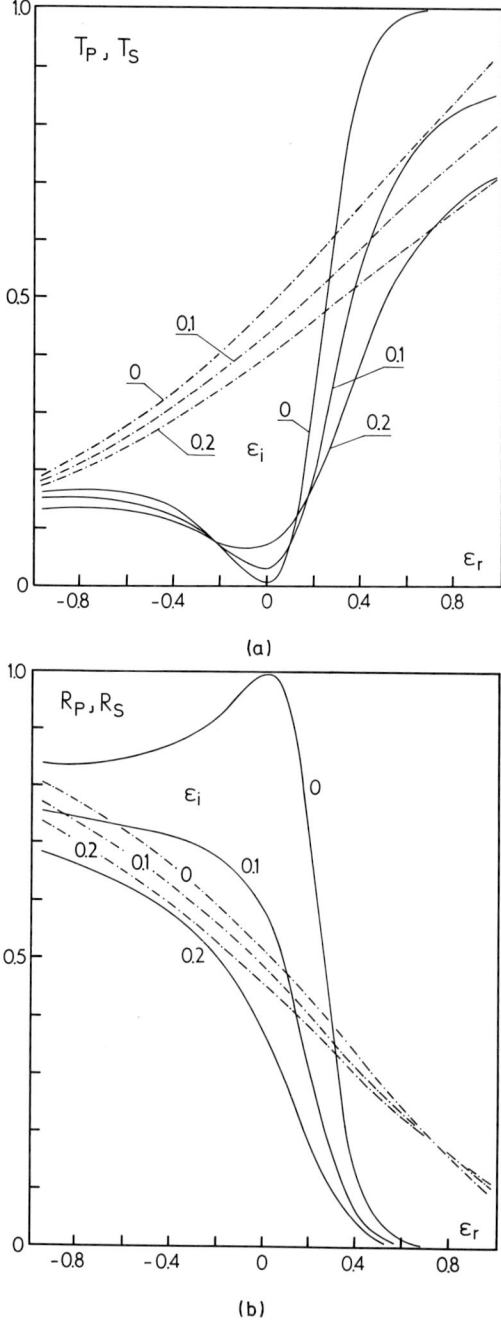

FIG. 25. The influence of ε_i on the values of T_p, R_p, and A_p as well as of T_s, R_s, and A_s plotted as a function of ε_r at an angle of incidence $\Theta = 45°$ and a film thickness divided by the light wavelength $d/\lambda = 0.182$.

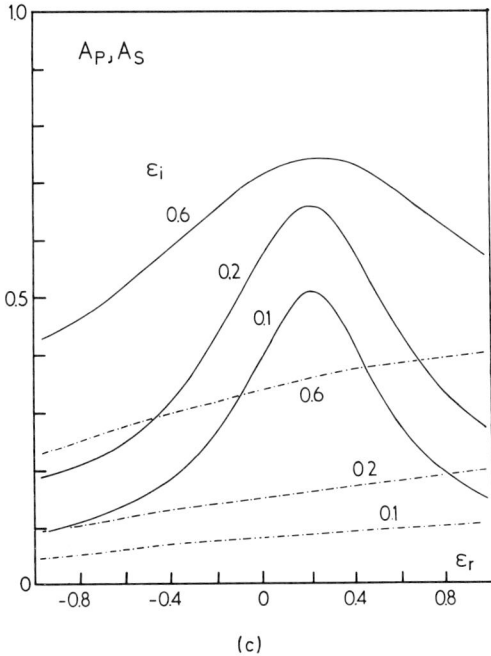

(c)

FIG. 25 (*continued*)

These relations are valid if the conditions of Eq. (90) and (91) are fulfilled. For comparison with the experiments, these equations have to be corrected since in the experiments the plasma film is deposited on a substrate such as quartz, sapphire, etc. The exact calculations are therefore made with the computer.

As we see, the denominator in Eqs. (85) and (86) is proportional to the dispersion relation D [Eq. (7)]. Thus it appears that $D = 0$ (or minimum) leads to the resonance process. However, one has to be careful, since the numerator goes to zero at the same time. A more careful consideration is necessary and it is necessary that Eqs. (85) and (86) be evaluated with a computer.

For a visualization of the arguments above, the values of T^p, A^p, and R^p are drawn for $\Theta = 45°$ and $d/\lambda = 0.182$ as function of ε_r with ε_i as parameter in Fig. 25 and are compared with the corresponding values for s-polarized light. The abscissa ε_r represents a frequency scale. The minimum in the transmitted intensity at $\varepsilon_r = 0$ is quite pronounced for small ε_i values indicating the excitation of the radiative mode. The reflectance R_p does not decrease as steeply to zero at $\varepsilon_r \sim 0$ as for thick films, even at small ε_i, since d/λ is not large enough.

The value $d/\lambda = 0.182$ does not fulfill the condition of Eq. (91), so that the curves of Fig. 25 cannot be described by Eqs. (95), (96), and (97).

At higher frequencies, there is a minimum in the R_p curve for low ε_i which is the well-known Brewster minimum determined by

$$\tan \theta = (\varepsilon_r)^{1/2} \qquad (\varepsilon_i \text{ small}) \qquad (98)$$

This happens at larger ε_r, which is not shown in Fig. 25.

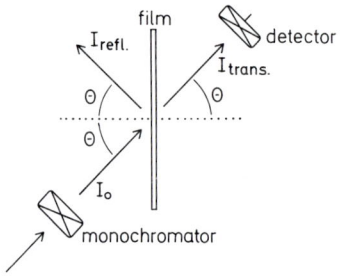

Fig. 26. Schematic diagram of an arrangement to observe the plasma resonance.

b. Experimental Results. The excitation of the radiative plasma oscillations by light has been demonstrated by the minimum transmittance at $\omega \sim \omega_p$ measured with p-polarized light. The first experiments were performed using silver *(62–64)*. This substance has a pronounced resonance frequency in the near ultraviolet ($\lambda \sim 3280$ Å) which is convenient for the experiments. Furthermore, the substance is not easily oxidized, so that the experiments can be made in air. Since interband transitions interfere with the plasmon in Ag, further experiments on substances with a quasi-free electron gas, such as K *(65–67)* Mg *(49, 50)*, and Al *(68, 69)*, have been made. The experimental arrangement is depicted schematically in Fig. 26. As the polarized, monochromatic light beam passes through the film, its transmitted intensity is measured as a function of the angle Θ and the light frequency. Figure 27 shows the phenomenon for a 300-Å thick silver film *(72)*. The curve (0°) represents the absorptance for normal incidence. At $\Theta > 0°$ the minimum is observed if p-polarized light is used. s-Polarized light (Fig. 27a) gives no effect. Values of the measured plasma resonance minima are listed in Table III.

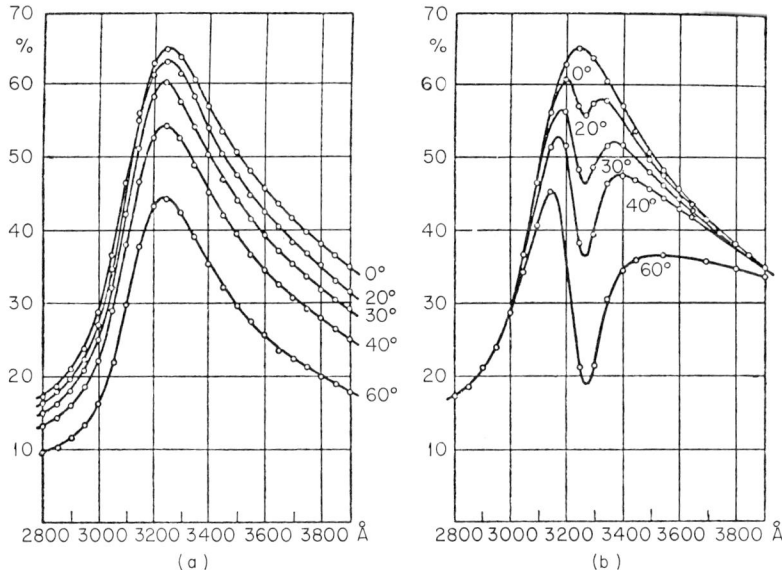

FIG. 27. Comparison of the transmission of s-polarized (T_s) (a) with that of p-polarized light (T_p) (b) of a silver film 300 Å thick for angles of incidence from 0° to 60°. The depth of the minimum is strongly dependent on the angle Θ [from Schulz and Zurheide (72)].

TABLE III

Observed Positions of the Plasma Resonance Minima (in eV) and their Half-Widths

K (67)	Mg (50)	Al (69)	Ag (71)	KCl (70)	KBr (70)
3.78	10.2	14.8	3.78	14.1	13.1
(250 meV	(750 meV		(75 meV		
at $-196°$C)	at $-80°$C)		at 20°C)		

This effect is strongly angle-dependent as follows from the condition that the electric field must have a component normal to the film ($E_z > 0$) to excite the mode ($\Theta = 0$ gives $\gamma_r = 0$ and thus $T_p = 1$), as Fig. 27 shows. Further detailed experiments on silver were performed with different film thicknesses and temperatures (71, 72). The evaluation of the observed minima with Eq. (85) makes it possible to determine $\varepsilon(\omega)$ around the plasma frequency rather exactly. Its temperature dependence can also be measured (73).

If one adds gallium to silver the minimum is displaced to lower energies shown in Fig. 28. At the same time, the minimum becomes less pronounced, which can be understood by the fact that the energy of the interband transition in silver of 4.1 eV is reduced by adding Ga and becomes more nearly equal to the plasma energy of 3.78 eV, leading to an increased damping of the oscillation (74). Similar observations have been made for silver alloyed with Pb, In, and Cd and for Ag–Au alloys (75, 76).

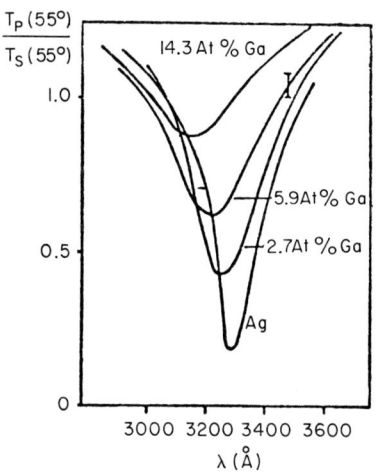

FIG. 28. The ratio T_p/T_s at $\Theta = 55°$ for an alloy (Ag + Ga) [from Bernert and Zacharias (74)].

Plasma resonance in reflection at thin films. The curves of Fig. 25 show a pronounced minimum in T_p even if $\varepsilon_i = 0$ (no interior absorption). This minimum occurs because the incident radiation has been reflected. For $\omega = \omega_p$, one has a higher reflectance $[R_p(\omega_p) = 1]$ than for $\omega < \omega_p$. This is valid only for thin films since in the region $\omega < \omega_p$ the reflectance of a thin film is smaller than 1, whereas in a thick material with $\varepsilon_i = 0$ it becomes 1. This increased reflectance has been observed on very thin films of Al (77) since its value of $\varepsilon_i(\omega_p)$ is rather small ($\varepsilon_i = 0.08$). This effect disappears with increasing ε_i so that for $\varepsilon_i > 0.03$ and $d/\lambda = 0.18$ (in Ag at $d = 600$ Å) it can no longer be observed. In transmission, however, the plasma resonance absorption remains observable up to $\varepsilon_i \sim 0.6$ due to the absorption in the plasma film, an advantage of this method.

Registration of the plasma resonance absorption by photoelectrons. Since the electric field E_z below the metal surface goes through a maximum if the frequency of the incoming light goes through ω_p ($D_z = \varepsilon E_z$ being continuous and thus E_z large at small ε), we expect, if the work function of the metal is smaller than $\hbar\omega_p$, that photoelectrons will be emitted. The number of the created photoelectrons per second (J_{ph}) by irradiating with p-polarized light will be approximately proportional to $|E_z|^2$ in the interior of the metal or

$$J_{ph} = \text{const}|E_0|^2 \cdot \frac{|1 + r_{01}^p|^2}{\varepsilon^2} \tag{99}$$

where r_{01}^p is the Fresnel coefficient at the boundary 0/1 [see Eq. (39)]. The constant factor will be difficult to determine since the escape depth of the photoelectrons comes into play, a quantity only approximately known.

Experiments of this kind have been made on very thin Al-films (thickness 50 Å) (78). In Fig. 29 the ratio of the photoelectrons produced by p-polarized to those produced by s-polarized light has been plotted as a function of the wavelength. The s-polarized field (E_y) which is continuous at $z = 0$ does not go through a resonance, so that the photoelectrons of the s-polarized light have been used as a normalizing factor. The ratio shows a maximum at $\lambda = 840$ Å (15 eV) as is expected because the value of $\hbar\omega_p$ of Al, obtained from plasma radiation experiments, is 14.9 eV (see Table I). The work function of Al is 4.2 eV.

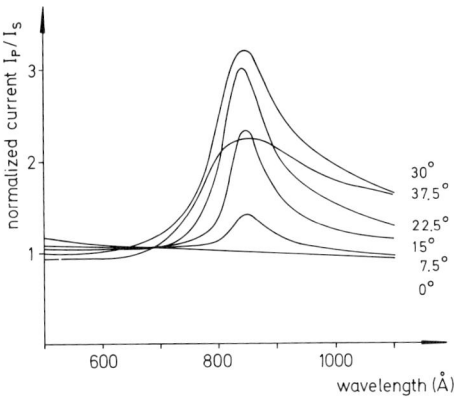

Fig. 29. Photoelectron current from a 50-Å Al-film produced by radiative plasmons. The ratio I_p/I_s is plotted [from Feuerbacher (78)].

In thicker magnesium films (~ 200 Å) a displacement of the peak position of the photoelectron intensity with varying angle Θ has been found which is explained by the dispersion of the radiative plasma oscillation (79). A linear dependence of the peak energy on $\tan^2 \Theta$ was observed, which agrees with the result of the experiments on plasma radiation excited by fast electrons in Mg-films (see Section II,2,b). If one identifies the energy of the photoelectron peak with $\hbar\omega$ of Eq. 49, the dispersion coefficient gives a thickness of ~ 200 Å in agreement with the film thickness measured independently. Experiments on very thin Mg-films (~ 80 Å) (80), however, showed a negative dispersion. Here the peak position of the photoelectron intensity is displaced to lower energies with increasing angle. This behavior has been ascribed to the structure of the film.

In experiments on silver films with $\hbar\omega_p = 3.78$ eV, whose work function of 4.6 eV is higher than the plasmon energy, photoemission is observed only by coating the silver surface with a monolayer of barium (81). p-Polarized light then produces a photoelectron peak at the expected energy $\hbar\omega_p$, as determined from other measurements. The thin Ba-layer did not change the energy of the radiative plasmons which is rather insensitive to contamination.

This photoelectric resonance phenomenon only occurs with the p-polarized component of the incoming light. This "vectorial" photoeffect is a well-known experimental fact and was reported a long time ago. One of the first observations was made on thin alkali films deposited on glass substrates (82). p-Polarized light gave a peak of photoemission at around 3.7 eV, whereas s-polarized light did not show any anomaly. The explanation of the resonance effect was left unanswered at that time.

Plasma resonance absorption in nonhomogeneous films. In general, metal films of a thickness of less than ~ 50 Å show an island structure and can no longer be regarded as plane-parallel slabs. Thus we expect them to have other optical properties which depend on different parameters, such as the shape of the island, their size, and their dielectric constants, which change in very small particles when the mean free pathlength of the conduction electrons in the bulk material becomes larger than the diameter of the particles (83). The interaction between the islands must also be considered (84).

The plasma resonance absorption of such thin films observed in silver (85) as well as in potassium (67), changes insofar as the minimum of transmittance is displaced to lower frequencies (see Fig. 35 on p. 197). In addition, further minima in the low frequency region appear partially only in the s-polarized light. Thus the experimental situation is no longer as simple as in the case of thin plane-parallel films (86).

The existing models do not explain the observations completely, however a number of simplifications in the calculations had been made which are perhaps the reason for the difference between the observed and calculated results.

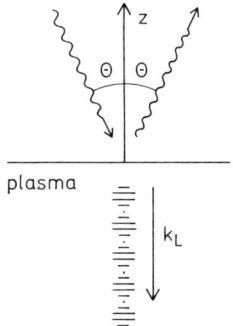

FIG. 30. Light of frequency $\omega \geq \omega_p$, falling at an angle Θ on the surface of a plasma (metal), can excite density oscillations or volume plasmons by the normal component of its electric field (E_z), which travel into the metal with the wave vector K_L. The vector K_L is much longer than ω/c so that K_L is practically perpendicular to the boundary.

Fine structure of the plasma resonance absorption by optical excitation of volume plasmons. If p-polarized light with its transverse electromagnetic field hits a boundary of a plasma at oblique incidence so that the electric field has a component along the surface normal (E_z), and if the light frequency ω becomes $\omega \gg \omega_p$, the electric field produces electron density oscillations at the surface which propagate into the interior as longitudinal plasma waves or volume plasmons (see Fig. 30).

If the light frequency ω exceeds ω_p the wave vector K_L of the longitudinal field in the plasma becomes real (see the dispersion relation Eq. (83)], and thus the resonant excitation of volume plasmons becomes possible. The electron density profile of Fig. 1 has to be supplemented in the interior of the metal by longitudinal oscillations, as Fig. 30 indicates. This idea has been verified experimentally. We treat this topic here because it is connected with the phenomenon of plasma resonance absorption. If we irradiate a plasma film with light to study the plasma resonance absorption, or excitation of the radiative surface plasmons, as described before, we must use light of frequency $\sim \omega_p$. On the other hand, the excitation of volume plasmons will occur for $\omega \geq \omega_p$, so that it will appear as a fine structure in the quantities T^p, R^p, and A^p on the high energy side (see Fig. 25). In order to treat this problem quantitatively, Maxwell's equation can be used. With the usual boundary conditions, however, one obtains the Fresnel coefficients and nothing new. To obtain effects as described above, the δ-like distribution of the charge in the surface, shown in Fig. 1, which considers E_z discontinuous as a boundary condition, has been replaced by a smooth profile. This results in a finite gradient of E_z which produces longitudinal density fluctuations propagating into the plasma interior. If one postulates the boundary condition E_z continuous from which follows the continuity of the current density j_z, one can calculate this process (*88*, *89*), see also (*87*). s-Polarized light does not show this effect. In this way the Fresnel coefficients are changed, but the corrected coefficient, e.g., the reflectance of light at a semi-infinite slab of a metal, changes by only a very small amount (less than 10^{-4}). This change in reflectance cannot be detected with the experimental means presently available.

This question has been treated carefully again (*90*), see also (*91*). To reinforce the effect it has been proposed to use a thin film so that half-wavelengths of the volume plasmons match with the thickness of the plasma film and standing waves result. A resonating system is now obtained with resonance at

$$\lambda_L = n(d/2) \qquad (100)$$

or

$$K_L = n(\pi/d)$$

The correlated frequencies are obtained from Eq. (83):

$$\omega_n^2 = \omega_p^2 + \tfrac{3}{5} v_F^2 (\pi/d)^2 \cdot n^2 \qquad (101)$$

Here n must be an odd number so that the thickness is equal to an odd number of half-wavelengths λ_L. The electric field of the transmitted light can thus transfer power into the resonating system of the volume plasmons, which is not possible if n has even values. As is evident, it is necessary that the wavelength of the transmitted light (λ_{tr}) in the metal be large compared to that of the volume plasma oscillations

$$\lambda_{tr} = \frac{\lambda_{vac}}{\varepsilon^{1/2}} \gg \lambda_L \qquad (102)$$

so that interferences of λ_{tr} do not disturb the standing wave configuration of λ_L. This can be realized since ε is small for $\omega \gtrsim \omega_p$ and λ_L has a value of some 10 Å. Furthermore the thickness of the film has to be small enough to avoid damping of the volume oscillations, which destroys the formation of the standing wave. In order to realize this condition, the thickness of the film has to be

$$d \sim v_F \tau \qquad (103)$$

where v_F is the Fermi velocity and τ the collision time. In potassium, with $\omega_p \tau \sim 20$, we have $d \sim 50$ Å. This is roughly the mean free pathlength of the electrons of the Fermi surface. Under these circumstances we expect dips in the reflectance due to the additional absorption of energy which the volume plasmons dissipate. Figure 31 demonstrates the additional maxima

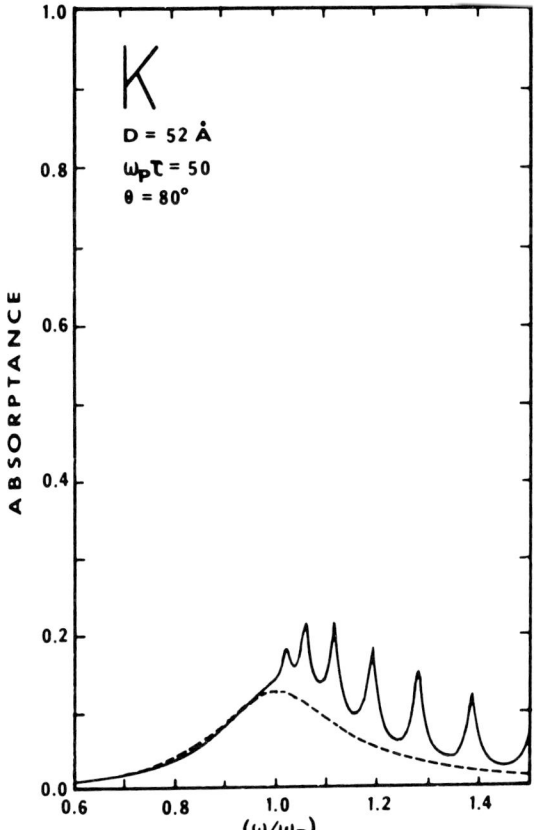

FIG. 31. Calculated absorptance of p-polarized light by 52-Å potassium foil. The maxima at $\omega > \omega_p$ are due to the excitation of volume plasma oscillations [from Melnyk and Harrison (*90*)].

in the absorptance of p-polarized light in a 52-Å potassium film at the frequencies ω_n.

Experiments have been performed on thin, very smooth silver (*92*) and potassium films (*93*) demonstrating the phenomena predicted above. Figure 32 shows the measured transmittance at 75° incidence of light through silver films of different thicknesses deposited on a very smooth quartz substrate. Minima with higher values than $n = 3$ and $n = 5$ could not be observed because the plasmon energy of silver of 3.8 eV at $n = 0$ is displaced with higher n values to higher energy values. This brings it nearer to an interband transition, leading to a strong damping of the plasma

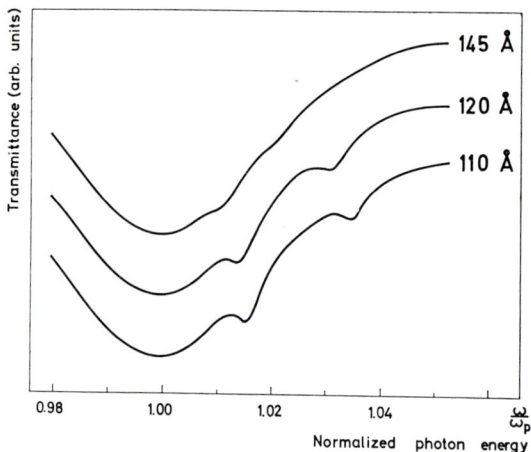

FIG. 32. Measured transmittance spectra of Ag for p-polarized light and for three different film thickness: 110, 120, and 145 Å. The angle of incidence is 75°. The minima indicate the excitation of volume plasmons [from Lindau and Nilsson (94)].

oscillation. The dependence of ω_L on K_L has been determined and did not give a linear relation between ω_L and K_L^2 as in the case of a free electron gas [see Eq. (83)]. Later experiments with electron energy losses verified this dispersion behavior (95).

Similar results have been obtained using thin potassium films. The method of detection was changed insofar as the energy absorbed in the film by the excitation of volume plasmons in the resonance case is partially transformed into photoelectrons, which can be detected by sensitive means. Resonance peaks up to $n = 10$ were observed. They are displaced with increasing n to higher energy values. The evaluation showed a linear dependence of ω_L^2 on K_L^2 as in the case of a free electron gas. Whereas in energy loss experiments K_L can be calculated from K_{el} and ϑ, the scattering angle [see Eq. (72)], here one needs the knowledge of n and the film thickness d to obtain K_L [see Eq. (100)]. The latter, however, is not exactly known so that an exact determination of the dispersion coefficient or the factor of K_L^2 is not possible.

Theoretical work to investigate this effect in other cases has been reported. The additional absorptance peaks are also obtained by looking at the absorptance of small metallic spheres. Besides the strong extinction at $\omega = \omega_p/\sqrt{3}$, maxima at $\omega \gtrsim \omega_p$ appear similar to those of Fig. 31 (96). The same phenomenon is obtained as a result of calculations on the fine structure of the transition radiation for λ smaller than λ_p (97). Both effects have not yet been found experimentally.

c. *Plasma Resonance Emission.* In transmission experiments with *p*-polarized light the excitation of the radiative surface plasmons is observed by the reduced intensity of light of $\omega \sim \omega_p$ in the direction of the incoming beam. No light should be observed in directions different from the reflected and transmitted beam. But, as already mentioned, exterior and interior roughness lead to light emission in nearly all directions. The physics of this phenomenon will be treated in Section III,4,d. Here we describe experiments demonstrating this phenomenon.

The first experiments (*98*) were done in an arrangement similar to that used to study plasma resonance absorption (see Fig. 26). The light intensity was measured with a photomultiplier in a direction Θ different from Θ_0, the angle of incidence, and at different azimuthal angles. The incoming light has to be *p*-polarized, so that the electric field has a component along the film normal. For silver, if the frequency of the incoming light is continuously changed, one obtains a strong maximum at $\hbar\omega = 3.79 \pm 0.01$ eV, with a half-width of 70 meV (*99*) (see Fig. 33), which equals the plasma wavelength as determined by plasma radiation experiments (see Table I). The light is mainly polarized in the plane of observation. Its intensity has a characteristic angular dependence $I_p(\Theta)$ as Fig. 34 depicts for a 400-Å thick silver film deposited by evaporation on a fused quartz plate and measured at an azimuth of $\psi = 90°$ and a fixed $\Theta_0 = 25°$ (*99, 100*). Looking at the film at $\psi = 90°$ with *p*-polarized light has the advantage of recording the E_z (or j_z) component and excluding the E_x (or j_x) component, which undergoes no resonance effect and contributes only to the background (*101*). The same dependency is obtained if Θ and Θ_0 are exchanged. This dependency is in

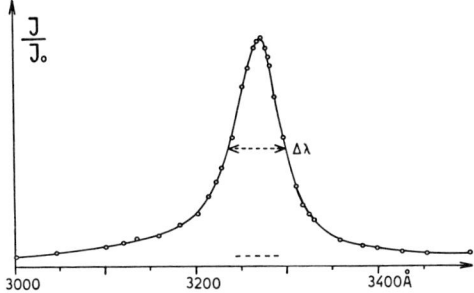

FIG. 33. Plasma resonance emission from silver (500 Å thick). Relative light intensity I/I_0, emitted at $\Theta = 40°$, is plotted as a function of the wavelength of the incoming light (I_0) incident at an angle of 23°. The half-width of ~ 60 Å corresponds to about 70 meV [from Schreiber and Raether (*99*)].

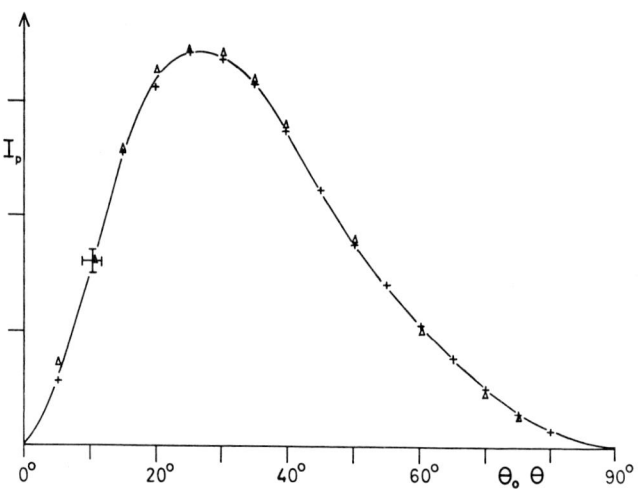

FIG. 34. Plasma resonance emission I_p of a silver film of 400 Å thickness as a function of Θ at $\Theta_0 = 25°$ and as a function of Θ_0 at $\Theta = 25°$. $\phi = 90°$ [from Schreiber (99)].

agreement with the picture that the scattered light emitted from the radiative surface plasmon, which has been excited by the incoming light of frequency ω_p, radiates approximately like a Hertz dipole on a metallic surface. [For more details, see Schreiber (99), Schröder (100), and Kretschmann and Raether (102).] The number of photons scattered out of the reflected and transmitted beam are about 5×10^{-4} of the incoming photons (99).

Looking ahead to the explanations presented in section III,4,d, we have to keep in mind that the excitation of the plasma electrons by light frequencies of $\omega \sim \omega_p$ ($\varepsilon \sim 0$) emphasizes the term contributing mainly to "interior roughness", in Eqs. (145) and (146) which contain surface and volume effects. Therefore, the effects described above can be explained by the theoretical considerations given in Section III,4,d. However, the evaluation of the relations concerning the roughness structure (correlation length, rms roughness height) in Schreiber (99), Schröder (100), and Steinmann et al. (101) must be revised, since it had been assumed in these papers that surface roughness is important. On the other hand, we have up to now no information on the structure of the interior roughness or how to describe it. It has been estimated that the observed effects can be described by varying the dielectric function ε_r of silver by a $\Delta\varepsilon_r$ of about 0.1, but this is a rough estimate (103).

Similar results were obtained using potassium films deposited on SiO_2 substrates (*104*). On the other hand, no resonance emission was observed if the K-film had been deposited on a sapphire substrate at a temperature less than $-165°C$ (*105*). This indicates that such a metal film consists of a rather homogeneous interior and has a smooth surface. If however the sapphire substrate reaches a temperature of $\gtrsim -165°C$, a strong light emission around ω_p is observed, and in addition, a weak emission maximum at $\lambda \sim 5400$ Å can be seen. This prominent peak can be observed up to temperatures of $\sim 0°C$. However it is displaced to longer wavelengths or temperatures above $-165°C$ (see Fig. 35). The transmittance minimum which

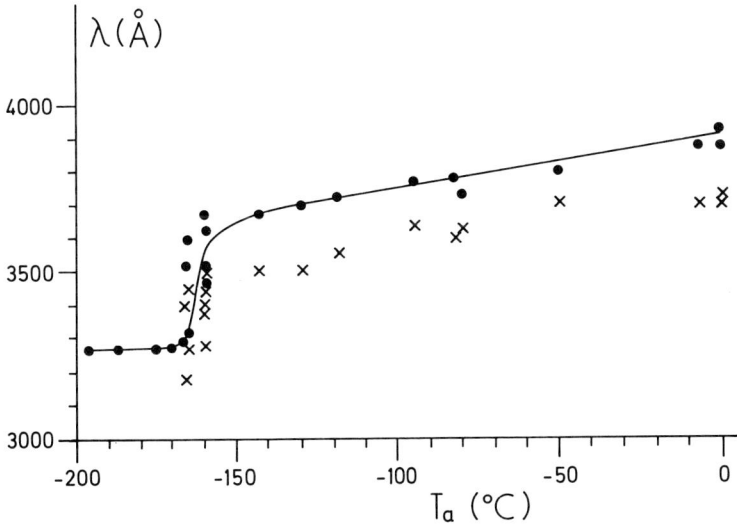

FIG. 35. Wavelength position (λ_{min}) of the transmittance minimum T_p in potassium (filled circled on the curve) as a function of the temperature of the substrate (sapphire). The X's show the same dependence for the maximum of emission (λ_{max}) [from Bösenberg (*105*)].

is observed at $T < -165°C$ also changes its position at $-165°C$ to wavelengths about 200 Å longer than that of the emission peak (see Fig. 35). These facts lead to the conclusion that the K-film transforms at $\sim -165°C$ into a film of an inhomogeneous structure, which causes the emission of light. This conclusion is supported by measuring the electrical resistance which rises to high values abruptly at $-165°C$, indicating a decay of the homogeneous film into island particles. A more detailed explanation of the light emission cannot be given at the present time.

Plasma resonance effects in small spheres. We shall report in Section III,1 on the excitation of surface plasmons on small spheres by electrons and X rays. Here the phenomena of plasma resonance absorption and emission shall be described in analogy to the same effects in thin films. Theoretically this question has been treated in Mie's calculations on light scattering by small spheres (*106*) which gives the absorption coefficient of photons incident on an ensemble of spheres of a dielectric function $\varepsilon_r + i\varepsilon_i$ embedded in a medium of a dielectric constant ε_0

$$\chi = \text{const Im} \frac{-1}{\varepsilon + 2\varepsilon_0} = \text{const} \frac{\varepsilon_i}{(\varepsilon_r + 2\varepsilon_0)^2 + \varepsilon_i^2} \tag{104}$$

The distance of the small spheres ($2R < \lambda$) has to be large compared to the wavelength of the radiation. This term has a resonant peak at

$$\varepsilon_1 = -2\varepsilon_0 \tag{105}$$

which in the case of a free electron gas and $\varepsilon_0 = 1$ lies at

$$\omega_1 = \omega_p/\sqrt{3}, \quad \text{(Mie mode)} \tag{106}$$

However the internal damping ε_i in Eq. (104) has to be corrected by including the radiation damping so that the total damping is $\gamma_{tot} = \gamma_i + \gamma_{rd}$ [see Crowell and Ritchie (*107*)] which is similar to that of Eqs. (78) and (79). The surface plasmon is radiative since it is coupled with photons by a curved boundary. The behavior of small cubic particles is discussed in Fuchs (*26*). The resonance absorption on spheres seems to have been explained as excitation of plasmons first by Doyle (*108*).

Experiments with light absorption by excitation of the surface mode $m = 1$ have been made on small spheres of silver and gold (diameter ~ 100 Å) embedded in gelatin of $\varepsilon_0 = 2.37$ or in glass with $\varepsilon_0 = 2.25$ (*109*). An absorption band was observed in silver spheres on glass at 4060 Å which agrees quite well with the calculated value of 3960 Å. The same is found true for the half-width. Similar good agreement was obtained in experiments with silver spheres in gelatin concerning the position of the peak, but poorer agreement concerning the half-width (*110*). Experiments on resonance emission of the $m = 1$ surface mode were made on spheres of sodium ($2R < 100$ Å) (*111*). Ultraviolet light scattered by these particles and analyzed with a monochromator shows a pronounced peak at $\lambda \sim 3200$ Å with a half-width of ~ 600 Å in poor agreement with the calculated value of 3760 Å using Eq. (104).

III. Excitation of Nonradiative Surface Plasmons

1. EXCITATION BY ELECTRONS: ELECTRON ENERGY LOSSES

If electrons of momentum $\hbar K_{el}$ and energy $\hbar^2 K_{el}^2/2m$ pass through a solid they transfer a momentum $\hbar q$ and an energy $\Delta E = \hbar\omega$ to the solid (see Fig. 36). This excitation of the plasma can be measured by analyzing the energy losses of the primary electrons with a suitable electrostatic or magnetic energy analyzer, which is schematically drawn in Fig. 36. Since the

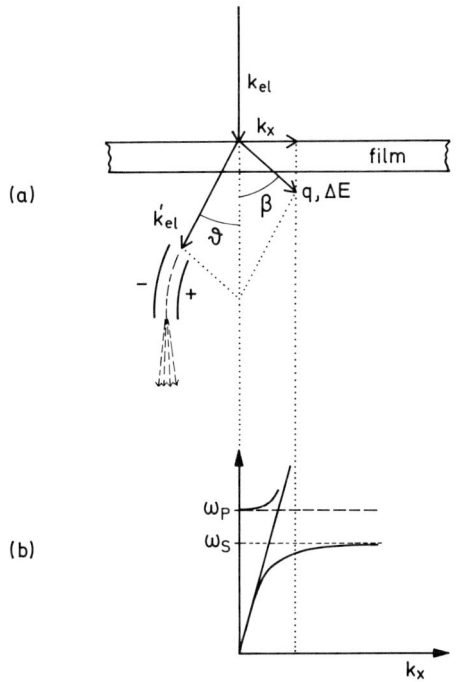

FIG. 36. (a) Scheme of the momentum configuration in the excitation process of surface plasmons by electrons. (b) The dispersion relation of nonradiative and radiative plasmons.

energy of the plasmons ΔE is of the order of a few eV and that of the primary electrons is ~ 50 KeV, one needs a system of rather high energy resolution to detect this fine structure of the energy loss spectrum (*112*). Measurements of the energy losses in a solid with a resolution of about 100 meV and with a suitable intensity is not difficult to realize.

Figure 36 shows that the projection of the vector q on the boundary equals the wave vector K_x of the plasmons in the boundary at normal incidence of the electron beam

$$\hbar g_x = \hbar K_{el} \vartheta = \hbar K_x \tag{107}$$

The angles β and ϑ are related by

$$\tan \beta = \vartheta/\vartheta_{\Delta E} \tag{108}$$

where $\vartheta_{\Delta E} = \Delta E/2E_0$ and E_0 is the kinetic energy of the primary electrons. If ϑ surpasses $\vartheta_{\Delta E}$ the angle β quickly approaches $90°$ so that large K_x values are easily realized. However, details in the dispersion relation around the light line require small K_x values and thus special high angular resolution of the electron detection. But here photons are a more suitable tool as we shall see later.

Before referring to detailed results on nonradiative plasmons derived from energy losses, we shall consider some theoretical relations.

Excitation probability of surface losses. The differential probability that one electron passing the boundary of a foil embedded in a medium ε_0 at normal incidence loses the energy ΔE and is scattered at an angle ϑ into the solid element $d\Omega$ is given, neglecting retardation, by

$$\frac{\partial^3 P_s}{\partial \Delta E \, \partial^2 \Omega} = \left(\frac{e}{\pi \hbar v}\right)^2 K_{el}^2 \, \text{Im} \left[\frac{(\varepsilon - \varepsilon_0)^2}{\varepsilon \varepsilon_0} \frac{2 K_x^2}{q^4} \cdot R \right] \tag{109}$$

$$R = \frac{\sin^2 \frac{\omega}{v} \frac{d}{2}}{iL^+} + \frac{\cos^2 \frac{\omega}{v} \frac{d}{2}}{iL^-} \tag{110}$$

where v is the electron velocity, and d the film thickness. The denominators in R are the expressions $L^\pm$ without retardation given in Eqs. (9) and (10). The splitting into two frequencies can be neglected if

$$\tfrac{1}{2} d K_{el} \vartheta \gg 1 \tag{111}$$

In the case of $\lambda = 0.05$ Å (50 keV electrons) this condition means

$$\vartheta d \gg 10^{-2} \tag{112}$$

(where d is in Å). Equation (110) can then be written as

$$R = \frac{1}{iK_x(\varepsilon + \varepsilon_0)} \quad (113)$$

which shows that for $\varepsilon_r \to -\varepsilon_0$ and small ε_i, a maximum excitation of the nonradiative plasmon is produced. This formula, derived by Ritchie (*113*), is valid for $K_x = K_{el}\vartheta \gg \omega_p/c$. Figure 36 shows that this is the region of the dispersion relation where the frequency has nearly reached its asymptotic value. The relativistic equation has been derived by Kröger (*114*) which gives a better description of the small angle scattering.

In the following we shall discuss the various properties of the nonradiative plasmons derived from the electron energy loss.

Position of the surface loss. If the plasma film is not too thin, but Eq. (111) is fulfilled, the inelastic electrons have a peak intensity at the maximum of the function $\text{Im}[1/(1 + \varepsilon)]$. This resonance behavior can easily be seen if we introduce the dielectric function of a free electron gas. We obtain ($\varepsilon_0 = 1$)

$$\frac{\partial^3 P}{\partial \Delta E\, \partial^2 \Omega} \sim \frac{A}{(\omega^2 - \omega_s^2)^2 + B} \quad (114)$$

with $\omega_s = \omega_p/\sqrt{2}$ and B measuring the half-width. In general, however, interband transitions change the dielectric function of the free electron gas. The surface mode will then be found at $\varepsilon_r = -1$ and the peak value of the surface loss function is given by

$$\left(-\text{Im}\,\frac{1}{1 + \varepsilon}\right)_{\omega=\omega_s} = \frac{1}{\varepsilon_i(\omega_s)} \quad (115)$$

Some measured values of $\Delta E_s = \hbar\omega_s$ are given in Table IV (*115*).

TABLE IV

SOME SURFACE AND VOLUME PLASMON ENERGIES

	Al	Mg	Li	Na	K
$\Delta E_s \begin{cases} \vartheta \sim 0 \\ \vartheta > 0 \end{cases}$	10.31 10.6	7.11 7.15	3.95 3.95	3.75 3.85	2.55 2.60
$\hbar\omega_p/\sqrt{2}$	10.58	7.34	5.01	3.96	2.62
$\Delta E_{s\,1/2}$	1.5–2	1.4–2	2–3	0.9–1.2	0.7–1
ΔE_p	14.95 ∓ 0.05	10.39 ∓ 0.05	7.08 ∓ 0.05	5.70 ∓ 0.05	3.70 ∓ 0.05
$\hbar\omega_p$	15.7	10.9	8.02	5.95	4.29

As Fig. 36 shows, the position of the surface loss peak depends on ϑ if we consider retardation. It changes rather quickly from 0 to $\sim \omega_s$ if ϑ varies between 0 and $\sim \vartheta_p = K_p/K_{el}$ where $K_p = \omega_p/c$ (see Fig. 4). Using 50-keV electrons ϑ_p becomes 0.5×10^{-4} rad in the case of Al with $\lambda_p = 830$ Å.

If the measurements are performed at $\vartheta = 0$ with a usual beam aperture of about 10^{-4} rad, we observe a value averaged over a large frequency range. Since the higher ω values contribute more to the intensity than the lower ones, the peak loss intensity almost approaches the asymptotic value as Table IV shows in the row $\vartheta \sim 0$. Table IV also lists values of ΔE_s for $\vartheta > 0$, which are of the order of several times 10^{-4} so that $K_x > K_p$. These values can, therefore, be regarded as asymptotic values. $\hbar\omega_p/\sqrt{2}$ represents data calculated using Eq. (15). Also the observed half-width $\Delta E_{s,1/2}$ is listed along with the observed volume loss ΔE_p.

For the quantitative comparison of the observed spectrum with the calculated one, the loss equation has to be folded with the angular distribution as well as with the energy distribution of the primary electrons. The $\varepsilon(\omega)$ dependence must be known (*116*). The exact relativistic form of the loss equation has to be used to describe the loss intensity, especially at low angles and energies.

The dependence of the loss value on its K_x value including the retardation part of the dispersion relation can be measured with electrons if the angular width of the primary beam is strongly reduced to apertures lower than 0.5×10^{-4}. One finds a decrease of the energy of the surface loss (see Figs. 37 and 39) (*7*), which becomes more pronounced if the film normal is inclined to the beam, as we shall discuss later (*141*). By applying special electron optical means, the angular width has been reduced to less than 10^{-5} so that the dispersion can be measured down to about 2 eV on thin Al-films (*117*). The region around the light line however is much more easily accessible nowadays by exciting these plasmons optically.

The influence of retardation becomes apparent not only in changes of the peak positions, but it is also important for calculating inelastic intensities as has been demonstrated with the silver loss spectrum at low energies (*118*).

For higher wave vectors one expects a different dispersion process which depends on the structure of the metal surface. This problem—difficult in theoretical and experimental respects—is still open. As the loss probability decreases quickly with the angle ϑ in the case of fast electrons [see Eq. (119)] the accuracy of measurement diminishes with ϑ, and a displacement, at least a small one, can be hidden. Experiments with fast electrons have been made in transmission of thin Al (*119*) and Mg-films (*119, 120*) under ultra high vacuum conditions, to observe the change in position of the surface losses, 10.6 and 7.1 eV, respectively, with increasing scattering angle ($K_x = K_{el}\vartheta$).

Measurements on polycrystalline films of Al with 50 kV electrons between $\vartheta = 2$ and 4×10^{-3} rad demonstrate an increase of the plasma frequency (*120a*). The theoretical evaluation has not yet been done (see Fig. 37).

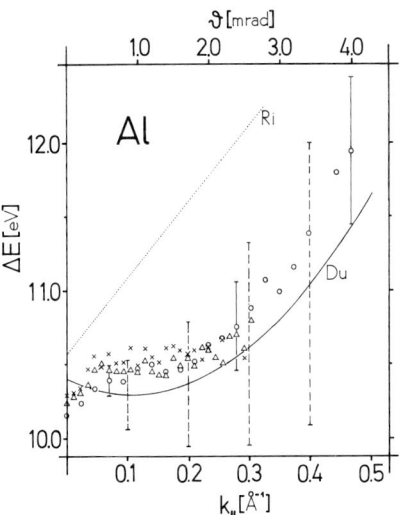

FIG. 37. Observed dependence of the surface loss (ΔE_s) of Al on the scattering angle ϑ or on $K_x = K_\parallel = K_{el}\vartheta$ where K_{el} is the wave vector of the primary electrons. The open circles are recent data obtained with 50 kV electrons (*120a*), which indicate an increase of ΔE with the wave vector. Earlier data from fast electron spectroscopy are ($\times$) (*120*) and (Δ) (*120b*). The solid line Du represents recent data from ILEED work (*123a*). Equation (43) is plotted in the dotted line (Ri).

At first glance, the possibilities of performing this type of experiment with slow electrons ($\sim 10^2$ eV) seems hopeful since the intensity problem is not so serious. The difficulty is that the strong diffraction maxima around the primary beam contribute inelastic scattering intensity in the same direction in which one observes the scattered electrons from the primary beam. Measurements have been made on Al (111) surfaces with electrons of 20–200 eV energy (*121*). New measurements on Al (111) surfaces (*122, 123*) showed better agreement insofar as the surface loss value at $K_x = 0$ of 10.5 ± 0.1 eV agrees with the value derived from fast electron work (10.6 ± 0.1 eV) by extrapolation to $K_x = 0$ neglecting retardation. Recent data (*123a*) yield for Al (100)

$$\omega_s(K_x) = 10.4(\pm 0.1) - 2(\pm 1)K_x + 9(\pm 3)K_x^2 \qquad (116)$$

K_x in Å^{-1} units. This curve is plotted in Fig. 37. The general trend of both dependences [Du and ($\bigcirc$)] agrees. In the (111) direction the coefficient of K_x^2 is with 0(+2) rather different from that in the (100) direction, whereas the coefficient of K_x is +2 (± 1), so that a nearly linear dependence results. The interpretation of these data is still open.

Angular dependence of the differential excitation probability of surface plasmons. The loss intensity as a function of ϑ for thicker films [see Eq. (111)] increases as

$$\vartheta/(\vartheta^2 + \vartheta_{\Delta E}^2)^2 \tag{117}$$

as can be seen from Eq. 109. The loss intensity goes from zero, at $\vartheta = 0$, to a maximum at

$$\vartheta = (1/\sqrt{3})\vartheta_{\Delta E} \tag{118}$$

Then it decreases as

$$\sim \vartheta^{-3} \tag{119}$$

for larger angles. Taking the numbers from above, one sees that the dependence for $\vartheta \lesssim \vartheta_{\Delta E}$ is experimentally rather difficult to verify. But the ϑ^{-3} dependence, which is in contrast to the ϑ^{-2} dependence of the volume loss intensity (*5*), has been measured for the Mg surface loss of 7 eV (*124, 125*). Similar results have been obtained on Al (*125*) and Ag (*125–127*) which support the theoretical result. For these measurements one needs a beam of small divergence because its intensity drop with angle has to be quicker than ϑ^{-3} to obtain a clear ϑ^{-3} decrease of the loss intensity.

Total probability for exciting surface losses. We see from Eq. (118) that the loss intensity is concentrated for fast electrons around the primary beam. With beam apertures larger than several 10^{-4} rad, one measures practically the integrated or total loss probability. For normal incidence, one obtains from Eq. (109) (free electron gas)

$$P_s = \frac{1}{1 + \varepsilon_0} \frac{\pi e^2}{4\pi \hbar} \frac{1}{v} \tag{120}$$

This is the probability that an electron excites a surface plasmon of energy $\hbar\omega_s$ on crossing one boundary of a plasma covered with a medium of ε_0. In the case of a thin film we have to double the value of Eq. (120). Its value depends only on the electron velocity and amounts to about 10^{-2} for 50 keV electrons, which means that a hundred electrons produce one nonradiative plasmon (see Table V).

TABLE V

EXPERIMENTAL AND CALCULATED TOTAL PROBABILITIES FOR ΔE_s ON THIN METAL FILMS DEPOSITED ON A CARBON SUBSTRATE (126).

	ΔE_s	$P_{s(exp.)}$	$P_{s(theor.)}$	Film thickness (Å)
Mg/C	7.3	$3-4 \times 10^{-2}$	3.3×10^{-2}	100
Al/C	10.3	$3-4 \times 10^{-2}$	3.3×10^{-2}	400
Na/C	3.9	$2-3 \times 10^{-2}$	3.3×10^{-2}	60

Influence of surface coatings. A characteristic property of surface plasmons is their sensitivity to thin surface coatings. Clean Al-surfaces have their surface loss, ΔE_s, at 10.3 eV. However, if the Al-film becomes oxidized it is covered with a homogeneous oxide film of about 30–40 Å whose dielectric constant in this energy region of ~ 10 eV is about 4. As we have seen in Eq. (109), such a system consisting of a plasma boundary (ε) covered with a thin medium of ε_0 has a surface loss function

$$\text{Im} \; \frac{1}{\varepsilon_1 \varepsilon_0} \frac{(\varepsilon_1 - \varepsilon_0)^2}{\varepsilon_1 + \varepsilon_0} \tag{121}$$

with a peak at

$$\varepsilon_r(\omega_s) = -\varepsilon_0 \tag{122}$$

This leads to a displacement from $\Delta E_s = 10.3$ for a clean surface to $\Delta E_s \simeq 7$ eV for an oxide-covered one (128). This displacement has been the first clear demonstration that the Al loss of about 10 eV is due to a collective oscillation of surface charges. In the above case it was assumed that ε_0 is constant and real. In general however, ε_0 contains an imaginary part which also changes the damping of the surface oscillation. Experiments with silver films covered on both sides with carbon films, demonstrate these effects on ΔE_s caused by such coatings, as shown in Fig. 38, where the observed and calculated curves show the same shape and the intensities agree rather well (129).

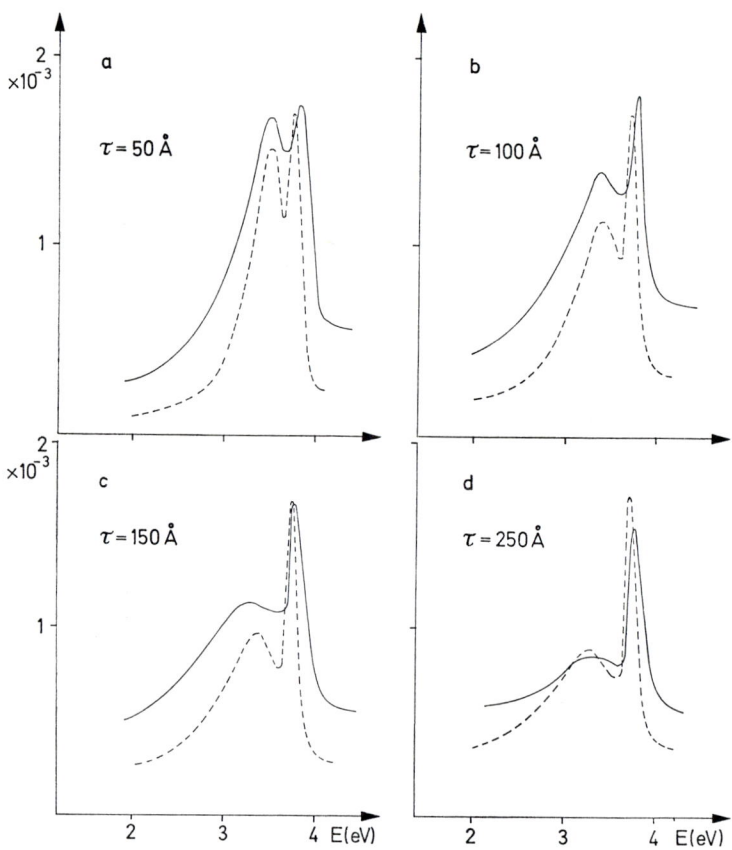

FIG. 38. Experimental spectra of a C/Ag/C layer (solid line) with different thicknesses τ of the carbon film. The volume loss remains at 3.8 eV, the surface loss (3.6 eV without carbon) changes in intensity and position. Dashed line: calculated spectrum from optical data with a dielectric constant for carbon of $\varepsilon_r = \varepsilon_i = 3.5$. The experimental angular width of the primary beam is 3.4×10^{-4} rad. Thickness of the silver film is 400 Å. The intensity is normalized with that of the no-loss peak [(from Daniels (*129*)].

Coupling dispersion of a symmetric layer system. As we have seen in Eq. (111), coupling dispersion exists in thin films $(dK_{el}9/2 \lesssim 1)$ (Fig. 4). This splitting of the surface losses and its angular dependence has been observed by Schmüser (*130*) on oxidized Al-films. The experimental difficulty lies in producing homogeneous unsupported metal films of about 100-Å thickness. A detailed study of the dependence of this splitting as function of the angle, the film thickness, and the electron energy confirmed the calculated behavior (*7, 124, 131*). Figure 39 demonstrates this splitting for an Al film 175-Å thick

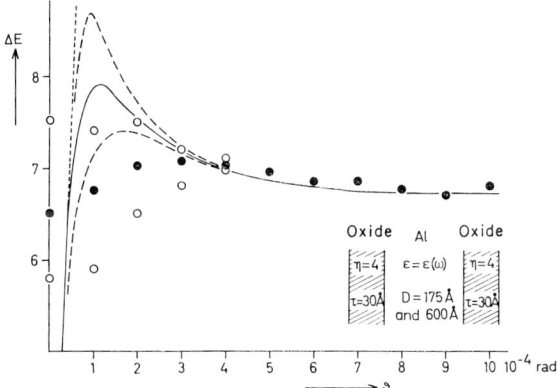

FIG. 39. Splitting of the Al surface loss (7 eV) by coupling of the two oscillations on both sides of an oxidized Al-film 175 Å thick (open circles). The dashed lines are calculated. The filled circles are measured values on a freely supported Al-film 600 Å thick. One recognizes the retardation effect at smaller angles by the decrease of the loss energy [from Kloos (7)].

covered on both sides with ~ 30 Å of oxide. This oxide film is inevitable on Al if one has to prepare a film that is freely supported. The open circles (lower curve) reproduce the ω^- behavior and show a strong ϑ-dependence of ω^-. The upper curve represents the ω^+ mode. A thick Al-film (600 Å) shows no splitting (filled circles). There is a slight increase of ω_s toward low ϑ values down to $\vartheta \sim 3 \times 10^{-4}$ rad due to thickness dispersion which is not negligible for ~ 30 Å thick oxide surface layers. The decrease of ω_s at low ϑ values, due to retardation, is clearly seen. For a quantitative comparison the angular distribution of the primary electron beam should have been folded with the excitation probability [see Eq. (109)]. This procedure leads to an agreement of observed and calculated values assuming a thickness of the oxide layer of about 30 Å [see (120)].

Asymmetric layer system. If the plasma film is deposited on a supporting membrane of carbon, SiO, or plastic, we have to do with an asymmetrically covered plasma film, considered in Section I,2,b. A typical example is the following (124). The loss spectrum of Al evaporated onto a Zapon substrate contains besides the 15-eV volume loss the 10-eV surface loss of the Al/vacuum boundary, and a 6.7-eV loss attributed to the Al/Zapon boundary, where ε_0 of Zapon ~ 5. This explanation is supported by Fig. 40, which displays the dispersion relation of such an asymetric system. Curve 1 (ω_+) and curve 1 (ω_-), for an oxide-free surface and for large ϑ values, represent the two surface losses just mentioned. Experimentally this interpretation

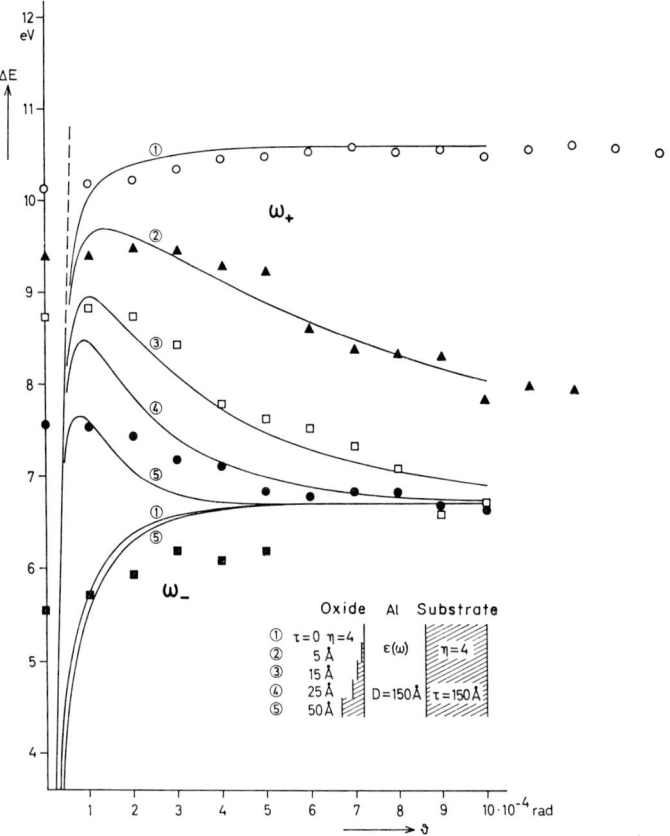

FIG. 40. Dispersion due to the varying thickness of the oxide film on an Al boundary. The open circles [curve (1) ω_+] and the filled squares [curve (1) ω_-] were measured for an Al film evaporated on an Al oxide substrate. With an increasing thickness of Al_2O_3 covering the free Al surface [curves (2)–(5)], the thickness dispersion is changing [from Kloos (7)].

can be demonstrated by covering the Al surface with a silver film. Then the 10-eV Al loss disappears and is replaced by the 3.6-eV Ag surface loss.

Dispersion due to finite thickness of the coating. As we have seen in Eq. (31), the extension of the electromagnetic field depends on its wave vector K_x. Therefore, plasmons with small K_x values or long wavelengths are rather insensitive to a thin layer of a different dielectric constant on the plasma surface, whereas those of large K_x values or short wavelengths are influenced much more. This means that the position of the surface loss produced on a plasma surface covered with a thin coating will be displaced more the higher the K_x values. This thickness dispersion has been observed on oxidized Mg- and Al-films (*7*, *126*) (Fig. 40). An Al film of $\simeq 150$-Å thickness is evaporated onto an Al oxide substrate. The two surface losses, 10 eV for the clean Al surface, and 6.5 eV for the Al/Al-oxide boundary are seen as a function of ϑ or $K_x = K_{el}\vartheta$. Now the clean Al surface is exposed to oxygen so that the oxide film grows in thickness and the thickness dispersion comes into play. Applying the relations calculated in Kloos (*7*) the measured curves are fitted to obtain the thickness of the oxide film given in the inset of Fig. 40. The ω_+ mode is much more sensitive to the oxide layer than the ω_- mode which remains nearly unchanged. Similar results have been obtained by recording the loss spectrum of a clean Mg surface in different stages of oxidation (*126*).

Coating with a plasma film. If the surface of a plasma film, e.g., Mg, is covered with a thin plasma film such as Al, the condition of Eq. (122), $\varepsilon = -\varepsilon_0$ yields a surface loss

$$\omega_s = \frac{1}{\sqrt{2}}(\omega_{p1}^2 + \omega_{p2}^2)^{1/2} \qquad (123)$$

assuming that both media resemble a free electron gas, and where ω_{p1} and ω_{p2} are the plasma frequencies of the two media.

Experiments have verified this result. A clean Al surface was covered by a thin Mg film and a clean Mg surface by a thin Al film. In both cases a new loss at 12.8 eV appeared in agreement with the value of 12.9 eV calculated from Eq. (123) (*126*, *132*).

Nonnormal incidence. These experiments are performed with a primary electron beam inclined at an angle α to the normal of the plasma film. At oblique incidence the momentum transferred to the boundary becomes

$$\begin{aligned}\hbar q_s &= \hbar q_\perp \cos \alpha \pm \hbar q_\parallel \sin \alpha \\ &= \hbar K_{el}(\vartheta \cos \alpha \pm \vartheta_{\Delta E} \sin \alpha) \\ &= \hbar K_{el}\vartheta \cos \alpha \left(1 \pm \frac{\vartheta_{\Delta E}}{\vartheta}\tan \alpha\right)\end{aligned} \qquad (124)$$

instead of $q_\perp = K_{el}\vartheta$. See Fig. 41. The plus sign is valid for the upper beam direction, the minus sign for the lower beam direction. The projection can become zero if

$$\vartheta^\circ = \vartheta_{\Delta E} \tan \alpha = (\Delta E_s/2E_0)\tan \alpha \qquad (125)$$

This value of ϑ^0 depends on ΔE. If one observes the scattered electrons on opposite sides of the primary beam, different values of q_s result which lead to an interesting asymmetry.

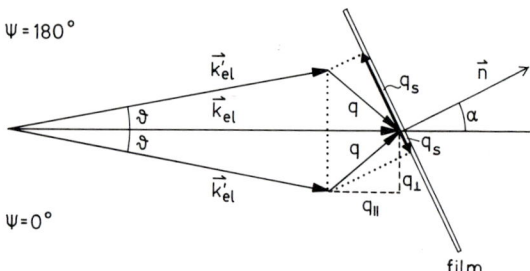

FIG. 41. Scheme of the momenta if the electron beam hits the film at oblique incidence. The projection of q on the boundary (q_s) of the film is different in the two scattering directions.

The scattered electrons can also be collected out of the plane ($\mathbf{n}, \mathbf{K}_{el}$). If the plane ($\mathbf{K}'_{el}, \mathbf{K}_{el}$) is turned around $\mathbf{K}_{el}$ by the azimuth angle ψ one obtains for the wave vector q_s in the boundary of the plasma (133)

$$q_s = K_{el} \cdot \vartheta \cdot \cos \alpha \cdot f \qquad (126)$$

where

$$f = \left[\frac{\sin^2 \psi}{\cos^2 \alpha} + \left(\cos \psi - \frac{\vartheta_{\Delta E}}{\vartheta} \tan \alpha \right)^2 \right]^{1/2} \qquad (127)$$

For $\psi = 0°$ ($-$ sign) or $\psi = 180°$ ($+$ sign) Eq. (124) is reproduced. This case is of interest if experiments at $\psi = 90°$ and $\psi = 270°$ are compared because here no asymmetry is to be expected.

Under these circumstances the excitation probability for a thin film becomes

$$\frac{\partial^3 P_s}{\partial \Delta E \, \partial^2 \Omega} = \left(\frac{e}{\hbar \pi v} \right)^2 K_{el}^2 \cdot \text{Im} \left[\frac{1}{\varepsilon} \frac{d}{\cos \alpha} \frac{1}{q^2} - \frac{2}{\cos \alpha} \frac{(\varepsilon - \varepsilon_0)^2}{i\varepsilon\varepsilon_0(\varepsilon + \varepsilon_0)} \frac{q_s}{q^4} \right] \qquad (128)$$

according to (134), (135), and (136).

The probability for volume excitations is proportional to $d/\cos \alpha$ which describes the increased path in the volume of the film due to the nonnormal incidence. The probability for surface losses becomes linearly dependent on q_s at constant q [we consider thicker films so we have replaced Eq. (110) by Eq. (113) in Eq. (128)] and proportional to $(\cos \alpha)^{-1}$. Two results follow from this: (1) the asymmetry of the surface loss intensity ($\sim q_s$), and (2) the predominance of the surface loss intensity in reflection experiments at grazing incidence ($1/\cos \alpha$).

Asymmetry of the surface loss intensity. The effect of tilting the foil on the intensity of the energy losses has been observed by Creuzburg (137) with the arrangement shown in Fig. 41. He studied Ge and Si losses excited with 40-keV electrons. Further experiments on the surface losses of oxidized Al at 6.3 eV, (*138*), of Ag at 3.6 eV (*139*), and oxidized Mg at 5.5 eV had similar results. Figure 42 demonstrates the result (*124*) for a magnesium film. The intensity has been recorded for values of ϑ from about 2×10^{-4} rad to -2×10^{-4} rad corresponding to an azimuth change from $\psi = 0°$ to $\psi = 180°$ in the plane ($\mathbf{K}_{el}, \mathbf{n}$) at an angle α of 20° to demonstrate the strong asymmetry. For a better understanding the calculations in Fig. 43 are shown. At normal incidence ($\alpha = 0°$), one obtains a dependence of the intensity of a surface loss on the scattering angle (as the dashed line of Fig. 43 shows) if the angular resolution is sufficient. The experimental resolution in these experiments, however, is not sufficient to show the minimum at $\vartheta = 0$ which is filled up (see the solid line in Fig. 43). By tilting the film, the asymmetric solid line is observed in the plane of incidence which demonstrates the proportionality of the loss intensity with q_s if $q = $ const. When the intensity is measured perpendicular to the plane ($\mathbf{K}_{el}, \mathbf{n}$) and $\psi = 90°$ and $\psi = 270°$, a symmetric intensity distribution occurs. The volume plasma loss has a symmetric distribution in both cases as is expected. If the film is inclined to the other side of the primary beam, the asymmetry changes its sign.

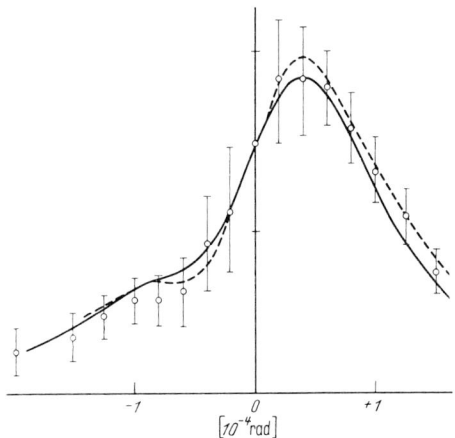

FIG. 42. Measured asymmetry of the 5.5 eV surface loss intensity (oxidized Mg film, several 100 Å thick), if the Mg film is inclined at 20° to the electron beam (dashed line) (40 keV electrons). The solid line is the calculated one. The two curves are set equal at $\vartheta = 0$ [from Schmüser (*124*)].

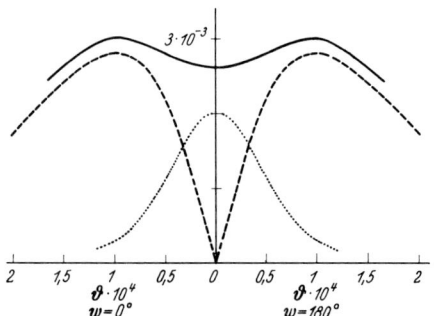

FIG. 43. Calculated angular intensity distribution of a 5-eV surface loss (dashed line) ($E_0 = 15$ KeV). Because of the large aperture of the primary beam (dotted line) one obtains the solid line for the loss intensity. The minimum at $\vartheta = 0$ is just not observable. Normal incidence [from Creuzburg (137)].

As mentioned above the values of ϑ^0 for zero intensity are a function of ΔE. For a given value of α the values of $\Delta E(\vartheta^0)$ lie on an oblique line (see Fig. 44). The K_x positions of the peak intensity of ΔE_s at normal incidence lie on the curves of Fig. 39. These relations, however, are changed in the case of oblique incidence, as indicated in Fig. 44. Here the peak positions are displaced as has been shown and discussed in (134, 135). This asymmetry of the peak positions has been observed (140) and studied in more detail with higher angular resolution (117). The quantitative evaluation of these curves is difficult since the optical constants of the surface film of aluminum oxide and its thickness are not exactly known.

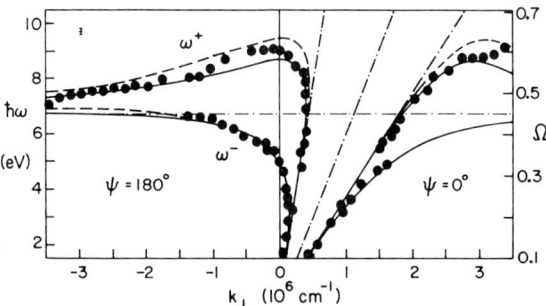

FIG. 44. Dispersion of the peak positions of the surface losses on an oxidized aluminum film (160 Å thick) tilted at 60°. (The two branches $\omega_\pm$ come from the coupling dispersion.) [From Pettit et al. (117).]

These experiments demonstrate that the wave vector K_x of these excitations is bound to the surface in contrast to the volume excitations.

Reflection of electrons at a surface. As mentioned above the experiments at nonnormal incidence have a second point of interest due to their strong dependence on $(\cos \alpha)^{-1}$ where α is the angle between the surface normal and the electron beam. This feature is due to the interaction of the Coulomb field of the primary electrons with the surface, which is more efficient, the more grazing the incidence of the electron beam.

To compare the observed with the theoretical intensities of surface losses in the reflection case, we chose the experimental conditions so that the integrated loss probability can be applied, and thus the problem of folding Eq. (128) with the angular and energy distribution of the no loss beam can be avoided. The intensity integrated over $d \, \Delta E$ (free electron gas) and $d\Omega$ becomes

$$P_s = \frac{\pi e^2}{hv(1 + \varepsilon_0)} \frac{1}{\cos \alpha} \qquad (129)$$

where v is the velocity of the primary electrons and ε_0 the dielectric constant of the bounding medium. A consequence of the relation given in Eq. (129) is that P_s grows with increasing $(\cos \alpha)^{-1}$ and the multiples of the surface loss $\hbar\omega_s$ become important. Theoretical considerations (*142, 143*) showed that the integrated intensity of the loss $n\hbar\omega_s$ is given by the Poisson distribution

$$P_n = \frac{1}{n!} P_s^n e^{-P_s} \qquad (130)$$

where P_s is given by Eq. (129) and n is a whole number.

One has to keep in mind that Eq. (129) has been derived for the passage of electrons through a boundary at nonnormal incidence. The application to reflection is made possible by assuming a break in the electron path either just at the surface or in the interior by a Bragg reflection on a lattice plane (see Fig. 45). If the penetration depth is not too small the electron beam crosses the boundary two times so to speak, at the entrance and at the exit, so we expect the double of the value of P_s in Eq. (129). In both cases the electron interacts with a system equivalent to an oscillator of frequency ω_s and its excited states $n\omega_s$. After the short interaction of the electron at the reflection with this oscillator, the different states are occupied corresponding to a Poisson distribution. This interaction becomes important at a distance less than ~ 10 Å from the surface because an electron running parallel to the boundary is accelerated by its image force to the surface and gains thus $\hbar\omega_s \sim 10$ eV in a distance of about 10 Å.

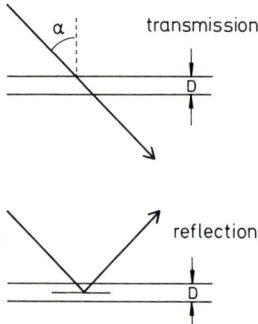

FIG. 45. Transmission and reflection of electrons.

Reflection experiments with slow electrons showed, at least qualitatively, the dependence of the loss intensity on the angle α (144). Similarly experiments with fast electrons indicated the $(\cos \alpha)^{-1}$ dependence, also qualitatively, by showing that the surface loss of Ge and Si is much more pronounced in the reflection loss spectrum than in the transmission spectrum (145, 146). Electron loss spectroscopy of the diffraction maxima of Al and Si cleavage surfaces confirmed that the surface loss intensity compared to the volume loss intensity increases more rapidly the lower the order of the diffraction maximum or the more grazing the incidence (146). Transmission experiments with fast electrons on thin silver films proved quantitatively the $(\cos \alpha)^{-1}$ relation when the foil was inclined up to $\alpha \sim 80°$ against the electron beam (4, 147). Experiments in reflection at grazing incidence on liquid Ga, In, and Al verified this relation up to $\alpha \sim 89°$, and due to the large value of P_s, recorded a high number of multiples of $\hbar\omega_s$. Si-Surfaces did not show these multiples due to the large half-width (6 eV) of this 10.5-eV loss (144, 148). The evaluation of the intensities of these loss series demonstrates quantitatively the Poisson distribution of the losses within limits of 10% using different experimental procedures (148). Furthermore, the value of P_s as a function of α has been obtained, and the quotient of $P_{s\,obs}$ and $P_{s\,theor}$ plotted as a function of $\cos \alpha$. Figure 46 displays the results. At angles of about $87°$ and larger, $P_{s\,obs}/P_{s\,calc}$ approaches ~ 1 which indicates that the electrons practically do not penetrate into the interior of the metals; they interact with the very first layers. At lesser angles of incidence, the ratio increases and tends to the value of two, indicating a certain penetration depth of the electrons into the metal, estimated to be roughly 10–15 Å in Si for 14 kV electrons at $\alpha = 85.6°$ (148).

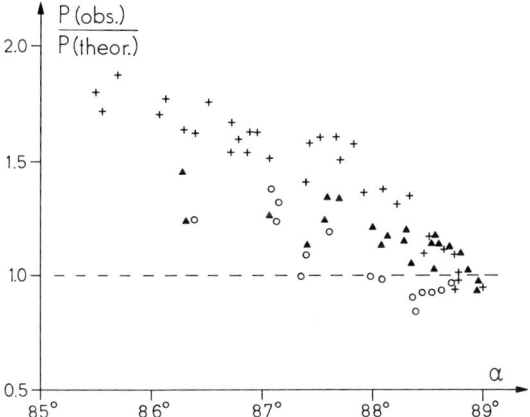

FIG. 46. The ratio of the (integrated) probability of the excitation of the surface plasmon by electrons (10 keV) as a function of the angle of incidence (α) in reflection. ○, Liquid indium; ▲, liquid gallium; +, liquid aluminum [from Schilling (148)].

Reflection experiments have been conducted to look at energy gains (149). Since the intensity of surface plasmons becomes rather large in reflection experiments with fast electrons at grazing incidence, it is possible that an interaction of surface plasmons ($\hbar\omega_s$) with the beam of incoming electrons, might lead to an energy gain of the incoming electrons. The interpretation of the experimental result is uncertain, however, since the probability of such an effect is very small.

Excitation by low energy electrons. Slow electrons can produce plasmons either in LEED or in electron emission experiments. In the first case, electrons of $\sim 10^2$ eV are shot onto the surface of a solid and reflected, particularly by the first upper crystal planes. Surface plasmons are excited and also volume plasmons. In the second case, photoelectrons are created in the interior of the crystal by light or X rays or secondary electrons, and leave through the surface, suffering collective scattering collisions among other types of collisions.

Only few LEED, or better ILEED, experiments on clean surfaces exist up to now. Studies on Cu (150), Ag (151), W (152), and Si (153) have been reported; see also Küppers (154). Recently the loss spectroscopy of low energy electrons has become a successful tool applied by Ibach and Rowe, who have used the high sensitivity of this method to study surface effects. On clean Si surfaces of orientation (100) and (111) (155), the losses due to collective and single excitation of bulk electrons have been found corresponding to those known from fast electron spectroscopy. However, further additional losses have been recorded which have been caused by surface effects, an interpretation which is supported by photoelectron experiments. Such surface effects are the "surface states," electronic states in the energy gap as well as in the energy band due to the boundary and its structural changes. These electronic levels can be seen in the loss spectrum, independently of whether they are filled or empty. Figure 47 displays such a spectrum of an ordered clean (111) 7 × 7 Si-surface (curve 1) (155). The second negative derivatives of the loss spectra ($-d^2N/dE^2$) are plotted as ordinates against the energy loss. One observes the collective excitations and the bulk interband transitions at E_1 (3.5 eV and E_2 (5 eV). Since the strongest curvature ($-d^2N/dE^2$) occurs in the original loss spectrum at the peaks of the losses,

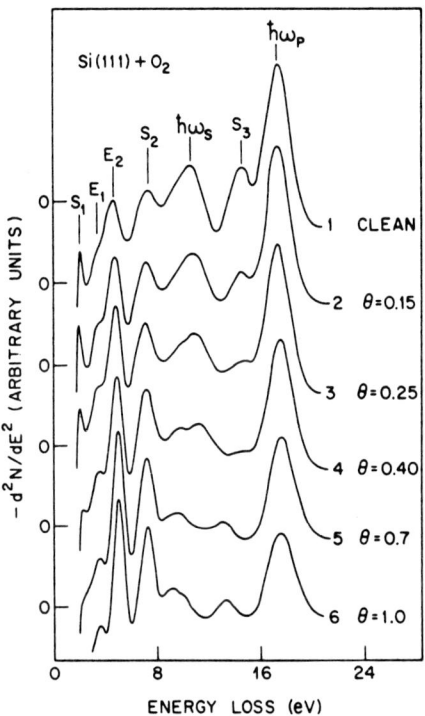

FIG. 47. Energy loss spectrum (negative second derivative) of a clean (111) (7 × 7) Si surface (curve 1) and spectra of increasing coverage with oxygen layer. Primary electron energy 100 eV [from Ibach and Rowe (155)].

one recognizes the loss peaks, e.g., $\hbar\omega_p = 17.3$ eV and $\hbar\omega_s = 10.6$ eV as maxima in the negative second derivative curve. The peaks S_1, S_2, and S_3 (2.0 eV, 7.4 eV, and 14.5 eV, respectively) are attributed to surface states which can be localized in the energy scheme of Si (155, 156). Such surface states are observed with the same method on GaAs, and Ge surfaces (158), and also on GeO_2 and SiO_2 surfaces (159).

A further point of interest is the sensitivity of the method against adsorbed atoms or molecules. In Fig. 47 the Si surface has been covered with different amounts of oxygen ($\Theta = 1$ signifies a monolayer) which produces strong changes, e.g., a splitting of the surface loss above $\Theta = 0.4$ [for more details, see Ibach and Rowe (155)]. Similarly, hydrogen adsorption on Si surfaces (157) and on Ni surfaces (153) has been studied using the same method. A further question might be, How are these molecules adsorbed at the surface? Profiting from the high energy resolution of this method, one has observed the mode of the vibration of the adsorbed oxygen which gives important information on its binding forces (160).

Typical electron emission experiments are performed in different ways. The interest is concentrated nowadays on photoelectrons produced by light or X rays. The appearance of plasmons on alkali metal surfaces has been reported in experiments with low energy (~ 10 eV) photons, (161, 162). The results agree in general, if the losses are compared with the peak positions obtained by fast electrons. In a detailed experiment on plasma excitation at liquid Hg surfaces with low energy photons (6.9–22 eV), a surface loss of 6.1 ± 0.15 eV and a volume loss of 11 ∓ 1 eV have been recorded (163). The excitation function of the volume plasmons has been measured and was found to be rather similar to that theoretically expected for surface plasmons (164).

Several papers report that if X rays are used, e.g., Al K_α-line of 1486 eV, a strong plasmon satellite structure is observed which accompanies the loss peaks of the core excitation in the photoelectron spectrum (165, 166). A similar plasmon structure is observed in "appearance potential spectroscopy" experiments in which photons produced by electrons having excited deeper shells of the atoms are detected. This structure consists of a series of volume peaks. The first ones are accompanied by surface losses (167–169). The interest in this process is the question of whether or not these losses are extrinsic wherein the free electron excites the plasmons on the way out of the metal, or intrinsic wherein the plasmon is created during the electron-hole creation by direct coupling with the collective oscillation of the plasma electrons. Information on the existence of intrinsic processes can be obtained by measuring the intensity distribution in the loss sequence, but it is difficult to separate the loss intensity from the background (169).

Inelastic collisions of very slow electrons are observed in the inelastic tunneling process of electrons. Besides molecular and phonon excitations, surface plasmons are observed in an n-type GaAs–Pb surface barrier tunnel junction. The observed peak ($\hbar\omega_s \sim 60$ meV) is attributed to collective surface excitations on the GaAs electrode which is consistent with their electron density of 5.4×10^{18} cm^{-3} (170, 171).

In an energy loss analysis of secondary electrons produced by primary electrons hitting a sodium film with an energy varying between 5 and 25 eV, the surface loss of Na at 3 eV could be identified. The excitation function, loss intensity as a function of the energy of the primary electrons, has been derived (172) which is in agreement with theoretical results (164).

Small spheres. Experiments with fast electrons have been made on small spherical particles to observe their oscillations, which we have mentioned in Section I,5. Spheres of aluminum (21, 173), silver (110), and gold (110) as well as of colloidal particles of alkali metals in alkali halogenides (174) have been investigated. Calculations to obtain the excitation probability in the case of a free electron gas (21) showed that besides the volume excitations, which are in general very weak due to the smallness of the spheres, surface plasmons of different m values are to be expected [see Eq. (57)]. The discussion of the probability shows that it depends on the value of m and on the product of qR where R is the radius of the sphere and q the transferred momentum [see Eq. (72)]. It passes a maximum for a fixed m or ω_m as a function of qR. This means that at a fixed radius the loss spectrum should consist of several peaks belonging to different m values. However, the rather large damping of the oscillations even in the case of the metals Al and Ag with their low damping rate smears out the structure of the spectrum, thus forming a broad loss peak. Using small values of $2R$, less than about 100 Å, the dipole oscillation ($m = 1$) is the main contributor to this loss peak so that the comparison of observed and calculated values becomes more meaningful. At larger values of $2R$ the maximum is displaced to higher energies since oscillations with higher m values participate. As an example we consider the results on silver. The experiments were made on an aqueous colloidal solution of Ag (110). The solution was stabilized with gelatin ($\varepsilon = 2.37$). The particles were fairly good spheres and their radii had a rather narrow size distribution. A thin specimen was obtained by drying the aqueous solution on a thin carbon film and an examination with the electron microscope showed well-separated particles.

The primary electron beam had an energy width of 60–100 meV and an angular aperture of 2–3×10^{-4} rad. At small values of $2R \lesssim 200$ Å a broad loss was observed at 2.99 eV, whereas the value calculated for $m = 1$ was determined to be 2.95 eV or 3.02 eV, depending on the dielectric function of Ag used for the calculation. On increasing $2R$ to about 800 Å the loss maximum is displaced to 3.19 eV, whereas the calculated value for $m = 2$ lies at 3.13 eV or 3.18 eV, respectively. [For comparison the loss peak for $m = \infty$ (slab) calculated with $\varepsilon = 2.37$, lies at 3.32 eV.] There is also satisfactory agreement in the peak positions in the case of gold.

The excitation of the surface modes is detected by the energy loss of the electrons. Since these modes have a dipole or multipole character, they have a rather strong radiation damping. It should be possible to detect the excitation for $m = 1$ on small spheres bombarded with fast electrons by radiation of a frequency $\omega_p/\sqrt{3}$. An experiment of this type which is the analog to that of plasma radiation of thin films has not yet been done. Calculations have been made to estimate the number of photons to be expected from such an experiment (175).

Instead of bombarding the small spheres with fast electrons, one can excite the surface modes by irradiating the small spheres with X rays. Such experiments have been performed using Ag spheres embedded in gelatin (176). In the particle picture, the high energetic photon suffers an inelastic collision and transfers the energy $\hbar\omega_p/\sqrt{3}$, together with a momentum $\hbar\omega_m/c$, to the electron collective. Instead of looking at the displaced "Compton line," which would indicate an energy loss of $\hbar\omega_m = \hbar\omega_p/\sqrt{3}$, the light is recorded with a monochromator. A light band with a peak at 4000 Å (3.10 ± 0.30 eV) and a half-energy width of ~ 750 Å or 600 meV has been found. Using Eq. (57) with $\varepsilon_a = 2.37$ for gelatin, a peak value of 3.0 eV was calculated in agreement with the observed value. Comparing these figures with those of the loss experiments made with fast electrons using similar targets, one obtains good agreement.

2. EXCITATION BY ELECTRONS: LIGHT EMISSION VIA ROUGHNESS

In the foregoing section the excitation of nonradiative plasmons has been recognized by the energy losses of electrons. But there is another possible way to detect these excitations; if the surface is not a smooth one the nonradiative waves can couple via roughness of the surface with light, so that radiation can show the production of these plasmons. Experiments of this type are described below. We discern two kinds of surface irregularities: the statistical roughness and the periodical perturbation of a grating with a sinusoidal profile.

a. Coupling by Statistical Surface Roughness. In the first experiments by Blanckenhagen *et al.* (177) a thick silver target was irradiated with 30-keV electrons and a radiation maximum was observed at about 3460 Å (3.6 eV). This energy corresponds to that of the surface plasmon at high K_x values, whereas the volume plasmon energy has a value of 3.8 eV (3280 Å). Stern (178) explained this observation as due to the coupling of plasmons with photons via the surface irregularities. In several further experiments (179) it has been shown that this interpretation is valid. The intensity of the light increases with increasing angle α between the electron beam and the surface normal [see Eq. (128)]. This is characteristic for the excitation probability of surface plasma waves. Furthermore, the roughness of the irradiated surface has a strong influence on the intensity of the emitted light insofar as the intensity is in general higher for rougher surfaces. In the same sense, a coating on the metal surface influences the emitted radiation; a dielectric thin film displaces the light peak to longer wavelengths as we know it from the surface energy losses [see Eqs. (17) and (122)].

There have been some discussions about whether or not this peak of the silver radiation can be explained by bremsstrahlung (180). Varying the energy E_0 of the primary electrons has a different influence on the radiation intensity in the case of bremsstrahlung ($\sim 1/E_0$) and plasma (or transition) radiation ($\sim E_0$). Calculations of the peak intensity of bremsstrahlung, however, showed that the plasma interpretation is right (179, 181). This has been further supported by experiments on Al surfaces (182) where, in contrast to silver, the energy of the volume plasmon (15 eV) and that of the

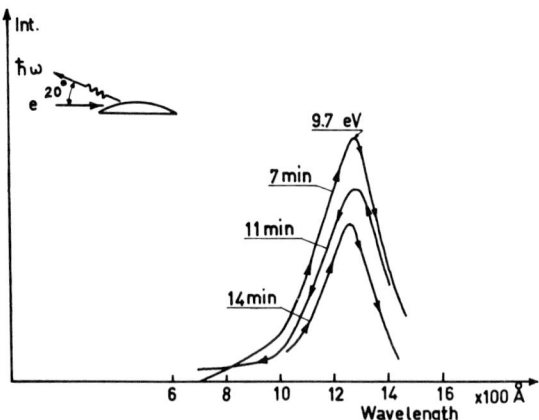

FIG. 48. Spectral distribution of the radiation emitted from a 500-Å thick Al film on a glass substrate, bombarded with 50-keV electrons at grazing incidence, at three different times after evaporation of the Al film [from Bürker and Steinmann (182)].

surface plasmon (10 eV) are well separated due to its nearly free electron behavior. Therefore experiments on Al surfaces under ultrahigh vacuum conditions were performed. To increase the intensity of the plasmons, the electron beam bombards a slightly curved Al surface at grazing incidence. The radiation was observed at an angle of 20° (see inset of Fig. 48). The result is depicted in Fig. 48 and demonstrates clearly, by the peak at about 9.7 eV, that the surface plasmon is responsible for this phenomenon. The light intensity at the 9.7 eV peak decreases with time due to the oxidation of the surface. If one measures at larger angles of incidence, the radiative plasmon (λ_p = 800 Å, 15 eV) is also observed (*183*). With these experiments it is demonstrated that the excitation of nonradiative plasma waves can be observed by photons which are coupled with the surface waves via the irregularities of the surface.

The calculation of these phenomena has been reported in several papers (*184–187*). The main difficulty in arriving at a quantitative comparison between the experimental and theoretical results comes from the lack of knowledge of the structure of the roughness, the same situation we meet in all experiments with statistical surface roughnesses.

b. Coupling by a Grating. The situation is much simpler if the coupling is done by a grating, thus restricting the "roughness" vector to only one value, $K_r = 2\pi/a$ (a grating constant), or to vK_r where v are whole numbers if higher orders come into play. The first experiments were done by Teng and Stern (*188*) who bombarded a stainless steel grating coated with an Al or Ag film of $\sim 1\mu$m thickness with electrons of ~ 10 keV at normal incidence. The grating had 1200 lines/mm. Light peaks were observed at certain angles Θ, which fulfilled the dispersion relation $\omega(|\mathbf{K}_x + \mathbf{K}_r|)$ of the nonradiative plasmons. The measurements had been performed approximately in the plane of incidence ($\varphi = 0$) as well as at higher values of φ obtained by turning the grating around its normal. In these experiments the dispersion relation of nonradiative plasmons could be verified on Al to values as low as 2 eV (7 eV is the asymptotic value of oxidized Al).

Similar, more detailed, experiments have been performed with holographic gratings which had a nearly sinusoidal profile (*189*). At oblique incidence an electron beam incident on a silver grating with a period of 8620 Å and a groove amplitude of 600 Å, excites nonradiative plasmons which couple back into the radiative region (see Fig. 13). Selecting a small wavelength band of the emitted radiation with a monochromator, one observes intensity maxima in certain directions, as shown in Fig. 49 for two wavelengths. Several peaks, the first and higher orders, are observed. As we have seen

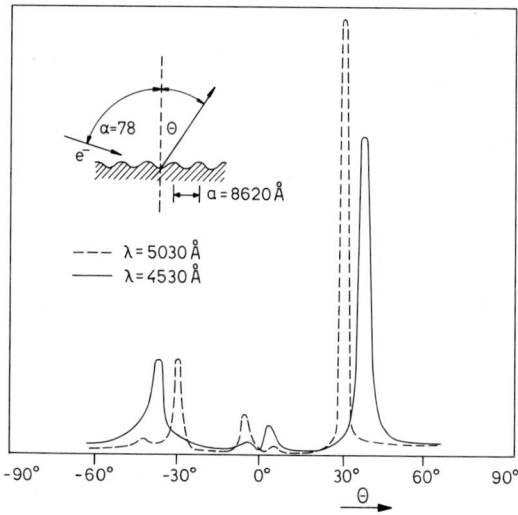

FIG. 49. The angular dependence of the intensity of the light excited by 80-keV electrons bombarding a grating [from Heitmann (*189*)].

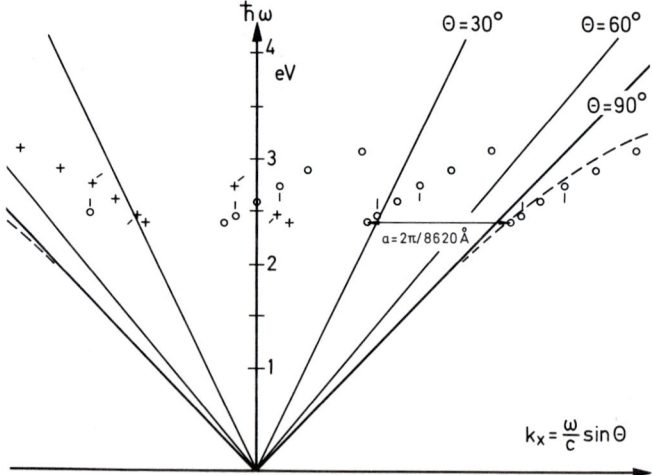

FIG. 50. Observed light peaks coupled into the radiative region by multiples of K_r. The spectrum of the 4530 Å line (solid line in Fig. 49) consists of a first-order peak to the right of $\Theta = 30°$ and of a second-order peak just to the right of $\Theta = 0°$ as Fig. 49 shows. These points are indicated in the dispersion curve by the sign ($\substack{\circ\\|}$). From the left dispersion curve the first order to the left of $-30°$, and the second order, left of $0°$ ($\substack{+\\|}$) are coupled into the radiative region by K_r and $2K_r$ [from Heitmann (*189*)].

in Eq. (108) the transferred momentum $\hbar q$ turns quickly into the direction perpendicular to that of the primary electron $\hbar K_{el}$ with increasing angle ϑ, so that, as Fig. 19 demonstrates, both wings of the dispersion curve are excited. Figure 50 shows the dispersion curve with the maxima observed in Fig. 49.

Experiments on gratings with sinusoidal profiles and variable amplitudes are of interest. They allow quantitative proof of the light scattering theory at nonsmooth surfaces since the roughness structure, here the sinusoidal profile, is known and the amplitude of the grating can be determined by other methods, e.g., from intensity measurements of the diffraction maxima (*189a*) or if possible by electron microscopy. A quantitative relation between the values $\Delta\Theta$ and $\Theta_{1/2}$ and the amplitudes of the grating (h) has been measured (*189b*). The experimental result is in rather good agreement with theoretical work (*189c*) as well as with a scattering theory of second order (*189d*) which takes into account the term δ^4 [$(\delta^2)^{1/2}$ is the rms height].

By varying the amplitude of the sinusoidal profile, the limits of the validity of the calculations become apparent for the intensity of the diffraction maxima (*189a*) as well as for the resonance position of the surface plasmons and its half-width on these sinusoidal surfaces (*189b*).

3. Excitation by Light

Light waves excite directly radiative surface plasmons at the boundary of a plasma, as we have seen in the case of the plasma resonance absorption. Nonradiative surface plasmons, however, cannot be produced by light in the same way. As we have already seen in Fig. 13, one must increase the momentum of the incoming photons to cross the light line $\omega = cK_x$ and to excite the nonradiative surface plasmons. There are two possible ways to change the momentum of the photons:

(a) If light travels along a surface with an irregular profile, such as a rough surface or in the simplest case a sinusoidal grating, the photons can pick up a "momentum" $\hbar K_r$ where $K_r = 2\pi/a$ and a is the period of the grating. In this way the phase velocity $\omega/(K_x + K_r)$ is reduced and can equal that of the nonradiative surface waves, thus leading to their excitation.

(b) If light enters a medium of refractive index n, its momentum normal to the boundary of this medium, $\hbar K$, increases to $n\hbar K$. This gain of momentum can be used to reduce the phase velocity of the incoming photons and thus to reach that of the nonradiative plasmons (ATR method—see Section III,4).

These physical principles have been applied in various ways. They have been employed to a large extent in the technique of exciting guided light modes, especially the grating coupling.

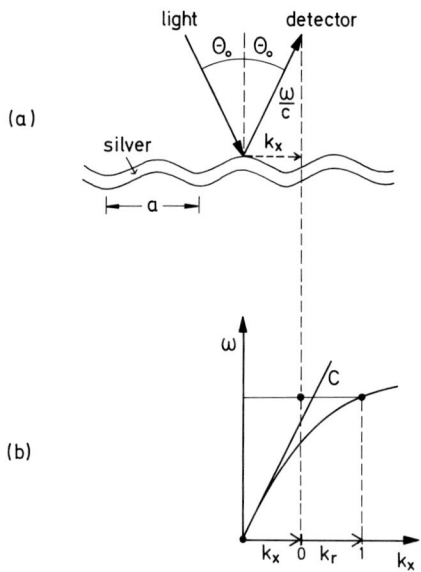

FIG. 51. Excitation of the nonradiative mode by light via a grating: (a) shows the experimental arrangement: at a certain Θ_0 a reflection minimum is observed; (b) explains this result with the help of the dispersion curve. K_r is the roughness vector $2\pi/a$.

a. Coupling by a Grating. A simple example for the coupling of plasmons with light passing along a rough boundary is the following. Light falls on a sinusoidally corrugated metal surface—the metal has to be regarded as an electron plasma—and the intensity of the reflected beam (zero order) is recorded (see Fig. 51). Changing the angle of incidence Θ_0 at a fixed wavelength, one observes at certain values Θ_0 a strong minimum in the reflected intensity. Figure 52 shows this result. Here *p*-polarized light of 5020 Å is observed after reflection at a silver film (~ 600 Å thick) evaporated onto a holographic grating with a grating constant $= 8050$ Å and groove amplitude of 430 Å. Figure 51b gives the explanation. If the tip of the wave vector $K_x + K_r$, $K_r = 2\pi/a$ touches the dispersion curve of the nonradiative plasmons, the latter are excited and power is absorbed through the imaginary part of ε_1 of silver (*190*).

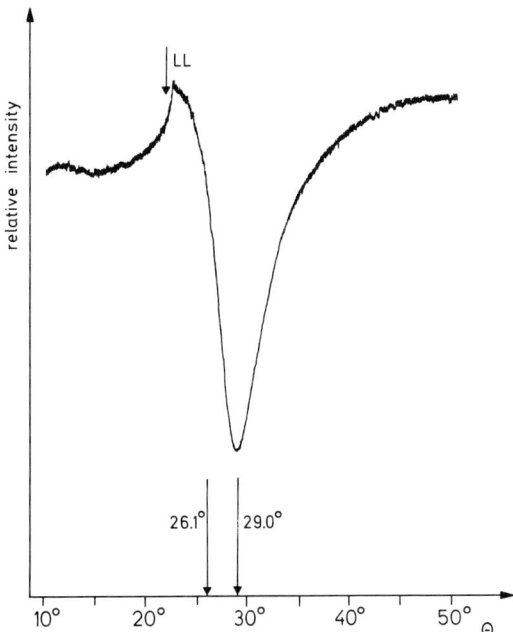

FIG. 52. Intensity reflected on a wavy silver film as a function of the angle of incidence Θ (see Fig. 51). The deep minimum indicates the excitation of the nonradiative plasmons. One sees a displacement from 26.1° (smooth surface) to 29.0° measured at the wavy surface. LL means the light line.

Another phenomenon can also be noticed. Before point 1 reaches the dispersion curve, by a suitable choice of K_x it crosses the light line. This means that the diffraction maximum present as long as point 1 lies left of the light line disappears. This is recognized by the change of intensity in the reflected beam. This intensity structure is of interest to the spectroscopist.

The experiment just described allows one to determine the dispersion curve by changing the frequency step by step. The deviation of the surface from a smooth one however leads to a displacement of the dispersion curve as we shall see in more detail in Section III,4,d. The same results are obtained by changing the frequency at a constant angle of incidence.

Experiments of this type have been performed by Teng and Stern (188) on blazed commercial gratings. Further experiments (191) used an optical grating, blazed for the first order, with $a = 15{,}000$ Å so that several diffraction maxima are observable. By looking at the first order intensity with p-polarized light in the range 7200–2700 Å, strong changes can be seen if

$K_x + (2\pi/a)n$ (n whole numbers) touches the dispersion curve so that $\omega[K_x + (2\pi/a)n]$ fulfills the dispersion relation. This indicates the strong coupling of these plasmons with light via the grating. s-Polarized light does not show this effect.

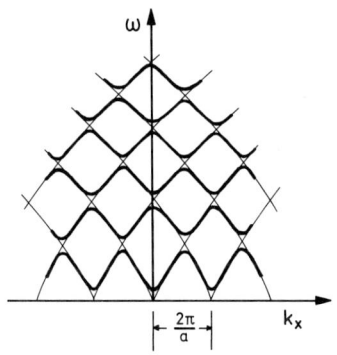

FIG. 53. The dispersion curve of nonradiative plasmons on a plane grating with period a. The strong coupling leads to a splitting of the dispersion curves at their crossing points which produces frequency gaps in which no plasma waves can exist.

The periodic structure of the surface produces a periodic repetition of the dispersion curve with $K_r = 2\pi/a$ (see Fig. 53). Thus it is possible that a plasmon of a wave vector K_x meets a plasmon of the same energy but with a wave vector of opposite direction $-K_x$ through a double scattering process, e.g., a forward and then a backward momentum change. This leads to standing waves and to the formation of a band gap (see Fig. 53). The evaluation of the reflection data on gold gives results in agreement with this concept (*191, 192*), and it demonstrates the strong coupling of the plasmons with each other through the grating. Further experiments supported this result (*193*). A somewhat similar coupling can take place if one uses a wavy thin film through which the plasma waves of both sides can couple. In this way a splitting can result under special conditions (*194*). The dispersion curves derived from these reflectance values showed slight deviations from those calculated with optical data using the relation (26). Experiments on the dispersion relation of rough and sinusoidally corrugated surfaces suggest that these differences are due to the deviation of the boundary from a smooth one and not to an oxide layer as was supposed in Cowan and Arakawa (*192*).

Further experiments of this type were undertaken with gratings coated with thin dielectric films as is often done with commercial gratings to increase their efficiency (*195*). The principal effect of such a coating on a metal grating can be demonstrated in the arrangement shown in Fig. 51 on a silver grating. The strong reflectance minimum produced by the excitation of nonradiative plasmons is displaced to larger angles by surface coatings of LiF of increasing thicknesses as a consequence of the thickness dispersion (see, e.g., Fig. 17). In the same way the results on commercial gratings are in general understandable considering surface plasmons and their dependence on such dielectric films.

These experiments on gratings have a practical aspect since this coupling contributes, to a certain extent, to the understanding of the Wood's anomaly. If one observes at a given wavelength, the intensity reflected by an optical grating into any one diffraction direction at various angles of incidence, a strong and rapid change of the intensity over a narrow range of angles can be recorded. The same effect can be obtained by varying the wavelength at a given angle of incidence. This variation is a function of the groove profile, of the value of $\varepsilon(\omega)$ of the metal and its coating, and the polarization of the light. The spectroscopist wants to avoid these strong intensity variations in his work. In general, these anomalies can be partly explained by taking into account the excitation of surface plasmons if the metals have negative values of the real part of ε in the wavelength region used. These intensity variations are also connected with the appearance and disappearance of diffraction maxima if K_x crosses the light line as mentioned above. This explanation has already been proposed by Rayleigh to explain these anomalies, see Fano (*196*). A discussion of this question has also been published recently (*197*). Theoretical work has lead to a better quantitative description of these complicated phenomena (*198*).

b. Coupling by Statistical Surface Roughness. Reflection experiments on plasma surfaces with a statistical roughness showed similar intensity anomalies. If a light beam is incident nearly normally on a plasma boundary and the intentisty reflected at $\sim 0°$ is recorded, one finds a dip in the reflected light intensity. This has been observed on silver (*199, 200*), aluminum (*201–204*) and magnesium (*205*) surfaces. Figure 54 shows this phenomenon on a thick silver surface. The reflected intensity on either side of the surface plasma frequency ω_s ($\lambda_s = 3450$ Å) is lower than that calculated with the known optical functions $\varepsilon_r(\omega) + i\varepsilon_i(\omega)$ on very smooth boundaries. The deficit ΔR of the reflectance extends over a whole frequency band from about λ_s up to longer wavelengths and its value increases with roughness (Fig. 54). The roughness has been changed by depositing films of CaF_2 of different thickness on the quartz substrate before depositing the metal film on it (*206*).

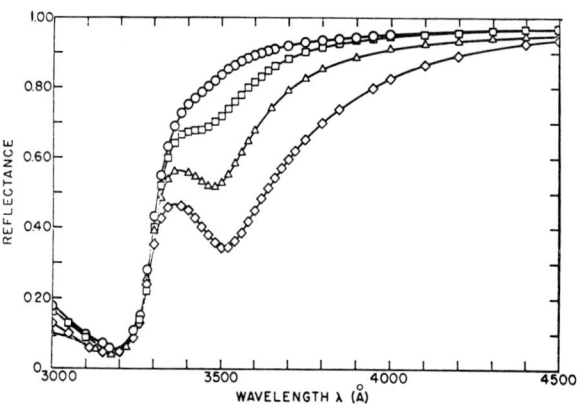

FIG. 54. Reflectance of silver surfaces of different roughnesses at normal incidence [from Stanford (206)].

Similar results were obtained in transmission (T_p) of rough silver films. Here T_p is reduced at wavelengths around λ_s and above (207). The explanation of these results is basically the same as that given above. The statistical roughness can be regarded as the sum of sinusoidal gratings of continuously varying values K_r so that the considerations related to Fig. 51 can be applied to each grating with the period a_r. The function which describes the probability of finding a value of $K_r = 2\pi/a_r$ describes the structure of the roughness and thus the form of the deficit ΔR as a function of the light frequency. Backscattering from the nonradiative region into the original direction of the radiative region (here $K_x = 0$) is of small influence. Coating of the reflecting surface with different thin films, such as CaF_2 (208), carbon (205), MgF_2 (199, 207), and LiF (203), proved by the calculable displacement of the reflectance dip that surface plasmons play the important role.

In addition to the plasmon effect, a further effect influences the dependence $\Delta R(\omega)$. This is the scattering of light by roughness within the radiative region due to the contributions of K_r values $<\omega/c$, which in addition, reduces the values of $R(\omega)$.

A further support to the interpretation of the effect as the excitation of surface plasmons is given by measuring the photoelectrons emitted by the strong electromagnetic fields of the plasmons if their quantum energy surpasses the work function of the metal. Such experiments were first done on roughened surfaces of Al (203). Figure 55 shows the reflectance deficit ΔR for a magnesium surface together with the photoelectric yield as function of the light frequency (205). Magnesium has a work function of about

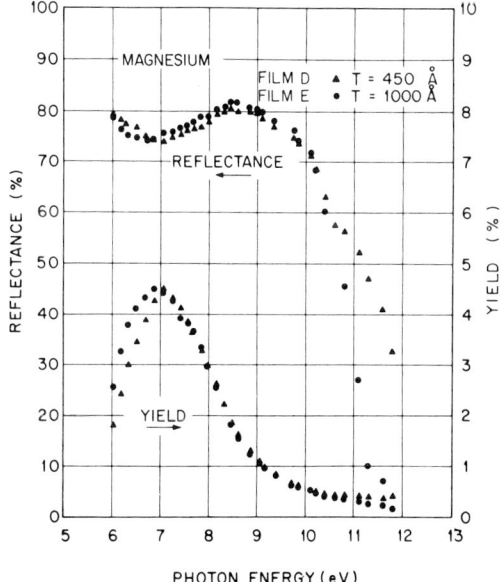

FIG. 55. Correlation of the deficit of the reflectance and the simultaneously measured yield of photoelectrons on Mg [from Gesell et al. (205)].

3.3 eV and a surface plasma energy of 7.1 eV. The number of photoelectrons has a peak in that frequency region where, among other factors, the high density of states of K_r values leads to an intense coupling between light and plasmons.

This reflectance deficit has been treated theoretically, using the ideas given above, by several authors (209, 210). In principal, the observed structure could be verified. However, the quantitative agreement depends on assumptions about the structure of the roughness. Assuming a Gaussian correlation, an analytical expression is obtained for ΔR. Adapting this relation to the experiments, one gets values of the correlation length and the rms height. However, the physical significance of these figures is still open to question (205). A number of experiments have shown that the assumption of a Gaussian correlation over the whole range of K_r values is not valid according to Endriz and Spicer (203), Dobberstein (211), and Kretschmann (212). A direct determination of the roughness and the correlation length from scattering experiments is possible over a limited region of K_r values and gives data that may provide certain information on the roughness structure.

4. Excitation by Evanescent Light Waves (ATR Method)

a. Introductory Remarks. We have seen that nonradiative surface waves cannot be excited on a metal/vacuum boundary by light reflected at this surface since the light wave vector is always smaller than that of the plasmon. The situation is different if the plasma is bounded by two media of different dielectric constants $\varepsilon_0 \neq \varepsilon_2$ (Fig. 56). We see immediately that in the case $\varepsilon_2 < \varepsilon_0$, a part of the dipersion relation of the boundary 1/2 lies left of the light line $c/\varepsilon_0^{1/2}$. Thus plasmons that lie on the dispersion curve designated by circles can be excited by photons coming from the medium ε_0 whose momenta have increased from $\hbar(\omega/c)$ to $\hbar\varepsilon_0^{1/2}(\omega/c)$ on entering the medium ε_0 from the vacuum (Fig. 56) (*213, 214*). If the angle θ, measured in the medium ε_0, is varied from zero to higher values at $\omega =$ const, with $\varepsilon_2 = 1$ (air or vacuum) total reflection starts at $\sin \Theta = (\varepsilon_0)^{-1/2}$ or $K_x = \omega/c$ and continues up to $\sin 0 = 1$ or $K_x = (\omega/c)(\varepsilon_0)^{1/2}$. Between these two limits, $1 \leq (\varepsilon_0)^{1/2} \sin \Theta \leq (\varepsilon_0)^{1/2}$, one meets the dispersion relation at a certain K_x value given by Eq. (26) (see Fig. 56).

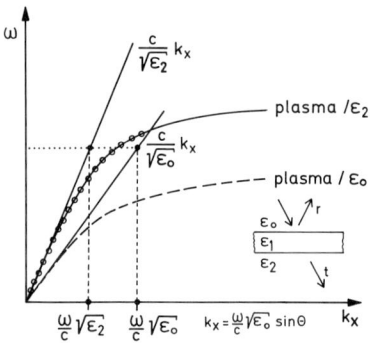

Fig. 56. A plasma slab (ε_1) bounded on the two sides by different media (ε_0 and ε_2). The dispersion curves of nonradiative plasma waves for both boundaries are depicted for $\varepsilon_0 > \varepsilon_2$ and $\exp(2|K_{z1}|d) \gg 1$.

If the dispersion relation is fulfilled, the reaction is seen in the intensity of the totally reflected light. Absorption takes place, and the reflectance drops from the value one to smaller values. This represents attenuated total reflection as we find it in all metals due to the finite value of their imaginary part.

The reflectance R of this two boundary system is calculated as the ratio of the reflected and the incoming electric field intensity by

$$R = \frac{J_r^p}{J_0^p} = \left|\frac{E_r^p}{E_0^p}\right|^2 = |r_{012}^p|^2 \qquad (131a)$$

where

$$r_{012}^p = \frac{r_{01}^p + r_{12}^p \exp 2iK_{z1}d}{1 + r_{01}^p r_{12}^p \exp 2iK_{z1}d} \qquad (131b)$$

for r_{vk}^p see Eq. (39).

For the discussion, $|r_{012}^p|^2$ can be developed at K_x^m, the position of the resonance peak (47). For values of $\exp(2|K_{z1}|d) \gg 1$ and assuming also that $|\varepsilon_{r1}| \gg \varepsilon_{i1}$ and $|\varepsilon_r| \gg 1$ we obtain

$$R = 1 - \frac{4\Gamma_i\Gamma_r}{(K_x - K_x^m)^2 + (\Gamma_i + \Gamma_r)^2} \qquad (132)$$

with the following abbreviations

$$\Gamma_i = \text{Im}(K_x) \quad [\text{see Eq. (28b)}] \qquad (133)$$

representing the interior damping, and

$$\Gamma_r = \text{Im}(\Delta K_x^m) \quad [\text{see Eq. (38)}] \qquad (134)$$

the radiation damping.

Equation (132) indicates a resonance minimum $R_{\min}$ at K_x^m given by Eq. (37) whose depth and width depends on Γ_i and Γ_r. Γ_i is determined by $\text{Im}(\varepsilon_1)$, whereas Γ_r is strongly dependent on the thickness d of the film which determines the exponential decay of the electromagnetic field between the boundaries 0/1 and 1/2 (see Fig. 60). In the medium ε_0, K_{z0} becomes real again so that radiation losses are the consequence. Therefore the reflectance minimum, as well as the scattering maximum, becomes broader with decreasing film thickness (see Section III,4,d). Under the special condition

$$\Gamma_i = \Gamma_r \qquad (135)$$

which fixes a certain film thickness, the reflected intensity becomes zero (complete adaption) (Fig. 60) (47). Variation of the wavelength λ of the incident light displaces the position of $R_{\min}$ and changes the width of the minimum (see Fig. 61).

This phenomenon represents the analogy to the plasma resonance absorption observed at radiative surface plasmons (215). When the reflectance has reached its minimum, the plasma oscillations will have a high amplitude, or the electromagnetic energy density on both sides of the plasma/vacuum boundary, $(\varepsilon_1/\varepsilon_2)$, will pass through a maximum at K_x^m. The energy density can be calculated with the following relation

$$|t^p_{012}|^2 = \left| \frac{t^p_{01} t^p_{12} \exp(iK_{z1}d)}{1 + r^p_{01} r^p_{12} \exp(2iK_{z1}d)} \right|^2$$

where $|t^p_{012}|^2$ measures the field energy just behind the plasma boundary, $z = -d$, in vacuum. It shows a maximum as function of Θ whereas R_p has a minimum. This peak height depends on the thickness of the plasma film like the value of R_p^{min} and has a maximum if $\Gamma_i \cong \Gamma_r$. The spatial distribution of the energy density $|H|^2$ is seen in Fig. 57. Outside the resonance, the field energy decays exponentially inside the plasma (ε_1) (see Fig. 58b) more rapidly than outside the plasma. Near the resonance case the field energy drops at first before reaching a high value at the boundary $\varepsilon_1/\varepsilon_2$. At resonance, there is no minimum inside the film and the value of

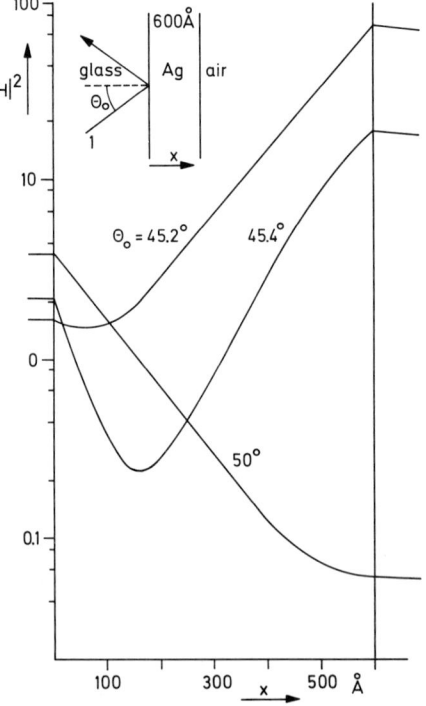

FIG. 57. The calculated electromagnetic energy density $|H(z)|^2$ in a 600-Å silver film [left medium (ε_0), right; air] with the angle of incidence as parameter $(\lambda = 6000$ Å$)$.

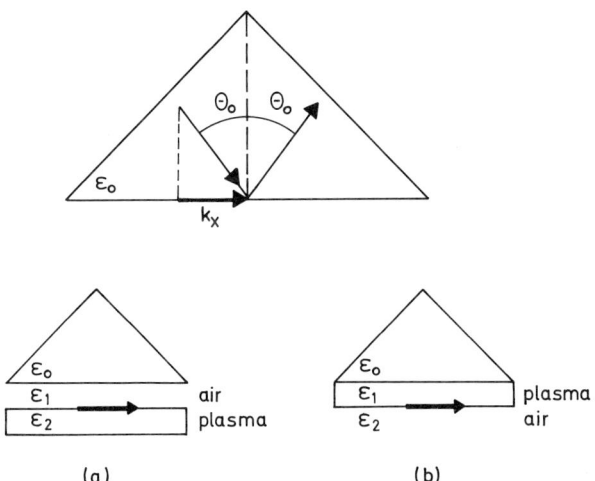

FIG. 58. ATR method of exciting nonradiative plasmons by light with different positions of the plasma film in relation to the prism.

the field energy at $z = -d$ is maximum. At small Im(ε), as in silver, it can reach about a factor 80 compared to the value without resonance (Fig. 57).

b. *ATR or Prism Method.* The physical consideration just explained is applied by using the total reflection in the following way, called the ATR method (attenuated total reflection). For example, a prism of quartz is irradiated with light of frequency ω that falls on the prism base at an angle Θ_0. This produces a tangential component of the wave vector

$$K_x = \varepsilon_0^{1/2}(\omega/c) \sin \Theta_0 \qquad (136)$$

which is longer by the factor $\varepsilon_0^{1/2}$ than the wave vector in vacuum, or, because $\varepsilon_0^{1/2} \sin \Theta_0 \geq 1$ an inhomogeneous wave propagates along the base with a phase velocity ω/K_x smaller than that of light. The plasma film can now be placed in two different positions. In the first, proposed by Otto (*213*), it is placed in a distance d from the base of the prism, where d is comparable with the wavelength of light in air. The inhomogeneous wave decays exponentially in this air gap and excites the surface waves on the plasma boundary (see Fig. 58a). In the second position proposed by Kretschmann and Raether (*215*), the plasma slab is placed directly on the base of the prism. Here the inhomogeneous wave has to penetrate the plasma film to excite the plasmons on the plasma/air boundary (see Fig. 58b).

Both methods have their advantages. In Fig. 58a there is an experimental difficulty in placing the plasma boundary at a distance of about 1000 Å parallel to the base. However, the air gap method is a suitable procedure to observe the surface plasmons on the boundary of a thick, smooth monocrystal. The arrangement of Fig. 58b allows to study properties of thin films that are deposited by evaporation or other techniques on the base of the prism, and the experimental procedure is much more convenient. A further advantage of case (b) is the possibility to observe easily processes in correlation with the resonance effect, such as, photoemission of electrons and emission of light due to surface roughness. Thus it is possible to detect the excitation, not by the reflectance minimum alone but also by other processes such as the emission peak of photoelectrons or light.

The observation of such a reflectance minimum of p-polarized light on thin films of Al of various thicknesses and other metals deposited on the base of a prism in the total reflection regime has been reported a long time ago by Turbadar (216) together with calculated results. However, his observations were not correlated with a resonating excitation process. The arrangements in Fig. 58a and b are physically the same and follow from each other mathematically by interchanging the media air and plasma, a fact which was overlooked at first (217).

c. *Applications. Determination of the Optical Constants or Dielectric Function of Metals.* By adapting the experimental reflectance curves with a least square fit procedure using the relation in Eq. (131), the values of ε_{r1}, ε_{i1}, and the thickness of the metal film are obtained. Because of the dispersion, the reflectance curves are displaced with the wavelength λ of the exciting light (see, e.g., Fig. 61), so that ε_1 as a function of ω is obtained. This accurate procedure allows one to record the variations of $\varepsilon_1(\omega)$ dependent on the conditions of the production of the film.

The experiments are shown in Fig. 59. The light beam enters a half-cylinder, generally of quartz, after having passed through a polarizer,

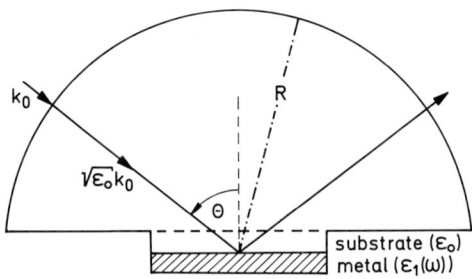

FIG. 59. Scheme of the arrangement to determine $\varepsilon_1(\omega)$ of thin metal films by exciting nonradiative plasmons on the metal/vacuum boundary.

diaphragms, and a cylindrical lens which reduces the divergence of the beam to offset the divergence that occurs on crossing the curved surface of the half-cylinder. The beam divergence is less than 0.02°. The metal to be investigated is deposited on a thin (~1 mm) plate of the same dielectric material ε_0. Since the axis of rotation for changing θ must lie in the center of the circle of radius R, the half-cylinder has to be reduced by the thickness of the plate of about 1 mm thickness. The optical contact of this plate with the half-cylinder is achieved by the use of a suitable oil. Thus a quick change of the plate is possible without a new adjustment. The surface of these quartz plates is fire-polished and, thus, rather smooth. The thickness of the metal film must be thin enough so that the intensity of the exciting field in the boundary does not become too weak, because the reflectance minimum decreases with increasing film thickness and disappears in silver at ~1000 Å, in gold at ~1500 Å, and in aluminum at ~1000 Å. Therefore the reflection method has a limited application for thicker films.

Figure 60 shows measured reflectance minima R_p in silver as a function of the thickness (10). It demonstrates that R_p becomes nearly zero at $d \sim$ 500 Å. The width increase at smaller d is due to the larger radiation damping by light emission into the medium ε_0. Measurements of the reflectance minima in silver deposited by evaporation onto a quartz substrate as a function of the wavelength of the incoming light yield $\varepsilon_1(\omega)$ values in good agreement with the mean values reported in the literature in this frequency

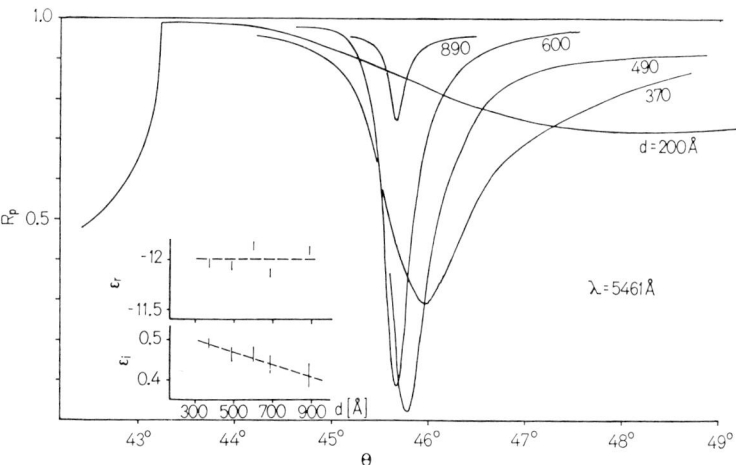

FIG. 60. Measured reflectivity of silver films of different thicknesses as a function of the angle of incidence at $\lambda = 5461$ Å. Results of $\varepsilon(\omega)$ are shown in the inset [from Kretschmann (61)].

region. ε_i increases slightly with decreasing thickness, probably due to the internal structure of the thin silver film (see inset of Fig. 60).

The same method has been applied to obtain data of $\varepsilon(\omega)$ of gold (218), however the accuracy of these measurements has not been optimal. Recent measurements on 500-Å gold films deposited on smooth quartz substrates agreed between 5000 Å $< \lambda <$ 7000 Å within 5% with literature data obtained by ellipsometric measurements (219, 219a).

If metals sensitive to oxidation, such as the alkali metals, have to be studied, ultrahigh vacuum conditions are necessary. A special prism has been used (220) to determine $\varepsilon(\omega)$ of sodium and lithium under these conditions. Figure 61 shows the reflectance of a Li film, 460-Å thick, at photon energies between 2 and 3.8 eV as a function of the angle of incidence. The

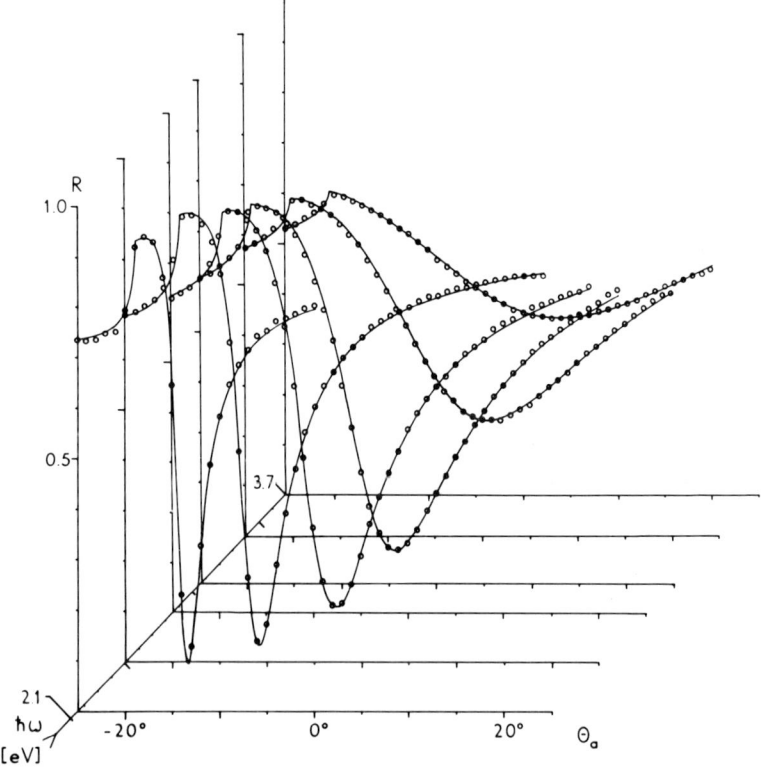

FIG. 61. Reflectivity of a lithium film, 460 Å thick, on a sapphire substrate as a function of the angle of incidence at different photon energies. ○, observed; solid line calculated with adapted values [from Bösenberg (221)].

minima become broader with decreasing wavelengths. From these curves $\varepsilon_r(\omega)$ and $\varepsilon_i(\omega)$ of sodium and lithium are obtained (221).

In Figure 60 the reflectance minima $R_{min}(\Theta)$, measured on Ag, are reproduced as a function of Θ at constant λ. Instead of varying Θ, measurements of $R_{min}(\lambda)$ at constant Θ and varying λ can be made. The curves $R_{min}(\lambda)$ and $R_{min}(\Theta)$ are not necessarily equal. The function $R_p(\lambda, \Theta)$ shows in the three-dimensional representation a more or less deep valley of R_{min} which becomes broader with decreasing wavelength. If one now cuts this valley in different directions (λ = const or Θ = const) it is not surprising that the different dependencies of $R_{min}(\lambda)$ and $R_{min}(\Theta)$ result. The larger $Im(\varepsilon_1)$ is, the more different the dependencies are. (222-225). This is displayed by the solid lines in Fig. 62. The lower curve is obtained from measurements at λ = const on a 600-Å silver film, whereas in the upper curve which shows a bending back at shorter wavelengths ($\lambda < \lambda_s$), Θ has remained constant. The dielectric function is the same in both cases. The bending back arises from cutting the "valley" at the high energy side. At still higher frequencies $\omega > \omega_p$, the deep valley of the interband absorption comes into play ($\lambda \lesssim$ 3200 Å) and R_{min} turns back again. Since one measures in the nonradiative region to the right of the light line (LL in Fig. 62), this structure of the attenuated total reflexion is due to the $Im(\varepsilon_1)$ determined by the interband absorption (transition from the Fermi level to the conduction band).

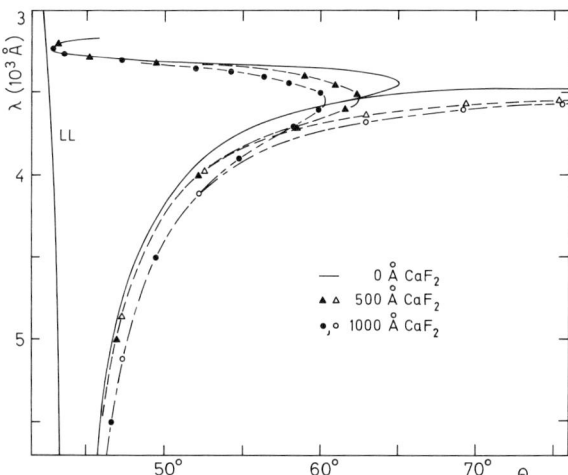

FIG. 62. Position of the reflection minima measured for a 600-Å silver film deposited on quartz. LL is the light line. A smooth Ag film (solid curve) is compared with Ag films of different roughnesses. The upper curves $R_{min}(\lambda)$ showing a bend back are observed at constant Θ, the lower curves $R_{min}(\Theta)$ at constant wavelength [from Orlowski and Raether (224)].

It should be emphasized that the R_{min} curves are not identical with the dispersion curve $\omega(K_x)$. These curves approach each other the less the plasmons are damped [see, e.g., (226)].

Figure 62 shows the results of the same types of measurements performed on silver films of different roughnesses, produced by underlying films of CaF_2 of varying thicknesses. The back bending position changes with growing roughness and is correlated with increased damping.

The discussions above for $R(\Theta)$ (λ = const) and $R(\lambda)$ (Θ = const) also hold for the dispersion relation. The dispersion of Eq. (7), $D(\omega, K) = D(\lambda, \Theta)$, can be plotted as a function of λ as well as of Θ. The minimum of D lie at different positions depending on which quantity has been kept constant, as can be seen for the example of gold (225). With decreasing damping this difference disappears.

Low energy surface plasmons in semiconductors. The ATR method can be applied in a photon energy region which can be extended to the far infrared region without serious limitation. This application is of interest for studying doped semiconductors whose density of free electrons is much smaller than that of metal electrons, thus leading to plasma frequencies in the infrared. Experiments have been made on different substances, for example, n-type germanium doped with 4×10^{19} electrons/cm^3 which shows a reflectance minimum at ~ 950 cm^{-1} or $\lambda \sim 50$ μm (227). In this energy region, experiments have also been made with a grating coupler inscribed on a InSb surface doped with 7×10^{18} electrons/cm^3 to excite surface plasmons of low frequency (228). The ATR method has also been used to excite surface phonons (229).

Very thin films on a plasma boundary. If the plasma surface is covered with a rather thin film of $\varepsilon_2(\omega)$ and a thickness d, the frequency of the surface oscillation is displaced. This asymmetric layer system, plasma $\varepsilon_1(\omega)$/film $\varepsilon_2(\omega)$/air (ε_3), has been considered already in Section I. Its dispersion relation is given by Eq. (7). Replacing the numbers 0 1 2 by 1 2 3 (zero shall be reserved for the prism), the resulting effects depend on the $\varepsilon_2(\omega)$ function. If $\varepsilon_2(\omega)$ is real, as in the case of a LiF layer, a displacement of the frequency results, which depends on the thickness of the film (thickness dispersion). This dependence can be seen in Fig. 17 as the lowest curve (surface mode). Figure 63 depicts a measurement of the reflectance curves of a silver film on which LiF films of different thicknesses are deposited (230). A slight broadening of the minima with increasing thickness is visible. This is due to the fact that the dispersion curve of the Ag/LiF system lowers with increasing thickness of the LiF, so that if the $R(\lambda, \Theta)$ valley is crossed by changing Θ, the

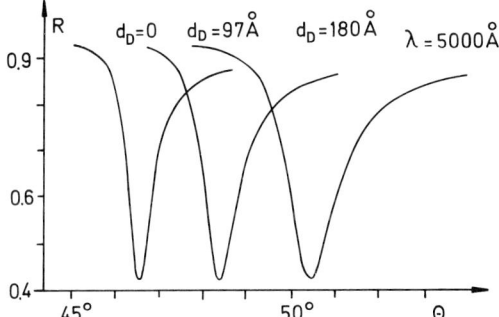

FIG. 63. Measured reflectivity of a silver film at $\lambda = 5000$ Å as a function of the angle of incidence. The silver film has been coated with Lif films of different thicknesses [from Twietmeyer (230)].

slope is less steep, the larger the LiF thickness. One notices already that very thin films, <5 Å, give rise to measurable displacements. If one keeps in mind that Θ can be determined within several 10^{-2} degree, this method is a rather sensitive one (231).

The displacement can be followed up to large thicknesses as long as absorption is negligible. Beyond a certain thickness, light modes can arise (see Fig. 17) excited by the incoming light through the silver film (231). If $\varepsilon_2(\omega)$ is complex as in a carbon, silver sulfide (231), or other metal coatings (230), damping also comes into play. A gold coating on a silver boundary (500 Å thick) leads to reflectance curves rapidly changing with the thickness of the gold film. On the other hand, a silver coating on a gold film (300 Å thick) leads to reflectance curves which look like the ones shown in Fig. 60.

The quantitative evaluation of these experimental results is difficult since the dielectric function $\varepsilon_2(\omega)$ of the bulk material, together with the thickness d given by the quartz oscillator, does not reproduce the observed curves accurately. The same is valid for the LiF films. It is probable that the structure of these very thin films is not that of the homogeneous bulk material, so that the dielectric functions of a very thin film seem to be changed. Therefore more knowledge of the structure of such films is needed.

Another point shows the difficulty of evaluation. If silver films of ~500-Å thickness are vapor-deposited in a good vacuum (~10^{-5} torr), one observes a slight displacement of the reflectance curves with time which leads to a continuous decrease of ε_r from $\varepsilon_r = -14.2$ to $\varepsilon_r = -14.9$ at $\lambda = 6000$ Å during several hours. The experiments show that after 4–6 hours the silver film has reached approximately a finite state. Certainly "structure" changes are responsible during which a lot of voids disappear, but the details of these changes are not well known. If these surfaces are exposed to air again, a small variation of the reflectance minima is visible which arises from gas adsorption and, partially, from surface reactions. The possibility that the ATR method contributes to such problems must be further studied (232, 232a).

In connection with these problems, calculations have been performed to determine the effect of thin "inhomogeneous" surface regions at the boundary of a plasma (233), e.g., at the boundary of a liquid Hg surface (234). A point of interest is the coupling of excited molecules to surface plasmons in very thin films (234a).

Photoelectron emission by nonradiative surface plasmons. The maximum excitation of surface plasmons is coupled with a maximum electric field at the plasma/vacuum boundary. If the work function of the metal has a value lower than $\hbar\omega$ of the surface plasmon, photoelectrons are emitted, as we have already seen. These electrons can be measured with a channeltron mounted just below the plasma film (as shown in Fig. 58b). By varying Θ

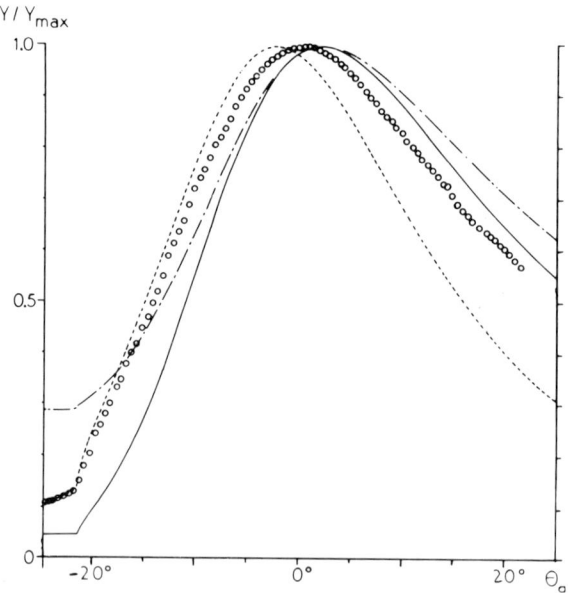

FIG. 64. Observed photoelectron yield (○) as a function of the angle Θ of a lithium film of 425 Å thickness, ($\lambda = 4047$ Å). Calculated with surface effect (----), with volume effect (———) $L = 1$ Å, (—·—) $L = 250$ Å [from Bösenberg (221)].

of the incoming p-polarized light, the photoelectron curve passes through a maximum, whereas the reflected light goes through a minimum. This allows one to derive the optical properties of the metal film *in situ*. This is important for the comparison with calculated values of the photoelectron emission assuming different mechanisms such as volume or surface effects. Such experiments were performed using cesium-coated Ag (*220*), Al (*235*), Li and Na (*221*). Figure 64 shows the comparison of observed photoelectron emission of a lithium film (425 Å thick) for the different angles Θ of the incoming light. These data are compared with calculated values assuming the two different mechanisms for the photoemission, surface effect and volume effect. For the calculation of the volume process, one needs the escape depth L of the photoelectrons, which is assumed to be 1 and 250 Å. A certain agreement is obtained, However, systematic deviations reveal that the right approach has not yet been attained.

Excitation of the guided light modes. In section II we have described several experiments in which the metal boundary has been covered with a thin dielectric film (ε_2). This leads to a displacement of the dispersion relation dependent on the thickness of the surface film (see Section III,1).

A new situation arises when the thickness of this dielectric film becomes so large that at least one light mode can exist in it. The number of these modes is regulated by the dispersion relation of Eq. (67). We regard then the dielectric film as a wave guide covered asymmetrically by the metal on one side and by air on the other side. By changing the angle Θ, or the value of K_x, one can fulfill the $\omega(K_x)$ condition for the different light modes of this system. The excitation of the light mode can be observed by the light scattered from the film surface, which has a certain roughness. Such experiments on silver covered with a LiF film have been reported (*231, 236*). In similar experiments on aluminum using the same arrangement these light modes have been observed in reflection (*237*).

Figure 65 shows the light intensity emitted in the direction of the film normal as measured with a photomultiplier as a function of the angle Θ. The constant frequency of the incoming He–Ne laser light corresponds to $\lambda = 6328$ Å. Besides the plasma mode at Θ ~ 57.5°, the different s- and p-light modes are observed whose number depends on the thickness, frequency, etc. The position of these light peaks can be measured very accurately, so that it is possible to derive the refractive index of the dielectric film. In this way it was observed that the refractive index of LiF films of some 10 μm thickness, deposited on Ag, has an anisotropy of about 1% comparing the normal and the tangential values. This can be recognized by an anomalous

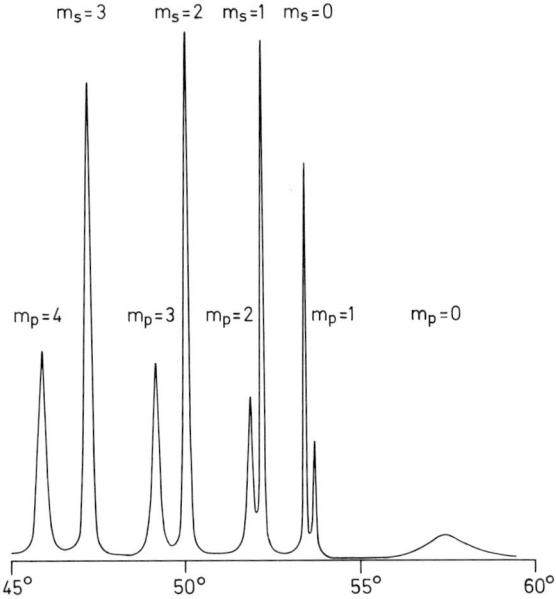

FIG. 65. Guided light modes s and p in a 20-μm LiF film deposited on a silver film of 500 Å thickness and excited by the prism method. $p = 0$ is the surface mode. $\lambda = 5500$ Å [from Hornauer and Raether (236)].

sequence of the s- and p-modes. [For more details, see reference (236).]

d. Roughness Effects. Influence of roughness on the properties of surface plasmons. The electromagnetic fields of the nonradiative plasma oscillations are concentrated in the boundary of the metal film. Therefore, changes of its composition which lead to a different dielectric function are easily observed as we have seen in the foregoing discussion. The sensitivity of the method of evanescent waves should also be sufficient to indicate geometric deviations of the boundary from a plane boundary. This deviation can be a statistical roughness or a periodically corrugated surface, such as a grating.

To measure such effects we can apply the reflection method just described. Another equivalent procedure uses the coupling of the nonradiative oscillations, via roughness, with the radiative region which produces the excitation of the plasmons by light emission (102, 9). If we look at the plasma boundary with a multiplier, in the direction of the surface normal, the emitted light passes through a maximum as a function of the angle of incidence Θ.

This way of detection differs from the reflection method because it is not limited by the thickness of the film. As mentioned in Section III,4,c, the minimum of R_p practically disappears for film thickness ~ 1000 Å whereas here there is no limit except that the intensity becomes too weak to be measured.

The experiments on surfaces of silver and gold films of different roughnesses with both methods revealed an appreciable dependence of the position of the reflectance minimum (scattering peak) and its half-width on the roughness (239, 240, 225). The roughness of a metal film can be varied by coating the smooth substrate with a layer of CaF_2, MgF_2, Ag, or other crystalline material before depositing the metal film by vacuum evaporation. In both cases the resonance peak (or minimum) is displaced to higher K_x values with increasing roughness, and the half-width of the resonance curves grows at the same time. Figure 66 shows an example of peaks of scattered light from 700-Å thick silver films of different roughnesses.

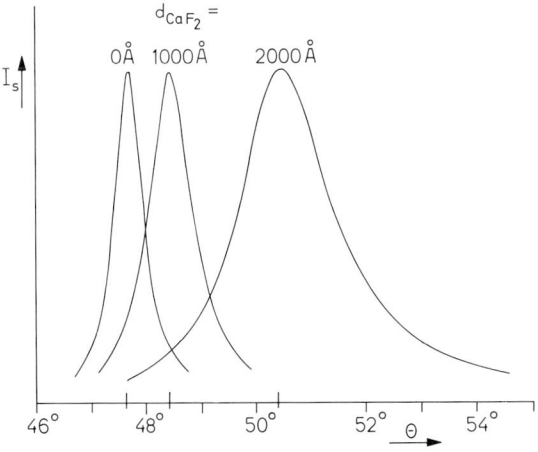

FIG. 66. Normalized intensity of the light (4579 Å) scattered at silver films 700 Å thick of different roughnesses as a function of the angle Θ.

There exists a correlation between the displacement $\Delta\Theta$ and the half-width of the peaks $\Theta_{1/2}$ which depends on the wavelength λ (see Fig. 67). Both methods, observation of the reflected and scattered light, give the same result (239, 225).

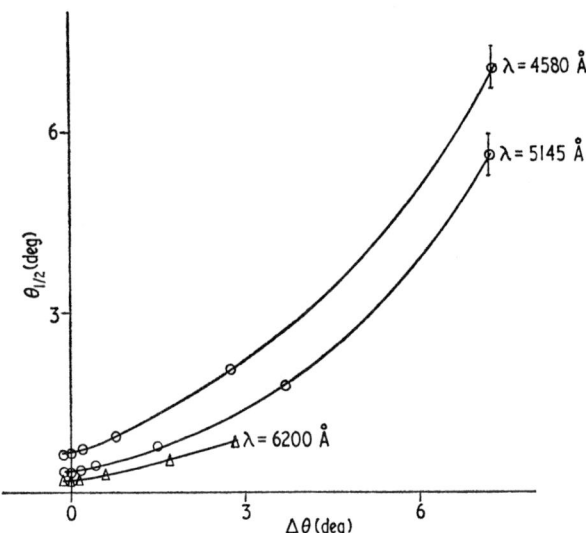

FIG. 67. The half-width ($\Theta_{1/2}$) as a function of the displacement of the plasma frequency ($\Delta\Theta$) on rough surfaces of silver at different wavelengths of the scattered light [from Hornauer et al. (239)].

Experiments on silver films with a sine profile (holographic grating) using the same arrangement had qualitatively the same result, which led to a very similar relation between $\Delta\Theta$ and $\Theta_{1/2}$ and the amplitude of the grating. The amplitude corresponds to the mean height of the statistical roughness (rms height) (241). It seems that experiments described in Bjork et al. (242) have to be interpreted in a similar way.

The interpretation of these results can be given very roughly. The increase of the peak width of the reflectance minimum or scattering maximum with greater deviation of the surface profile from a smooth surface is caused by the stronger light emission leading to a higher damping of the oscillations. The displacement of the resonance frequency to higher K_x values, or the decrease of the phase velocity of the surface waves, has its analogy in the open structure of an electric circuit where the phase velocity is less than the light velocity. The electric fields have to travel along the real and longer boundary, so that the velocity along the fictive smooth surface becomes smaller. A theoretical treatment of this problem showed qualitatively the displacement of the resonance peak as observed (243). In the case of a rough surface it is better to say that the resultant electromagnetic field—the unperturbated together with the waves scattered at the irregularities—move with a lower phase velocity.

Light scattering by rough surfaces assisted by surface plasmons. Theoretical considerations. The interaction of plasmons via roughness has some interesting consequences with regard to light scattering. Photons hitting a rough metal surface are in general diffusely scattered. If the frequency of these photons is chosen so that plasmons are excited, as discussed above, the emitted resonant intensity is rather strong, allowing one to study this interaction process in detail. One hopes in this way to obtain information on the structure of the roughness and to test existing scattering theories at the same time [see, e.g. *(244)*].

In the following we sketch in simplified form a theory of light scattering of Stern *(248)* for small roughnesses *(103, 245)*. If light falls from the vacuum on a smooth boundary of a solid, it produces in the interior below the surface a polarization current density

$$j_1 = \frac{dP_1}{dt} = \frac{d}{dt}(\varepsilon_1 - 1)E_1 \qquad (137)$$

where E_1 is the field in the interior. If we have a rough surface whose profile $S(x, y) = Z$ is given as indicated in Fig. 68, a fictive smooth boundary is defined by $\overline{Z} = S(x, y) = 0$.

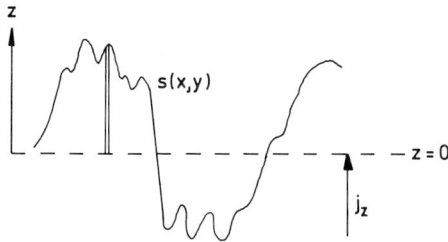

FIG. 68. Schematic profile of a rough surface.

For simplicity we neglect the components j_y and j_x and regard j_z alone. Each section of the roughness profile now acts as an antenna of height $S(x, y)$, fed by j_{z1}, which radiates like a Hertz dipole on a surface; in this case a metallic surface.

The height of these antennas is assumed to be small compared with the wavelength λ of the light, $|S| < \lambda$, so that phase differences along $S(x, y)$ can be neglected. The valleys contribute with negative amplitudes. A plane electromagnetic wave with the intensity J_0 may fall on the surface at an angle of incidence Θ_0. Then the questions arise, Which is the emitted intensity observed at an angle Θ and an azimuth Φ for both polarizations? and How does it depend on the mean height and on the spatial distribution of the peaks of the roughness?

The result of this calculation (*212, 246, 247*) can be written as a product of two factors. One describes essentially the angular distribution of the emission: $|W_z(\Theta_0, \Theta)|^2$ and the other the distribution of the roughness, $|S(\Delta K)|^2$ which will be discussed later. Thus we write for the intensity scattered into the solid angle element $d\Omega$

$$dJ = \frac{1}{4}\left(\frac{\omega}{c}\right)^4 \cdot J_0 \frac{(\varepsilon_0)^{1/2}}{\cos \Theta_0} |S(\Delta K)|^2 |W_z(\Theta_0, \Theta)|^2 \, d\Omega \qquad (138)$$

where J_0 is the intensity of the incoming light, ε_0 the dielectric constant of the medium from which the light falls on the plasma boundary. In the case of the ATR method, $\varepsilon_0^{1/2}$ represents the refractive index of the prism. If we have to deal only with one dipole of length δ, the factor $|S|^2$ is replaced by δ^2 and the factor $|W_z|^2$ by $\sin^2 \Theta_0 \sin^2 \Theta$ or

$$dJ = \text{const } J_0 \cdot \delta^2 \cdot \sin^2 \Theta_0 \cdot \sin^2 \Theta \, d\Omega \qquad (139)$$

In our case such a dipole is placed on a metallic surface.

Now if p-polarized light falls from the ε_0, or the vacuum side, onto the surface we obtain regarding the z component of the angular factor $W(\Theta_0, \Theta)$

$$W_z(\Theta_0, \Theta) = A(\Theta_0) \cdot B(\Theta) \qquad (140)$$

$A(\Theta_0)$ represents the quantities describing the excitation of the Hertz dipole and depends therefore only on Θ_0, whereas $B(\Theta)$ describes the emission at

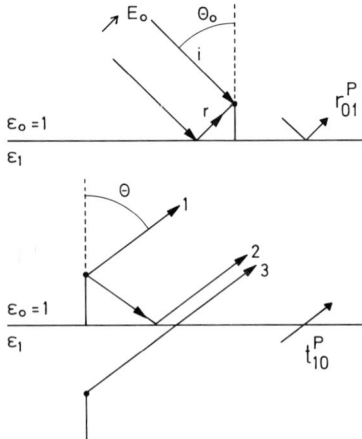

FIG. 69. Excitation of and emission by a dipole.

the angle Θ (see Fig. 69). The $A(\Theta_0)$ factor becomes

$$|A(\Theta_0)|^2 = |\sin \Theta_0|^2 \left|\frac{1 + r^p_{01}(\Theta_0)}{\varepsilon_1}\right|^2 \tag{141}$$

where r^p_{01} is the Fresnel coefficient of Eq. (39). It is derived from the current density j_z using Eq. (137)

$$j_z = \frac{dP_z}{dt} = i\omega(\varepsilon_1 - 1)\{E_0 \sin \Theta_0 + E_0 \sin \Theta_0 r^p_{01}(\Theta_0)\} \frac{1}{\varepsilon_1} \tag{142}$$

The resultant acting field outside the metal, contained in the braces, is composed of the incoming plus the reflected z component of the p-polarized electric field amplitude at the boundary 0/1. Multiplication with $1/\varepsilon_1$ gives the z component in the interior.

The second factor, $B(\Theta)$, in Eq. (140), describes the emission at the angle (Θ), which is different depending on whether the dipole radiates just above or below the surface in the metal. In the first attempts to explain the results of the resonance emission experiments, Stern assumed (*248*, see also *249*) that both possibilities are realized with the same probability. A detailed calculation however had the result (*103*) that in the case of a purely geometric deviation of the surface profile from a plane one ("exterior roughness", see Fig. 68) the radiating dipole has to be placed just above the surface. If $\varepsilon_0 \neq 1$ one has to place the dipole in a small vacuum slit between the media ε_0 and ε_1. If we have inhomogeneities in the interior, such as deviations of the dielectric constant from the mean value due to voids at grain boundaries, it has to be placed just below the surface ("interior roughness"). The consequence is that the emission factor $B(\Theta)$ changes if the roughness is "exterior" or "interior" and is

for exterior roughness
(beam 1 + 2 in Fig. 69)
$$|B(\Theta)|^2 = \sin^2 \Theta |1 + r^p_{01}(\Theta)|^2 \tag{143}$$

and for interior roughness
(beam 3 in Fig. 69)
$$|B(\Theta)|^2 = \sin^2 \Theta |1 + r^p_{01}(\Theta)|^2 \cdot \frac{1}{|\varepsilon_1|^2} \tag{144}$$

Now the expressions (143) and (144) together with the excitation term of Eq. (141) can be written as

exterior roughness:

$$|W_{z,\text{ex}}|^2 = \sin^2 \Theta_0 \sin^2 \Theta \left|\frac{2\varepsilon_1 K_{z0}}{L^\infty}\right|^2_{\Theta_0} \cdot \left|\frac{2\varepsilon_1 K_{z0}}{L^\infty}\right|^2_{\Theta} \cdot \frac{1}{\varepsilon_1^2} \tag{145}$$

and interior roughness

$$|W_{z,\text{int}}|^2 = \frac{1}{\varepsilon_1{}^2} \cdot |W_{z,\text{ex}}|^2 \qquad (146)$$

with

$$L^\infty = \varepsilon_1 K_{z0} + \varepsilon_0 K_{z1} \qquad (147)$$

See Eqs. (9) and (10) for $d \to \infty$.

If now the frequency of the incoming light fulfills the dispersion relation, either for radiative or nonradiative plasma oscillations, the oscillations participate in the light scattering.

(i) *Participation of nonradiative plasmons*. In this case

$$\omega \leq \omega_s \qquad (\varepsilon_1 \leq -1) \qquad (148)$$

and either ω and $(K_x)_{\Theta_0}$, or ω and $(K_x)_\Theta$, or both, fulfill the relations $L^\infty = 0$ or a minimum (see Fig. 14, lower part). In the first case the nonradiative plasmons are excited, e.g., by ATR, and coupled with radiative plasmons or photons by roughness ($\mathbf{K}_r$) (see Fig. 14b: 2–3); in the second case the nonradiative plasmons are coupled by roughness with other nonradiative plasmons (Fig. 14b: 2–4) which can be detected by ATR.

The expressions $1/L^\infty$ become large so that the scattered intensity for these special angles (Θ_0, Θ) becomes strong for exterior as well as for interior roughness.

From Eqs. (145) and (146) one concludes that the volume contribution is strongly reduced by $1/\varepsilon_1{}^2(|\varepsilon_1| > 1)$ if both types of roughness are present in the film with comparable probability. This is valid for even smaller percentages of surface roughness contribution. One needs in this case another experiment to obtain information on the roughness type which is possible by measuring the angular dependence of the different polarization components (see p. 252).

(ii) *Participation of radiative plasmons*. In this case we have

$$\omega \sim \omega_p \qquad (\varepsilon_1 \sim 0) \qquad (149)$$

so that the value of the internal contribution to the scattering intensity generally predominates. For thin films we have to replace the Fresnel coefficients r_{ik} by r_{ikl}; thus the relations become more difficult to handle but give similar results.

In the above examples we concentrated on the z components. At oblique incidence, as usual, the vector $\mathbf{j}$ has j_x and j_y components. E_x and E_y being continuous, there is no difference between external and internal roughness, so that the total dipole function W becomes, instead of W_z

$$|W|^2 = |W_x + W_y + W_z|^2 \tag{150}$$

where W_z alone depends on the kind of roughness as just mentioned. We see that the sum of the three terms is strongly influenced by their phase relations and by the value of ε_1. Thus it is possible to derive from angular measurements the character of the roughness.

We now come to a brief discussion of the term $|S(\Delta K_x)|^2$ in Eq. (138). The considerations just described are concerned with the radiation of one dipole on a metallic or dielectric boundary. Now we have a whole assembly of dipoles, so that integration over the irradiated surface F is needed. $K_x(\Theta)$ and $K_x(\Theta_0)$ are the wave vectors of the scattered and the incident light beam *in* the surface so that $\Delta K = K_x(\Theta) - K_x(\Theta_0)$. We introduce r, the distance of the radiating point (x, y) from a zero point. The intensity factor due to the distribution of the dipoles in the surface becomes, after averaging over the surface F

$$|S(\Delta K_x)|^2 = \frac{1}{F} \left| \int S(x, y) \exp i(\Delta K_x \cdot r) \, dx \, dy \right|^2 \tag{151}$$

We neglect the phase difference due to the heights $S(x, y)$ since we assume $|S| < \lambda$ the wavelength of the light.

$dJ/d\Omega$ of Eq. (138) represents the additional diffuse scattering as compared with the smooth surface if the reflected and the transmitted beam are excluded. In the special case where the antennas are periodically distributed on the surface (linear grating), intensity maxima of scattered light appear at

$$\Delta K = v(2\pi/a) \tag{152}$$

where v are whole numbers. $v \neq 0$ produces light beams in directions different from the reflected and transmitted beam; here a is the period of the grating.

In the general case of a rough surface one introduces the correlation function

$$G(x, y) = \frac{1}{F} \int S(x, y) S(x' - x, y' - y) \, dx' \, dy' \tag{153}$$

to describe the spatial distribution of the "antennas." As we see $|S(\Delta K)|^2$ is the Fourier transform of the correlation function. $G(0, 0)$ represents the mean square height of the peaks S^2. In the literature one frequently finds the assumption that G can be written as

$$G(x, y) = G(0, 0) \exp\left(-\frac{x^2 + y^2}{\sigma^2}\right) \tag{154}$$

where σ is the correlation length. This relation is valid on real surfaces only in certain regions of ΔK. With this relation it is possible to simplify Eq. (151) and one obtains

$$|S(\Delta K)|^2 = \frac{1}{4\pi} \sigma^2 \overline{S^2} \exp(-|\Delta K|^2 \sigma^2 \cdot \tfrac{1}{4}) \qquad (155)$$

which is often used to evaluate experiments.

Since the experimental results have shown [e.g., *(247)*] that $\ln|S(\Delta K)|^2$ as a function of $|\Delta K|^2$ gives a linear dependence in certain ΔK regions (see Fig. 74), one can apply Eq. (155) and evaluate the observed curve. If we wish to apply similar considerations to the interior roughness we meet the difficulty of having no concept of this interior roughness. We can think of a smooth film with an interior roughness produced by a statistical variation of the dielectric constant (*103, 250*)

$$\varepsilon = \overline{\varepsilon} + \delta\varepsilon(x, y, z) \qquad (156)$$

The polarization current becomes then

$$j = \delta\varepsilon \cdot \dot{D}_0 \qquad (157)$$

It is rather difficult to fix details of the function $\delta\varepsilon(x, y, z)$ its origin being due to different physical phenomena, such as, grains which do not join intimately so that a density variation results. This structure may also determine the surface roughness, so that a correlation between the exterior and the interior roughness may be the result. Therefore, the roughness function cannot be more detailed, since it depends on the unknown structure of the ε variations.

This is the reason why the evaluation of the plasma resonance emission experiments, with regard to the derived roughness structure, need reconsideration since it has become probable that volume scattering is the decisive factor in these experiments and not surface roughness.

Measurements of light scattering on rough silver surfaces. As we have seen, light is emitted from a nearly smooth surface of Ag or Au if nonradiative plasmons are excited, e.g., by the prism method. The observation has been made in the direction of the film normal ($\Theta = 0°$) in the resonance case, i.e., Θ_0 has always been chosen to produce maximum excitation of the surface mode. Now the variation of the angle Θ resulted in a characteristic intensity distribution of the emitted light (*9*). A detailed theoretical and experimental analysis of this scattered light for frequencies $\omega < \omega_s$ has been reported by Kretschmann (*246, 247*).

In order to obtain the roughness structure $|S(\Delta K)|^2$ from the measured light intensity as a function of the angle Θ, one has to know which of the equations, (145) or (146), has to be applied, dependent on the type of the roughness structure. This is possible by measuring the ratio of the intensities of the p-polarized to the s-polarized light, J_p/J_s, as a function of Θ in the x–z plane or the plane of observation perpendicular to the plane of incidence, or $\phi = 90°$. Figure 70 shows the experimental arrangement. This ratio is

$$\frac{J_p}{J_s} = \frac{J(\psi = 90°)}{J(\psi = 0°)} \sim \tan^2 \Theta \qquad \text{for surface scattering} \qquad (158)$$

$$\frac{J_p}{J_s} \sim \frac{1}{|\varepsilon_1|^2} \cdot \tan^2 \Theta \qquad \text{for volume scattering} \qquad (159)$$

The different behavior is the consequence of the fact that both polarization current components j_x and j_z interfere and contribute with comparable values, if exterior scattering takes place, whereas j_z is strongly reduced by $1/|\varepsilon_1|$ in the case of volume scattering. The experimental result (Fig. 71) decides this question in favor of exterior roughness for silver irradiated with light of the wavelengths indicated in Fig. 71. The roughness $|S(\Delta K)|^2$ of this silver film can now be measured.

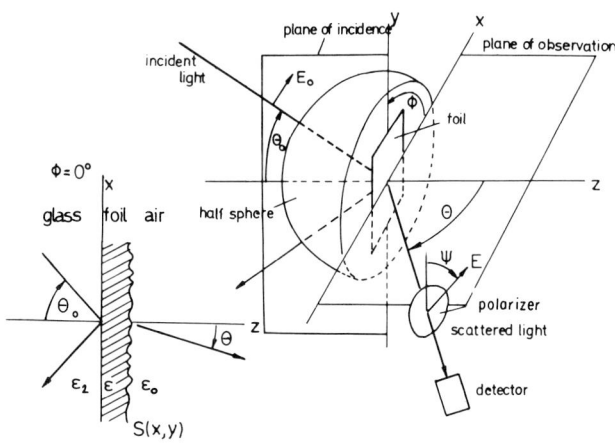

FIG. 70. Scheme of the experiment to analyze the light intensity scattered by a rough plasma surface due to nonradiative surface plasmons [from Kretschmann (247)].

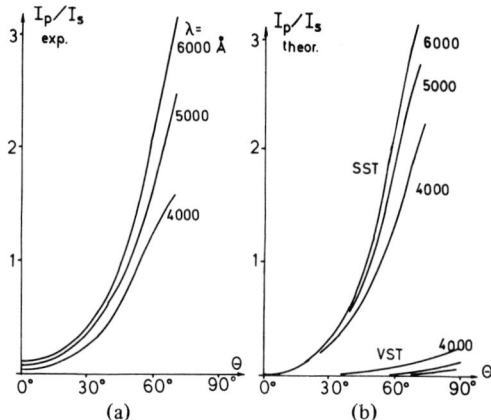

FIG. 71. Experimental (a) and theoretical (b) ratio $I_p(\Theta)/I_s(\Theta)$, for a 500-Å thick Ag foil at three wavelengths (6000 Å, 5000 Å, 4000 Å). SST and VSP are surface and volume scattering, respectively [from Kretschmann (247)].

Figure 71 shows further that the intensity ratio does not go to zero for $\Theta \to 0$, but retains finite values which are larger the longer the wavelength. This is probably due to multiple scattering effects (247) since the intensity ratio at $\Theta = 0$ increases with roughness (247a). First-order theory is no more sufficient to explain this result.

With this result we now know the angular distribution of the radiating dipole or the function $|W|^2$. Figure 72 reproduces this dipole function calcu-

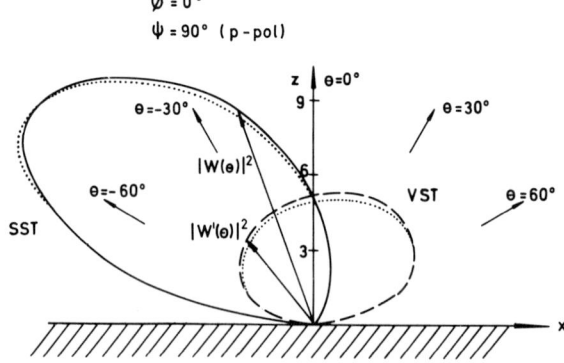

FIG. 72. Calculated polar diagram of the radiation pattern $|W(\Theta)|^2$ of a silver film 700 Å thick at $\lambda = 5000$ Å. SST is surface scattering and VST is volume scattering. The K_x vector of the incident light coming from below lies in the x direction. SST represents the distribution that has to be used for the evaluation of the scattered intensity [from Kretschmann (246)].

lated for ($\phi = 0°$) and for p-polarized light ($\psi = 90°$, see Fig. 70). It has a maximum in the backward direction, a consequence of the interference of the two components of the polarization current, j_z and j_x which have comparable values for surface scattering.

In the case of volume scattering, the z component of the polarization current is suppressed ($|\varepsilon_1| \gg 1$) so that the characteristic dipole function of j_x(VST) remains (see Fig. 72).

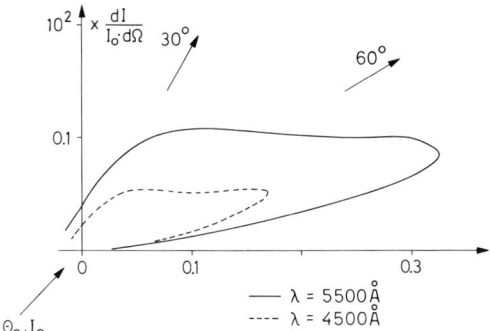

Fig. 73. Polar diagram of the light intensity $dI/d\Omega$ scattered by a silver film 550 Å thick deposited on a smooth quartz substrate for three wavelengths of light [from Hornauer (252)].

Measurements of the distribution of the scattered radiation at $\lambda > \lambda_s$ on a metal film allows one to derive $|S(\Delta K)|^2$ using Eq. (138). Figure 73 shows an observed angular distribution which has its peak intensity in the forward direction in contrast to Fig. 72. This is apparently due to the influence of the function $|S(\Delta K)|^2$, which, derived with the help of Eq. (138), is shown in Fig. 74. Here ΔK is given by

$$\Delta K = (\omega/c)\sin\Theta - (\omega/c)n_0 \sin\Theta_0 \qquad (160)$$

$\Theta = 0$ is marked in Fig. 74 by an arrow. Increasing Θ values (right of the foil normal) lead to smaller ΔK, whereas decreasing or negative Θ values (left of the foil normal) produce larger ΔK values. The result that the function $|S|^2$ takes on larger values for small ΔK and decreases at larger ΔK leads to the change of the peak position from the left to the right side (cf. Fig. 72 with Fig. 73). In other words, the probability of finding long wavelength components in the spatial roughness spectrum (distances of $\sim 10^4$ Å) is higher than that of short wavelength components. This spectrum is probably essentially due to the polished quartz substrate. In the region of higher ΔK_x values (backward direction in Fig. 73), we obtain a linear dependence on $|\Delta K|^2$

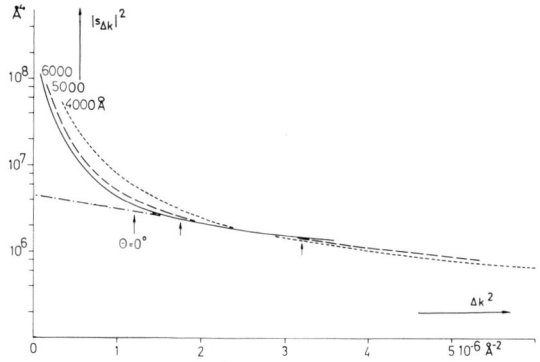

FIG. 74. Roughness spectrum $|S(\Delta k)|^2$ as function of $|\Delta k|^2$ derived from Fig. 73. The angle $\Theta = 0$ (foil normal) is indicated by an arrow for the three wavelengths used in the experiments [from Kretschmann (212)].

(logarithmic plot). Assuming a Gaussian relation for the correlation function, one obtains 7–10 Å for the value of $(S^2)^{1/2}$, for values of ΔK between 2×10^{-3} Å^{-1} and 5×10^{-3} Å^{-1}. These figures have been verified for silver films under comparable experimental conditions (251). If the surface becomes rougher, the peak turns back to the region of negative Θ values. This indicates that the probability of finding long wavelength roughness components is reduced in favor of shorter wavelength components, an observation which is expected (252, 253). Certainly this method allows one to obtain detailed information on the roughness structure of metal films, but the difficulty is that we do not know the distribution of roughness vectors (K_r) outside the light circle, a region which is not accessible to measurements with visible light. Here the distribution is certainly determined by the crystalline structure of the surface (grain size, crystal form) as revealed by electron microscope photographs. The evaluation has not yet been extended into this region.

Roughness questions also play an important role in leakage problems of guided light waves since deviations of the waveguide surface from a smooth one lead to radiation losses in addition to absorption. Detailed calculations have been made by Marcuse (254). These calculations, however, differ not only by the method but also in that they are applicable only to s modes. In this case the continuity of the tangential field components do not raise the question of where to place the polarization current. For the application the s mode—especially the waveguide mode $s = 0$ (mono-mode waveguide)—is important.

The p modes can also be treated in a rough approximation if the difference of the refractive indices of the two neighboring media is small. In the case of a metal film these calculations cannot be used; here the problem has to be reconsidered as has been done in (*103*) and (*210*).

5. Closing Remarks

Surface plasmons are polarization waves guided along the surface of a metal whose electrons behave like quasi-free electrons, i.e., the electrons are bound to the ions with an eigenfrequency which is small compared to the plasma frequency ω_p. The concept of surface plasmons is, as we have seen, a useful means of describing different physical phenomena.

Similar polarization waves, called phonons, are produced, e.g., by the displacement of positive and negative ions of an ionic crystal. In an infinite volume there exist longitudinal and transverse phonons with the frequencies ω_L and ω_T; if the crystal is bounded, surface phonons result as eigensolutions of the Maxwell's equations, as has been shown by Kliewer and Fuchs (*238*) by applying the same mathematical formalism. These solutions have very similar properties except that their energies are about 10^3 times smaller and the lowest energy equals ω_T rather than zero. The dispersion relations have nearly the same shape, and the frequency approaches asymptotically $\omega_L/\sqrt{2}$.

The experimental development has proceeded similarly by application of the same methods as used for surface plasmons: excitation by fast or slow electrons as well as by light via the grating coupler or the ATR method.

Acknowledgment

I wish to thank Mr. D. Heitmann for many helpful discussions.

References

1. R. H. Ritchie, *Phys. Rev.* **106**, 874 (1957).
2. E. A. Stern and R. A. Ferrell, *Phys. Rev.* **111**, 1214 (1958).
3. R. H. Ritchie, *Surface Sci.* **34**, 1 (1973).
4. H. Raether, *Surface Sci.* **8**, 233 (1967).
4a. E. N. Economu and K. L. Ngai, *Adv. Chem. Phys.* **27**, 265 (1974).
5. For more details, see H. Raether, *Springer Tracts Mod. Phys.* **38**, 85 (1965).
6. N. D. Lang, *Solid State Commun.* **7**, 1047 (1969); N. D. Lang and W. Kohn, *Phys. Rev. B* **1**, 4555 (1970).
7. T. Kloos, *Z. Phys.* **208**, 77 (1968).
8. E. A. Stern, as quoted in reference 2.
9. R. Bruns and H. Raether, *Z. Phys.* **237**, 98 (1970).

10. E. Kretschmann, *Z. Phys.* **241**, 313 (1971).
11. J. Schönwald, E. Burstein and J. M. Elson, *Solid State Commun.* **12**, 185 (1973).
12. A. S. Barker, *Phys. Rev. B* **8**, 5418 (1973).
13. E. Kretschmann, Thesis, Univ. of Hamburg, Hamburg, 1973.
14. R. H. Ritchie, *Prog. Theor. Phys.* **29**, 607 (1963).
14a. H. Kanazawa, *Progr. Theor. Phys.* **26**, 851 (1961).
14b. D. Wagner, *Z. Naturforsch. Teil A* **21**, 634 (1966).
14c. K. Sturm, *Z. Phys.* **209**, 329 (1968).
15. A. J. Bennet, *Phys. Rev. B* **1**, 203 (1970).
16. A. Eguiluz, S. C. Ying, and J. J. Quinn, *Phys. Rev. B* **11**, 2118 (1975).
17. J. Harris and A. Griffin, *Phys. Lett. A* **34**, 51 (1971); *Phys. Rev. B* **3**, 749 (1971); P. J. Feibelman, *Phys. Rev. Lett.* **30**, 975 (1973); A. D. Boardman, B. V. Paranjape, and R. Teshima, *Surface Sci.* **49**, 275 (1975), (a brief guide to the current theoretical position is given here).
18. P. J. Feibelman, *Surface Sci.* **40**, 102 (1973).
19. K. L. Kliewer and R. Fuchs, *Phys. Rev.* **153**, 498 (1967).
20. V. E. Pafomov, *Zh. Eksp. Teor. Fiz.* **6**, 829 (1958) [the denominator of his equation is identical with Eq. (7)].
21. F. Fujimoto and K. Komaki, *J. Phys. Soc. Jpn.* **25**, 1679 (1968).
22. M. Natta, *Solid State Commun.* **7**, 823 (1969); *J. Phys. (Paris)* **32**, 639 (1971).
23. A. A. Lucas, *Phys. Rev. B* **7**, 3527 (1973).
24. C. Cawthorne and E. J. Fulton, *Nature (London)* **216**, 576 (1967).
25. J. H. Evans, *Nature (London)* **229**, 403 (1971); G. I. Kulcinski, J. L. Brimhall, and H. E. Kissinger, *J. Nucl. Mater.* **40**, 166 (1971); R. Bullough and A. B. Lidiard, *Comments Solid State* 4(3), 69 (1972).
26. R. Fuchs, *Phys. Rev.* **11**, 1732 (1975).
27. J. C. Ashley and L. C. Emerson, *Surface Sci.* **41**, 615 (1974).
28. P. K. Tien, R. Ulrich, and R. J. Martin, *Appl. Phys. Lett.* **14**, 291 (1969).
29. "Integrated Optics, Applied Physics," (T. Tamir, ed.), Vol. 7. Springer-Verlag, Berlin and New York, 1975.
30. R. A. Ferrell, *Phys. Rev.* **111**, 1214 (1958).
31. W. Ginzburg and I. Frank, *Zh. Eksp. Teor. Fiz.* **16**, 15 (1946).
32. H. Boersch, C. Radeloff, and G. Sauerbrey, *Z. Phys.* **165**, 464 (1961); H. Boersch, P. Dobberstein, D. Fritzsche, and G. Sauerbrey, *Z. Phys.* **187**, 97 (1965); J. E. Lilienfeld, *Phys. Z.* **20**, 280 (1919).
33. R. H. Ritchie and H. B. Eldrige, *Phys. Rev.* **126**, 1935 (1962).
34. J. C. Ashley, *Phys. Rev.* **155**, 208 (1967).
35. E. Kröger, *Z. Phys.* **235**, 403 (1970).
36. E. A. Stern, *Phys. Lett.* **8**, 7 (1962).
37. W. Steinmann, *Phys. Status Solidi* **28**, 437 (1968).
38. W. Steinmann, *Phys. Rev. Lett.* **5**, 470 (1960); *Z. Phys.* **163**, 92 (1961).
39. W. R. Brown, P. Wessel, and E. P. Trounson, *Phys. Rev. Lett.* **5**, 472 (1960).
40. R. J. Herickhoff, W. F. Hanson, E. T. Arakawa, and R. D. Birkhoff, *Phys. Rev.* **139**, 1455A (1965).
41. A. J. Braundmeier, E. T. Arakawa, and M. W. Williams, *Phys. Lett. A* **32**, 241 (1970).
42. A. J. Braundmeier and E. T. Arakawa, *Opt. Commun.* **2**, 257 (1970).
43. E. T. Arakawa, N. O. Davis, and R. D. Birkhoff, *Phys. Rev.* **135**, 224A (1964).
44. E. T. Arakawa, L. C. Emerson, D. C. Hammer, and R. D. Birkhoff, *Phys. Rev.* **131**, 719 (1963).

45. R. J. Herickhoff, E. T. Arakawa, and R. D. Birkhoff, *Phys. Rev.* **137**, 1433A (1965).
46. H. D. Hattendorff, *Phys. Status Solidi* in press (1977).
47. E. Kretschmann, *Z. Phys.* **241**, 313 (1971).
48. M. Schlüter, *Z. Phys.* **250**, 87 (1972).
49. H. O. Tittel, *Phys. Lett. A* **26**, 145 (1969).
50. B. Schmalfeld, unpublished work (1969).
51. G. Binias and K. Kuhnert, *Phys. Lett. A* **38**, 467 (1972).
52. U. Bürker and W. Steinmann, *Phys. Status Solidi* **12**, 853 (1965).
53. W. Steinmann and H. Wille, *Phys. Status Solidi* **15**, 507 (1966).
54. R. H. Ritchie, J. C. Ashley, and L. C. Emerson, *Phys. Rev. A* **135**, 759 (1964).
55. R. P. DiNardo and A. W. Goland, *Phys. Rev. B* **4**, 1700 (1971).
56. D. Heitmann, *Z. Phys.* **249**, 356 (1972).
57. C. von Festenberg, *Z. Phys.* **227**, 453 (1969).
58. D. Heitmann, *Z. Phys.* **245**, 154 (1971).
59. R. Vincent and J. Silcox, *Phys. Rev. Lett.* **31**, 1487 (1973).
60. A. Daudé, A. Savary, G. Jezequel, and S. Robin, *Opt. Commun.* **1**, 237 (1969).
61. E. Kretschmann, *Z. Phys.* **241**, 313 (1971).
62. A. J. McAlister and E. A. Stern, *Phys. Rev.* **132**, 1959 (1963).
63. M. Hattori, K. Yamada, and H. Suzuki, *J. Phys. Soc. Jpn.* **18**, 203 (1963).
64. S. Yamaguchi, *J. Phys. Soc. Jpn.* **17**, 1172 (1962).
65. J. Brambring and H. Raether, *Z. Naturforsch. A* **21**(9), 1527 (1966).
66. J. Brambring, *Z. Phys.* **200**, 186 (1967).
67. J. Bösenberg, *Z. Phys.* **218**, 282 (1969).
68. A. Ejiri and T. Sasaki, *J. Phys. Soc. Jpn.* **20**, 876 (1965).
69. M. Skibowski, B. Feuerbacher, and W. Steinmann, *Z. Phys.* **211**, 329 (1968).
70. A. Ejiri, *J. Phys. Soc. Jpn.* **23**, 901 (1967).
71. E. Kretschmann, *Z. Phys.* **221**, 357 (1969).
72. D. Schulz and M. Zurheide, *Z. Phys.* **211**, 165 (1968).
73. E. Kretschmann, *Z. Phys.* **221**, 346 (1969).
74. B. Bernert and P. Zacharias, *Z. Phys.* **241**, 205 (1971).
75. P. O. Nilsson, I. Lindau, and S. B. M. Hagström, *Phys. Rev. B* **1**, 498 (1970).
76. K. Ripken, *Z. Phys.* **250**, 228 (1972).
77. B. Feuerbacher, R. P. Godwin, and M. Skibowski, *Phys. Lett. A* **26**, 595 (1968).
78. B. Feuerbacher, Thesis, Univ. of Munich, Munich, 1968.
79. T. F. Gesell and E. T. Arakawa, *Phys. Lett. A* **36**, 79 (1971).
80. B. Feuerbacher and B. Fitton, *Phys. Rev. Lett.* **24**, 499 (1970).
81. I. Hoffmann and W. Steinmann, *Phys. Status Solidi* **30**, K53 (1968).
82. See, e.g., O. Klemperer, "Elektronik," p. 122. Springer-Verlag, Berlin, 1933.
83. U. Kreibig and C. von Fragstein, *Z. Phys.* **224**, 307 (1969).
84. E.g., S. Yoshida, T. Yamaguchi, and A. Kimbara, *J. Opt. Soc. Am.* **62**, 1415 (1972).
85. N. Emeric and A. Emeric, *Thin Solid Films* **1**, 363 (1968).
86. O. Hunderi and H. P. Myers, *J. Phys. F* **3**, 683 (1973).
87. V. P. Silin and E. P. Fetisow, *Phys. Rev. Lett.* **7**, 374 (1961).
88. F. Sauter, *Z. Phys.* **203**, 488 (1967).
89. F. Forstmann, *Z. Phys.* **203**, 495 (1967).
90. A. R. Melnyk and M. J. Harrison, *Phys. Rev. Lett.* **21**, 85 (1968); *Phys. Rev. B* **2**, 835 (1970).
91. K. Kliewer and R. Fuchs, *Phys. Rev. B* **185**, 905 (1969).
92. I. Lindau and P. O. Nilsson, *Phys. Lett. A* **31**, 352 (1970).
93. M. Anderegg, B. Feuerbacher, and B. Fitton, *Phys. Rev. Lett.* **27**, 1565 (1971).

94. I. Lindau and P. O. Nilsson, *Phys. Scr.* **3**, 87 (1971).
95. P. Zacharias, *Vac. Ultraviolet Radiat. Phys., Proc. Int. Conf., 4th, Hamburg* p. 593 (1974).
96. R. Ruppin, *Phys. Rev. B* **11**, 2871 (1975).
97. T. K. Bolland, *Phys. Rev. B* **2**, 798 (1970).
98. J. Brambring and H. Raether, *Phys. Rev. Lett.* **15**, 882 (1965); *Z. Phys.* **199**, 118 (1967).
99. P. Schreiber, *Z. Phys.* **211**, 257 (1968); P. Schreiber and H. Raether, *Z. Naturforsch. Teil A* **21**, 2116 (1966).
100. E. Schröder, *Z. Phys.* **225**, 26 (1969).
101. W. Steinmann, I. Hoffmann, and K. Stettmaier, *Phys. Lett.* **23**, 234 (1966).
102. E. Kretschmann and H. Raether, *Z. Naturforsch. A* **23**, 615 (1968).
103. E. Kröger and E. Kretschmann, *Z. Phys.* **237**, 1 (1970).
104. J. Bösenberg and H. Raether, *Phys. Rev. Lett.* **18**, 397 (1967).
105. J. Bösenberg, *Phys. Lett. A* **26**, 74 (1967).
106. G. Mie, *Ann. Phys. (Leipzig)* **25**, 377 (1908).
107. J. Crowell and R. H. Ritchie, *Phys. Rev.* p. 436 (1968).
108. W. T. Doyle, *Proc. Phys. Soc., London* **75**, 649 (1960).
109. R. H. Doremus, *J. Chem. Phys.* **40**, 2389 (1964); **42**, 414 (1965).
110. U. Kreibig and P. Zacharias, *Z. Phys.* **231**, 128 (1970).
111. C. J. Duthler, S. E. Johnson, and H. P. Broida, *Phys. Rev. Lett.* **26**, 1236 (1971).
112. For more details, see K. D. Sevier, "Low Energy Electron Spectrometry." Wiley, New York, 1972.
113. R. H. Ritchie, *Phys. Rev.* **106**, 874 (1957).
114. E. Kröger, *Z. Phys.* **216**, 115 (1968).
115. T. Kloos, *Z. Phys.* **265**, 225 (1973).
116. J. Daniels, C. von Festenberg, H. Raether, and K. Zeppenfeld, *Springer Tracts Mod. Phys.* **54**, 77 (1970).
117. R. B. Pettit, J. Silcox, and R. Vincent, *Phys. Rev. B* **11**, 3116 (1975).
118. P. Zacharias, *Z. Phys.* **238**, 172 (1970).
119. C. Kunz, *Z. Phys.* **196**, 311 (1966).
120. T. Kloos and H. Raether, *Phys. Lett. A* **44**, 157 (1973).
120a. K. J. Krane and H. Raether, *Phys. Rev. Lett.* **37**, 1355 (1976).
120b. J. Langkowski, in H. Raether, *Vac. Ultraviolet Radiat. Phys., Proc. Int. Conf., 4th,* Hamburg (1974).
121. A. Bagchi, C. B. Duke, P. J. Feibelman, and J. O. Porteus, *Phys. Rev. Lett.* **27**, 998 (1971).
122. J. O. Porteus and W. N. Faith, *Phys. Rev. B* **8**, 491 (1973).
123. C. B. Duke and U. Landmann, *Phys. Rev. B* **8**, 505 (1973).
123a. C. B. Duke, L. Pietronero, I. O. Porteus, and I. F. Wendelken, *Phys. Rev. B* **12**, 4059 (1975).
124. P. Schmüser, *Z. Phys.* **180**, 105 (1964).
125. C. Kunz, *Z. Phys.* **167**, 53 (1962).
126. C. Kunz, *Z. Phys.* **196**, 311 (1966).
127. J. Geiger, *Z. Phys.* **161**, 243 (1961).
128. C. J. Powell and J. B. Swan, *Phys. Rev.* **115**, 869 (1959); **118**, 640 (1960).
129. J. Daniels, *Z. Phys.* **213**, 227 (1968).
130. P. Schmüser, *Solid State Commun.* **2**, 41 (1964).
131. A. Otto, *Z. Phys.* **185**, 232 (1965).
132. W. R. Miller, Jr. and N. N. Axelrod, *Solid State Commun.* **3**, 133 (1965).
133. E. A. Stern and R. A. Ferrell, *Phys. Rev.* **120**, 130 (1960).
134. J. Geiger, *Phys. Status Solidi* **24**, 457 (1967).

135. A. Otto, *Phys. Status Solidi* **22**, 401 (1967).
136. E. Kröger, *Z. Phys.* **235**, 403 (1970).
137. M. Creuzburg, *Z. Naturforsch. A* **18**, 101 (1963); *Z. Phys.* **174**, 511 (1963).
138. C. Kunz and H. Raether, *Solid State Commun.* **1**, 214 (1963).
139. C. Kunz, *Z. Phys.* **180**, 127 (1964).
140. A. Otto and J. B. Swan, *Z. Phys.* **206**, 277 (1967).
141. J. B. Swan, A. Otto, and H. Fellenzer, *Phys. Status Solidi* **23**, 171 (1967).
142. A. A. Lucas and M. Sunjic, *Phys. Rev. Lett.* **26**, 229 (1971).
143. M. Sunjic and A. A. Lucas, *Phys. Rev. B* **3**, 719 (1971).
144. E. J. Powell, *Phys. Rev.* **175**, 972 (1968).
145. H. Raether, *Electron Microsc., Proc. Int. Congr., 5th, Philadelphia* A A 3 (1962).
146. M. Creuzburg and H. Raether, *Z. Phys.* **171**, 436 (1963); J. Lohff, *Z. Phys.* **171**, 442 (1963).
147. J. Daniels, *Z. Phys.* **203**, 235 (1967).
148. J. Schilling, *Z. Phys. B* **25**, 61 (1976).
149. J. Schilling and H. Raether, *J. Phys. C* **6**, L358 (1973).
150. L. K. Jordan and E. J. Scheibner, *Surface Sci.* **10**, 373 (1968).
151. M. P. Seah, *Surface Sci.* **17**, 161 (1969).
152. J. O. Porteus and W. N. Faith, *Phys. Rev. B* **2**, 1532 (1970).
153. E. Bauer, *Z. Phys.* **224**, 19 (1969).
154. J. Küppers, *Surface Sci.* **36**, 53 (1973).
155. H. Ibach and J. E. Rowe, *Phys. Rev. B* **9**, 1951 (1974).
156. J. E. Rowe and H. Ibach, *Phys. Rev. Lett.* **31**, 102 (1973).
157. H. Ibach and J. E. Rowe, *Surface Sci.* **43**, 481 (1974).
158. R. Ludeke and L. Isaki, *Phys. Rev. Lett.* **33**, 653 (1974).
159. J. E. Rowe, *Appl. Phys. Lett.* **25**, 576 (1974).
160. H. Ibach, K. Horn, R. Dorn, and H. Lüth, *Surface Sci.* **38**, 433 (1973).
161. N. V. Smith and W. E. Spicer, *Phys. Rev. Lett.* **23**, 769 (1969); *Phys. Rev.* **188**, 593 (1969).
162. R. G. Oswald and T. A. Callott, *Phys. Rev. B* **4**, 4122 (1971).
163. P. Oelhafen, *Surface Sci.* **47**, 422 (1975).
164. Y. U. Romanov, *Radio Phys.* **7**, 67 (1964).
165. R. A. Pollak, L. Ley, F. R. McFeely, S. P. Kowalczyk, and D. A. Shirley, *J. Electron Spectrosc. Relat. Phenom.* **3**, 381 (1974).
166. Y. Baer and G. Busch, *Phys. Rev. Lett.* **30**, 280 (1973).
167. A. M. Bradshaw and D. Menzel, *Phys. Status Solidi* **56**, 135 (1973).
168. A. M. Bradshaw and W. Wyrobisch, *Electron Spectrosc. Relat. Phenom.* **7**, 45 (1975).
169. W. J. Pardee, G. D. Mahan, D. E. Eastman, R. A. Pollak, L. Ley, F. R. McFeely, S. P. Kowalczyk, and D. A. Shirley, *Phys. Rev. B* **11**, 3614 (1975).
170. D. C. Tsui, *Phys. Rev. Lett.* **22**, 293 1969).
171. K. H. Gundlach, *Festkoerperprobleme* **11**, 237 (1971).
172. J. Hölzl, H. Mayer, and K. W. Hoffmann, *Surface Sci.* **17**, 232 (1969).
173. F. Fujimoto, K. Komaki, and K. Ishida, *J. Phys. Soc. Jpn.* **23**, 1186 (1967).
174. M. Creuzburg, *Z. Phys.* **194**, 211 (1966).
175. J. Crowell and R. H. Ritchie, *Phys. Rev.* **172**, 436 (1968).
176. T. Kokkinakis and K. Alexopoulos, *Phys. Rev. Lett.* **28**, 1632 (1972).
177. P. von Blanckenhagen, H. Boersch, D. Fritsche, H. G. Seifert, and G. Sauerbrey, *Phys. Lett.* **11**, 4 (1964).
178. E. A. Stern, in "Optical Properties and Electronic Structure" (F. Abelès, ed.), p. 396. North-Holland Publ. Amsterdam, 1966.
179. H. Boersch, P. Dobberstein, D. Fritzsche, and G. Sauerbrey, *Z. Phys.* **187**, 97 (1965).

180. L. S. Cram and E. T. Arakawa, *Phys. Rev.* **153**, 455 (1967).
181. P. Dobberstein and G. Sauerbrey, *Phys. Lett. A* **31**, 6 (1970).
182. U. Bürker and W. Steinmann, *Phys. Rev. Lett.* **21**, 3 (1968).
183. A. J. Braundmeier, Jr., M. W. Williams, E. T. Arakawa, and R. H. Ritchie, *Phys. Rev. B* **5**, 2754 (1972).
184. R. H. Ritchie, *Phys. Lett. A* **27**, 660 (1968).
185. P. Dobberstein, *Phys. Status Solidi* **38**, 649 (1970).
186. R. H. Ritchie, *Phys. Status Solidi* **39**, 297 (1970).
187. G. Sauerbrey, E. Woeckel, and P. Dobberstein, *Phys. Status Solidi B* **60**, 665 (1973).
188. Y. Teng and E. A. Stern, *Phys. Rev. Lett.* **19**, 511 (1967).
189. D. Heitmann, *J. Phys. C* **9** (1976) in press.
189a. D. Heitmann and H. Raether, *Surface Sci.* **59**, 17 (1976).
189b. I. Pockrand and H. Raether, *Opt. Commun.* **18**, 395 (1976).
189c. Petit, R. *Nouv. Rev. Optique* **6**, 129 (1975).
189d. E. Kröger and E. Kretschmann, *Phys. Status Solidi.* **76**, 515 (1976).
190. I. Pockrand, *J. Phys. D* **9**, 2423 (1976).
191. R. H. Ritchie, E. T. Arakawa, J. J. Cowan, and R. N. Hamm, *Phys. Rev. Lett.* **21**, 1530 (1968).
192. J. J. Cowan and E. T. Arakawa, *Z. Phys.* **235**, 97 (1970).
193. D. Beaglehole, *Phys. Rev. Lett.* **22**, 708 (1969).
194. I. Pockrand, *Opt. Commun.* **13**, 311 (1975).
195. G. Hass and R. Tousey, *J. Opt. Soc. Am.* **49**, 593 (1959); L. R. Canfield, G. Hass, and J. E. Waylonis, *Appl. Opt.* **5**, 45 (1966).
196. U. Fano, *J. Opt. Soc. Am.* **31**, 213 (1941).
197. M. C. Hutley and V. M. Bird, *Opt. Acta* **20**, 771 (1973); M. C. Hutley, *Opt. Acta* **20**, 607 (1973); J. E. Stewart and W. S. Gallaway, *Appl. Opt.* **1**, 421 (1962).
198. R. C. McPhedran and D. Maystre, *Opt. Acta* **21**, 413 (1974).
199. S. N. Jasperson and S. E. Schnatterly, *Phys. Rev.* **188**, 759 (1969); *Bull. Am. Phys. Soc.* **12**, 399 (1967).
200. P. Dobberstein, A. Hampe, and G. Sauerbrey, *Phys. Lett. A* **27**, 256 (1968)..
201. J. L. Stanford and H. E. Bennett, *Appl Opt.* **8**, 2556 (1969).
202. B. P. Feuerbacher and W. Steinmann, *Opt. Commun.* **1**, 81 (1969).
203. J. G. Endriz and W. E. Spicer, *Phys. Rev. Lett.* **24**, 64 (1970); *Phys. Rev. B* **4**, 4144 (1971); *Phys. Rev. B* **4**, 4159 (1971).
204. A. Daudé, A. Savary, and S. Robin, *J. Opt. Soc. Am.* **62**, 1 (1972).
205. T. F. Gesell, E. T. Arakawa, M. W. Williams, and R. N. Hamm, *Phys. Rev. B* **7**, 5141 (1973).
206. By courtesy of Dr. J. L. Stanford, published in H. Raether, *J. Phys. (Paris), Colloq.* **31**, C1 (1970).
207. E. Schröder, *Opt. Commun.* **1**, 13 (1969).
208. J. L. Stanford, *J. Opt. Soc. Am.* **60**, 49 (1970).
209. J. Crowell and R. H. Ritchie, *J. Opt. Soc. Am.* **60**, 794 (1970).
210. E. Kretschmann and E. Kröger, *J. Opt. Soc. Am.* **65**, 150 (1975).
211. P. Dobberstein, *Phys. Lett. A* **31**, 307 (1970).
212. E. Kretschmann, *Opt. Commun.* **10**, 353 (1971).
213. A. Otto, *Phys. Status Solidi* **26**, K99 (1968); *Z. Phys.* **216**, 398 (1968).
214. A. Otto, *Z. Phys.* **219**, 227 (1969).
215. E. Kretschmann and H. Raether, *Z. Naturforsch. A* **23**, 2135, (1968).
216. T. Turbadar, *Proc. Phys. Soc., London* **73**, 40 (1959); *Opt. Acta* **11**, 207 (1964).

217. A. Otto, *Z. Angew. Phys.* **27**, 207 (1969).
218. A. S. Barker, Jr., *Phys. Rev. B* **8**, 5418 (1973).
219. P. B. Johnson and R. W. Christy, *Phys. Rev. B* **6**, 4370 (1972).
19a. W. H. Weber and S. L. McCarthy, *Phys. Rev. B* **15**, 5643 (1975).
220. J. Bösenberg, *Phys. Lett. A* **37**, 439 (1971).
221. J. Bösenberg, *Z. Phys. B* **22**, 267 (1975).
222. E. T. Arakawa, M. W. Williams, R. N. Hamm and R. H. Ritchie, *Phys. Rev. Lett.* **31**, 1127 (1973).
223. R. W. Alexander, G. S. Kovener, and R. J. Bell, *Phys. Rev. Lett.* **32**, 154 (1974).
224. R. Orlowski and H. Raether, *Surface Sci.* **54**, 303 (1976).
225. H. Kapitza, *Opt. Commun.* **16**, 73 (1976).
226. A. S. Barker, Jr., *Surface Sci.* **34**, 62 (1973).
227. A. S. Barker, Jr., *Phys. Rev. Lett.* **28**, 892 (1972).
228. N. Marshall, B. Fischer, and H. J. Queisser, *Phys. Rev. Lett.* **27**, 95 (1971).
229. N. Marshall and B. Fischer, *Phys. Rev. Lett.* **28**, 811 (1972).
230. H. Twietmeyer, unpublished work (1975).
231. K. Holst and H. Raether, *Opt. Commun.* **2**, 312 (1970).
232. E. Burstein, W. P. Chen, Y. J. Chen, and A. Hartstein, *J. Vac. Sci. Technol.* **11**, 1004 (1974).
232a. F. Abelès and T. Lopez-Rios, *Opt. Commun.* **11**, 89 (1974).
233. H. L. Lemberg, S. A. Rice, P. H. Naylor, and A. N. Bloch, *Solid State Commun.* **14**, 1097 (1974).
234. E. M. Conwell, *Phys. Rev. B* **11**, 1508 (1975).
34a. M. R. Philpott, *J. Chem. Phys.* **62**, 1812 (1975).
235. C. Maczek, A. Otto, and W. Steinmann, *Phys. Status Solidi B* **51**, K59 (1972).
236. D. Hornauer and H. Raether, *Opt. Commun.* **7**, 297 (1973).
237. A. Otto and W. Sohler, *Opt. Commun.* **3**, 254 (1971).
238. R. Fuchs and K. L. Kliewer, *Phys. Rev. A* **140**, 2076 (1965); K. L. Kliewer and R. Fuchs, *Phys. Rev.* **144**, 495 (1966).
239. D. Hornauer, H. Kapitza, and H. Raether, *J. Phys. D* **7**, L100 (1974).
240. A. J. Braundmeier, Jr. and E. T. Arakawa, *J. Phys. Chem. Solids* **35**, 517 (1974).
241. I. Pockrand, *Phys. Lett. A* **49**, 259 (1974).
242. R. H. Bjork, A. S. Karakashian, and Y. Teng, *Phys. Rev. B* **9**, 1394 (1974).
243. P. M. van den Berg and J. C. M. Borburgh, *Appl. Phys.* **3**, 55 (1974).
244. H. Raether, *Thin Solid Films* **28**, 119 (1975).
245. E. Kretschmann, *Z. Phys.* **227**, 412 (1969); H. J. Juranek, *Z. Phys.* **233**, 324 (1970); J. M. Elson and R. H. Ritchie, *Phys. Rev. B* **4**, 4129 (1971); V. Celli, A. Marvin, and A. Toigo, *Phys. Rev. B* **11**, 2777 (1975).
246. E. Kretschmann, *Opt. Commun.* **5**, 331 (1972).
247. E. Kretschmann, *Opt. Commun.* **6**, 185 (1972).
47a. Chr. Horstmann, *Opt. Commun.* in press (1977).
248. E. A. Stern, *Phys. Rev. Lett.* **19**, 1321 (1967).
249. E. Kretschmann and H. Raether, *Z. Naturforsch. Teil A* **22**, 1623 (1967).
250. R. E. Wilems and R. H. Ritchie, *Phys. Rev. Lett.* **19**, 1325 (1967).
251. J. Bodesheim and A. Otto, *Surface Sci.* **45**, 441 (1974).
252. D. Hornauer, *Opt. Commun.* **16**, 76 (1976).
253. A. J. Braundmeir, Jr. and D. G. Hall, *Surface Sci.* **49**, 376 (1975).
254. D. Marcuse, "Integrated Optics." Institute of Electrical and Electronics Engineers, New York, 1972.

Magnetic Bubble Films

P. CHAUDHARI, J. J. CUOMO, R. J. GAMBINO, AND E. A. GIESS

IBM Thomas J. Watson Research Center
Yorktown Heights, New York

I. Introduction.	263
Criteria for Materials Selection	265
II. General Magnetics.	266
1. Static Properties.	266
2. Domain Dynamics.	269
3. Summary.	272
III. Materials for Magnetic Bubbles.	272
1. Amorphous Materials.	273
2. Crystalline Materials.	275
IV. Growth of Magnetic Bubble Materials.	276
1. Preparation of Amorphous Films.	277
2. Preparation of Single-Crystal Garnet Films.	281
V. Anisotropy in Magnetic Bubble Films.	284
1. General Considerations.	284
2. Amorphous Materials.	285
3. Anisotropy in Garnets.	286
VI. Defects in Films.	288
1. Defects in Amorphous Films.	289
2. Defects in Garnets.	289
VII. Summary and Conclusions.	293
References	294

I. Introduction

Magnetic bubbles are cylindrical magnetic domains found in thin films. The large interest in these domains arises principally from their potential as cheap, relatively high performance memory or storage components in computers. Bubble devices can supplant discs or tapes and offer the additional advantage of solid state storage. Mechanical motion which leads to wear and physical breakdown in discs and tapes is not present. Their relatively low cost stems from very high density of information packing and simplicity of processing. Magnetic bubbles ranging in diameter from 0.08 to 80 μm have been observed. At the smaller end of the size range the packing density is of the order of 10^9 bits/cm^2; and while this density has yet to be realized in hardware, shift registers with 0.8-μm bubbles have been built and operated (Hu et al., 1973).

Information is stored in magnetic bubble devices by the presence or absence of a bubble in a shift register. The information content can be moved around the shift register by applying a rotating in-plane magnetic field. Although magnetic bubbles had been reported by Kooy and Enz (1960), the ability to move and manipulate them was not demonstrated until later by Bobeck (1967). An example of a typical shift register made of permalloy is shown in Fig. 1. The in-plane rotating field causes the polarization of the magnetic field to move from one tip of the permalloy to an adjacent one, and the magnetic bubble follows the field. A cycle of operation is shown in Fig. 2 (Bobeck and Scovil, 1971).

FIG. 1. A portion of a typical shift register consisting of a pattern of T's and I's made of permalloy. The permalloy circuit elements are about one micron wide.

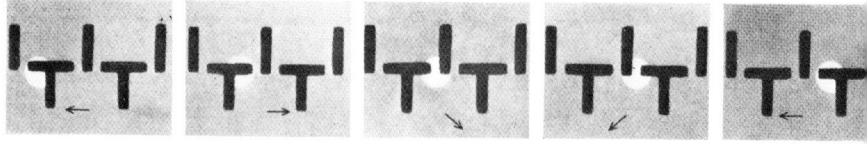

FIG. 2. The cycle of operation of a T- and I-bar type shift register. As the external magnetic field is rotated clockwise, the magnetic bubble follows the induced magnetic polarization of the permalloy and propagates to the right. From Bobeck and Scovil, "Magnetic Bubbles." Copyright © 1971 by Scientific American, Inc. All rights reserved.

Information is transferred along the shift register at a rate (the so-called data rate) given by dividing the velocity of a bubble by the period of the shift register, which is typically four times the bubble diameter. Since higher data rates are preferred, the mobility of a magnetic domain plays a significant role in the selection of optimum materials for devices. Typically, the mobility of a reasonable quality bubble material is of the order of few hundred centimeters per second oersted (cm/sec-Oe). If we assume that the bubble moves at velocities of 10^3 cm/sec, then a 1-μm bubble yields a data rate of 2.5 MHz. These numbers suggest that high performance devices are possible with magnetic bubbles. The repulsive interaction between magnetic bubbles also offers the possibility of carrying out certain logic functions. Several schemes have been proposed (Sandfort and Burke, 1971; Bobeck and Scovil, 1971; Chang et al., 1973).

In virtually all magnetic bubble technology, the materials that contain magnetic bubbles have played a key role. Not only must the materials have suitable static magnetic properties, but the dynamic properties of bubbles are also very important. The purpose of this chapter is to provide a summary of the selection and growth of two classes of materials that are currently candidates for the bubble technology: various single-crystal garnets and the more recently discovered amorphous alloys of Gd–Co. We shall also be concerned with the perfection of materials, the structure of domain walls, and their relationship to the motion of a bubble.

CRITERIA FOR MATERIALS SELECTION

The two most important criteria for the selection of a material for any device are performance and cost. Magnetic bubbles are no exception to this axiom. Performance is determined by systems considerations such as device design and, within a given device design, by how well the device operates. Of these two we shall be concerned only with the second and the factors that affect it. Similar considerations apply to cost. For example, processing and packaging of a device play a significant role in cost estimates along with materials cost. We shall be concerned only with materials here.

The operational performance of a device is related to its speed and stability. In magnetic bubble technology, speed is directly proportional to the mobility of a magnetic bubble and inversely proportional to its diameter. The stability of the device is determined by its reliability, which involves defects in the materials, structure of domain walls, and the temperature sensitivity of the material. Performance is also gated by drive circuits which provide the rotating or switching magnetic fields (drive fields). The requirement of a large magnetic drive field implies large currents through the circuit to supply the magnetic field, which at high frequencies lead to substantial eddy current losses. High performance devices therefore require low drive fields and these in turn demand materials with low magnetization (Kryder et al., 1974).

The cost of materials involves processing costs such as the manufacturing of crystals or sputtering equipment. It also involves the diameter of a magnetic bubble; the cost per bit of information stored in a material increases with the square of the bubble diameter. It is readily apparent that the smaller the bubble, the better the cost performance. Also, size of the material, i.e., area of a film, determines cost. Here again this is related to materials questions such as perfection, uniformity, etc. Selection of a suitable bubble material, therefore, involves understanding how to optimize magnetic properties of a material—saturation magnetization, the exchange constant, anisotropy energy, mobility, and coercivity. In addition we have the usual considerations of imperfection or more generally the fabrication of a magnetic bubble film given a set of optimum magnetic parameters. In the following section, we shall, therefore, first introduce the elementary basics of magnetics relevant to this paper and then proceed to describe the various trade-offs that have to be made to arrive at an optimum set of magnetic properties. In the subsequent sections we then consider the two classes of materials—the amorphous Gd–Co alloys and the single-crystal garnets in some detail.

II. General Magnetics

1. STATIC PROPERTIES

In a magnetic film that has its easy axis of magnetization perpendicular to the plane of the film and whose anisotropy field is greater than the demagnetizing field, the observed domain distribution is called a stripe domain pattern. A schematic example is shown in Fig. 3a. This domain pattern with its resultant width of domain arises from a balance between the magnetostatic energy and domain wall energy (see, e.g., de Jonge and Druyvesteyn, 1972). The magnetostatic energy is lowered by the formation of domains and favors narrow stripe domains. However, with an increase in wall area associated with an increasing number of stripes, the wall energy increases. The resultant domain width is obtained by minimizing the total film energy with respect to width. If an external magnetic field (usually called the bias field) is applied perpendicular to the film plane, then those domains which have their magnetization pointing in the direction of the external field grow at the expense of the others. At some critical bias field the stripe domain pattern is no longer energetically favored but rather the cylindrical domain

  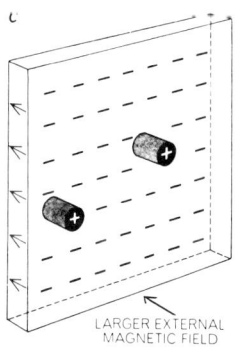

FIG. 3. The zero external field domain pattern of a bubble domain material (a), in a small perpendicular field (b), and in a sufficient field to collapse the stripe domains into bubbles (c). From Bobeck and Scovil, "Magnetic Bubbles." Copyright © 1971 by Scientific American, Inc. All rights reserved.

(magnetic bubble). A schematic illustration from stripes to bubbles is shown in Fig. 3a–c. Thiele (1970) has worked out a theory of static domain configuration, and we present one of his figures (Fig. 4) showing the range of stability of stripe and bubble domains plotted on a coordinate system where the stripe width (approximately equivalent to a bubble diameter) and the film thickness are normalized with respect to a quantity known as the fundamental length, l, defined as

$$l = \frac{\sigma_w}{4\pi M_s^2} = \frac{(AK_u)^{1/2}}{\pi M_s^2} \qquad (1)$$

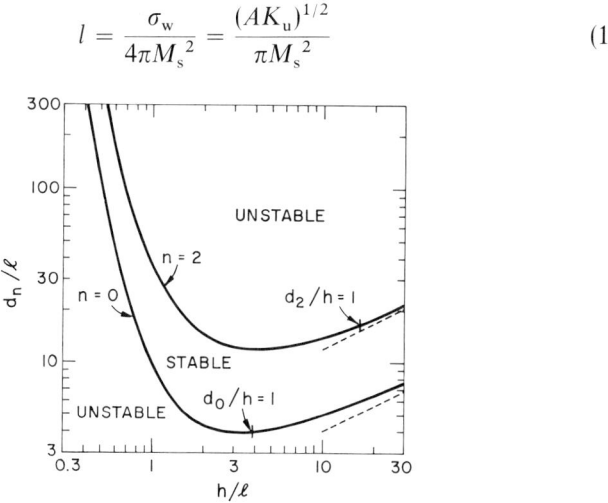

FIG. 4. Bubble diameter d_n vs film thickness, both normalized to the characteristic length l. The stable bubble size range is broadest when h/l is 4 (from Thiele, 1970).

where σ_w is the wall energy per unit area, $4\pi M_s$ the saturation magnetization, A the exchange stiffness constant (units of energy per unit length), and K_u the uniaxial anisotropy energy per unit volume. We note that the range of stability of a bubble with respect to stripe formation (usually called run out) and bubble collapse (annhilation of bubble) is maximum when the film thickness is $4l$. Therefore, this is usually the preferred thickness. The bubble diameter in the center of that stability region is $8l$. This implies that the ratio of bubble size to film thickness is 2, and this leads to film thicknesses that range from 0.1 to 2.5 μm.

The uniaxial anisotropy energy required to obtain stable bubbles has to be larger than the demagnetizing energy $2\pi M_s^2$. The ratio of anisotropy energy to the demagnetizing energy per unit volume is called the quality factor of the film and is given by

$$Q = K_u/2\pi M_s^2 \tag{2}$$

and for stability $Q > 1$. Using Eq. (2), Eq. (1) can be rewritten in the form

$$l = (2/\pi)^{1/2}(AQ)^{1/2}/M_s \tag{3}$$

and we note that the bubble diameter, $d = 8l$, is directly proportional to the square root of the exchange constant and Q, and inversely proportional to M_s.

The required magnitude of the exchange constant is determined by the temperature sensitivity that the device can tolerate. Operating a device close to the Curie or Néel point of a magnetic material is undesirable since the temperature dependence of the magnetization is generally too large and l changes rapidly, leading to a range of bubble diameters which may cause the bubble to stripe out or collapse at a given external bias field. In general a large temperature sensitivity is undesirable, and it is therefore preferable to choose materials that are well away from a Néel or Curie point relative to ambient temperature. Because the exchange constant decreases rapidly in the vicinity of the Curie temperature, the value of A plays an important role in selecting materials with low temperature sensitivity. Too large a value is generally accompanied by undesirably high magnetization, whereas too low a value produces sensitivity to temperature changes. These arguments also show the utility of a ferrimagnetic bubble material. In such a material a low magnetization and a high exchange constant A are achievable and, hence, offer good temperature stability. Both the amorphous and garnet materials are in this class.

Earlier it was noted that the value of Q has to be greater than unity from stability considerations. However, it cannot be made arbitrarily large since

l increases for a given A and M_s. The choice of Q also influences the dynamic properties of a magnetic bubble. Experimental evidence indicates that large Q ($Q > 6$) is associated with malfunction in the device which is related to erratic behavior of the bubble caused by the generation of Bloch lines in the domain wall. This is discussed in detail in Sections II, 2 and 3. When the bubble diameter is small, the value of Q close to or somewhat larger than two appears to be favorable. However, when the bubble diameter is relatively large and the magnetization is small, the in-plane rotating drive field is a significant fraction of the saturation magnetization. In this case, the stability of operation of a device is poor for values of Q close to or slightly greater than two. Hence the choice of Q depends on the size of the bubble.

2. Domain Dynamics

The velocity of a magnetic bubble is given by (Thiele, 1970)

$$V = \frac{\mu_w}{2}\left(\Delta H - \frac{8}{\pi} H_c\right) \quad (4)$$

where μ_w is the domain wall mobility, ΔH the product of the drive field gradient and the bubble diameter, and H_c the coercivity. It was noted earlier that the data rate is directly proportional to the velocity and inversely proportional to the diameter of a bubble. It is obvious that we need high velocities and small bubble diameters for high performance devices. For a given value of the drive field Eq. (4) shows that high mobility and low coercivity material is required.

We therefore consider material parameters that effect μ_w and H_c. The structure of a domain wall in a typical film suitable for magnetic bubbles can be characterized as that of a Bloch wall. This holds true for most of the wall except in the vicinity of the two surfaces where the structure is better described in terms of a Néel wall (Slonczewski, 1973). During the motion of such a domain wall, the Néel wall extends increasingly into the film and at some critical velocity the wall structure is unstable and a Bloch line is generated in the wall at or near one surface of the film. This Bloch line, since it lies parallel to the plane of the film, is called a horizontal Bloch line (HBL). These lines move through the wall and annihilate at the opposite surface. A new HBL is generated at this surface and this then proceeds back to the first surface. This cyclic generation of HBL continues as long as the drive field exceeds the critical drive field required for this process. Slonczewski (1973) has shown that the critical velocity V_s, associated with this phenomena is related to material parameters of a film by an expression

$$V_s = 24\gamma A/hK_u^{1/2} \tag{5}$$

where γ is the gyromagnetic ratio and h the film thickness.

When the velocity is less than this critical velocity the mobility of the wall is given by

$$\mu_w = \Delta\gamma/\alpha \tag{6}$$

where Δ is the width of the Bloch wall and is proportional to $(A/K_u)^{1/2}$, and α is a Landau–Lifshitz (Landau and Lifshitz, 1935) damping parameter.

In selecting materials for bubble devices it would appear that the highest possible value of μ_w is desirable. However, if the drive fields available in a typical bubble device exceed the fields required to reach a critical velocity given by Eq. (5), then it is believed that the device can malfunction due to erratic behavior of a bubble. This behavior has been attributed to the presence of vertical Bloch lines (VBL) discussed below (Hagedorn, 1974). Using Eqs. (5) and (6) and setting the local field in a device equal to the net drive field H_d, we find that

$$H_d \leq 24\alpha A^{1/2}/h \tag{7}$$

We note that this inequality favors a large value of α and A, and a small value of film thickness h. The range over which h can be varied is limited by bubble stability (or in a device—the device margin) consideration. The preferred value is $4l$. As discussed previously, a large exchange constant is desirable. However, as l depends on A also [see Eq. (3)], the effect of increasing A for a given value of l is to increase magnetization. This in turn increases the drive field for the operation of a device. As mentioned earlier, a large drive field is undesirable from the standpoint of providing large currents at high frequencies. We note, therefore, that A and h can be varied over rather narrow ranges. This leaves the remaining parameter α, the damping constant.

Equation (7) can also be recast into a slightly different but instructive formulation by using Eq. (3):

$$H_d \leq 6(\pi/2)^{1/2}\alpha Q^{-1/2}M_s(l/4h) \tag{8}$$

So, in order for the inequality to hold, M_s should be large or for a given value of M_s the quality factor Q must be low. Taking a $4\pi M_s$ of 200 G, $\alpha = 0.1$, and $h = 4l$, we find that $H_d \lesssim 24$ for $Q = 1$ and $H_d \lesssim 9.8$ for $Q = 6$. Drive fields are typically of the order of 10 Oe so that a material with a $Q = 1$ meets the inequality of Eq. (8), whereas $Q = 6$ may not and hence following Hagedorn's (1974) suggestion the device is likely to malfunction. Smith et al. (1974) find that low-Q materials have less tendency to exhibit properties associated with Bloch lines.

In the preceding paragraph we remarked on the presence of vertical Bloch lines (VBL) in a magnetic bubble. It has been shown that VBL cause the bubble to move in a direction which is at an angle to the gradient drive field (see Tabor et al., 1972; also Vella Coleiro and Tabor, 1972). If a large number of VBL are present the bubble moves almost 90° to the drive field. The effective mobility of a bubble in the direction of the drive field is therefore reduced. Bubbles with a large number of VBL are frequently called hard bubbles. This nomenclature is descriptive of the observation that the external bias field required to collapse a bubble with VBL is greater than that for a bubble with no VBL (Malozemoff, 1972). When the number of VBL is small—say, four—the bubble does not deflect to large angles and hence has a mobility approaching normal bubbles. The deflection of a bubble containing VBL arises from a gyrotropic force exerted on them by a drive field. This force is at right angles to the perpendicular of the film plane and the drive field. The force is proportional to the number of pairs of VBL; the larger the number, the greater the deflection. VBL also occur in integer pairs; hence the deflection is quantized as has been demonstrated by Slonczewski et al. (1973). Hard bubble formation can be suppressed by providing a capping layer (Bobeck et al., 1972) to the bubble so that a pair of Bloch lines are joined and can therefore annihilate. The capping layer can be provided by ion implantation (Wolfe and North, 1972) or by an additional film which may either be of the same material as the bubble materials and differ from it in magnetization or it can be a different magnetic material (Takahashi et al., 1973; Lin and Keefe, 1973).

We now consider the final and possibly the least understood quantity in Eq. (3), which is coercivity. There is almost unanimous opinion that H_c should be as low as possible. Two limiting criteria have been considered. Coercivity should be higher than any stray magnetic field that a device may encounter, for example, the Earth's magnetic field. However, this limit is hard to define since magnetic shielding can readily be incorporated as part of a device. The other limiting value was proposed by Thiele (1970) who pointed out that for a satisfactory operation of a device $H_c < 0.01\,(4\pi M_s)$. This criterion is useful for materials with a low $4\pi M_s$ but not so important for those with larger $4\pi M_s$ where it is invariably satisfied.

The origin of coercivity can be diverse, and it is therefore difficult to generate a satisfactory theory. However, for domain wall motion where the domain wall experiences spatially varying fluctuations of magnetic properties it can be shown that (Gambino et al., 1974).

$$H_c = \frac{2K_u}{M_s} \sum_i \frac{\Delta x_i}{x_i} = 4\pi M_s Q \sum \frac{\Delta x_i}{x_i} \tag{9}$$

Minimum coercivity is therefore obtained by lowering $4\pi M_s$ and/or Q. Clearly minimizing fluctuations or perturbations of the magnetic properties also reduces H_c.

3. SUMMARY

A quantity that plays a central role in magnetic bubble technology is the fundamental length l. The preferred film thickness is approximately four times and the bubble diameter eight times this fundamental length. These two conditions maximize the range of stability of a magnetic bubble with respect to an external magnetic field. For given values of the quality factor Q and the exchange constant A, the value of the fundamental length and therefore the preferred bubble diameter decreases with increasing saturation magnetization. For small bubble materials, in order to avoid excessively high drive fields which increase with the magnetization, both Q and A must be kept low. The quality factor cannot be made less than one for stability reasons; therefore Q is chosen to be larger than one but close to it for small bubbles (of the order of a micron or less) and greater than one when the bubble diameter is several microns. An additional constraint on Q arises from dynamic considerations which suggest that materials with a high Q ($Q > 6$) (unless hard bubble suppression is provided) may show device malfunction due to the generation of Bloch lines. The value of the exchange constant is selected on the basis of two considerations. Low A allows lower magnetization and lower H_d. However, large A is preferred if the magnetic properties of the material are required to be relatively temperature insensitive.

III. Materials for Magnetic Bubbles

Magnetic bubble materials, which are suitable for devices, can be classified into two broad categories: (1) amorphous materials of which Gd–Co is a prototype and (2) single-crystal materials of which the garnet is most favored. Apart from obvious structural differences between these two classes of materials there are also other differences. For example, the present amorphous materials are metallic, while garnets are insulators. Amorphous materials are characterized as indirect exchange materials in which the interaction is via conduction electrons. Crystalline materials such as the orthoferrites, hexaferrites, and garnets are superexchange materials; hence the interaction is among the ordered cations through intervening oxygen ions in the crystal lattice. It is interesting to note that when these superexchange materials are prepared in the amorphous state they are paramagnetic or superparamagnetic. There are also differences in preparation of the two materials (as discussed in detail in Section IV). Both classes of materials, however, have striking similarities. Amorphous and garnet materials are ferrimagnetic, have high mobilities, and their compositions can be varied to give a range of bubble sizes. Both rely on a growth-induced anisotropy rather than on structure as in the case of ortho- and hexaferrites. We consider now the general properties of each class of materials.

1. Amorphous Materials

The magnetic properties of crystalline rare earth–transition metal intermetallic compounds have been studied extensively (Nesbitt and Wernick, 1973). It has been shown that the heavy rare earths (Gd to Lu) couple antiparallel to the 3d transition metals Fe, Co, and Ni. It has also been shown that the Co and Ni moments decrease as the rare earth concentration in the intermetallic compound increases. This effect has been attributed to the partial filling of the transition metal d band by electrons contributed by the rare earth. These two features of the magnetic behavior of the crystalline compounds, ferrimagnetism and d bands filling, have been observed in the analogous amorphous systems (Orehotsky and Schroeder, 1972; Rhyne et al., 1972; Chaudhari et al., 1973a; Tao et al., 1974.)

The ferrimagnetic nature of these materials is apparent from magnetization when plotted as a function of composition (see for example the curve of the Gd–Co system plotted in Fig. 5). The spontaneous magnetization goes through a minimum indicating compensation at approximately 85 at. %

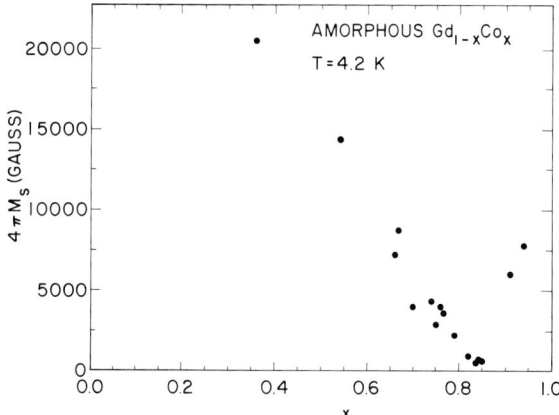

Fig. 5. Magnetization at 4.2 K as a function of composition in the Gd–Co binary system (Tao et al., 1974).

cobalt. The cobalt moment in crystalline materials drops from 1.7 μB in pure cobalt to 0.65 μB in Gd_4Co_3. A decreasing cobalt moment with increasing Gd concentration is also observed in the amorphous alloys but is somewhat less extreme (Tao et al., 1974). The d-band filling effects are apparently weaker in the amorphous alloys because they are less dense. The magnetic properties of the amorphous rare earth–transition metal alloys are a smooth function of composition because continuous solid solutions are produced in the amorphous state.

The temperature dependence of magnetization is shown in Fig. 6. The compensation temperature as a function of composition is shown in Fig. 7. This behavior is clearly consistent with the ferrimagnetic nature of these materials. Polarized neutron diffraction studies (Rhyne et al., 1972) of

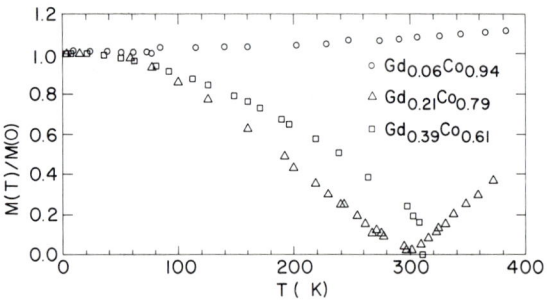

FIG. 6. Temperature dependence of the magnetization of three compositions in the Gd–Co binary system. Note the compensation point of 290 K in the $Gd_{0.21}Co_{0.79}$ composition (Tao et al., 1974).

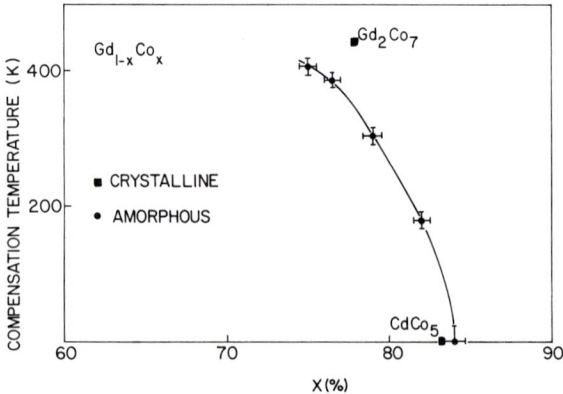

FIG. 7. Composition dependence of the compensation temperature in the Gd–Co binary system.

amorphous TbFe$_2$ have also been interpreted on a ferrimagnetic model as well as the results of Lorentz electron microscopy on Gd–Co films (Herd and Chaudhari, 1973).

2. Crystalline Materials

Historically, magnetic bubble devices were built first with single-crystal orthoferrites. These have been rapidly supplanted by the versatile garnets. When the bubble size decreases below 0.4 μm the garnets are no longer satisfactory, and the hexaferrites may be useful below this diameter. We shall therefore describe all three classes of materials but pay particular attention to garnets which are currently the strongest bubble device candidates along with the amorphous materials. Typical orthoferrites, hexaferrites, and garnets all exist in the PbO–Fe$_2$O$_3$–Y$_2$O$_3$ pseudo-ternary phase diagram studied extensively by Nielsen and Dearborn (1958).

a. Orthoferrites. This class of orthorhombic ferrites has the general formula RE FeO$_3$ where RE can be any of the "rare earth" series of elements La to Lu wherein the 4f atomic shell is being filled with electrons. Bobeck *et al.* 1969; Bobeck, 1967) used orthoferrite single-crystal platelets to demonstrate the utility of bubbles for magnetic storage devices. The orthoferrites have relatively low saturation magnetization $4\pi M_s$ (Goodenough and Longo, 1970) and consequently will only support large (>20 μm) diameter bubbles (Gianola *et al.*, 1969), which is a disadvantage when high bit densities are required. The orthoferrites do, however, exhibit high domain wall mobility; for example, Rossol (1970) measured a mobility of 6000 cm/sec-Oe for YFeO$_3$.

b. Hexaferrites. The commercially important ferrites with the hexagonal magnetoplumbite structure have the general formula MFe$_{12}$O$_{19}$, where M can be a divalent cation Pb, Sr, or Ba. Wijn (1970) has recently reviewed the literature on magnetoplumbites. These materials have received relatively little attention in the bubbles area because of the low domain wall velocities observed by Van Uitert *et al.* (1970) in bulk grown magnetoplumbite platelets.

c. Garnets. Cubic garnets have three sublattices with rare earths, singly or in combinations, occupying the largest dodecahedral sites and with nonmagnetic Ga, Al, or Ge ions substituting for Fe ions on the smaller octahedral and tetrahedral sites. [See, e.g., Huber (1970) or Geller (1967) for more details of garnet crystal chemistry.] A 5 μm diameter bubble film might be a garnet composed of two rare earths and iron oxide with gallium oxide, e.g., Eu$_{0.6}$Y$_{2.4}$Fe$_{3.8}$Ga$_{1.2}$O$_{12}$ (Giess *et al.*, 1971). It is advantageous to keep the composition as simple as possible, generally only three to four cations are employed but as many as six cations have been used. Substituting for iron with a nonmagnetic ion generally has a greater effect on $4\pi M_s$ than does the choice of rare earths. Both Ga and Al have about 90% preference for tetrahedral sites while Ge has a 99% preference (Geller, 1967). Consequently, it takes less Ge^{4+} (charge compensated by Ca^{2+}) to make a 5 μm bubble garnet than it does Ga or Al, and the Ge garnet has properties with better temperature stability by virtue of having a higher Curie point (Bonner et al., 1973). For instance, $Y_{1.88}Lu_{0.2}Ca_{.92}Ge_{.92}Fe_{4.08}O_{12}$ has Curie point T_c of about 190°C, while $Eu_{0.6}Y_{2.4}Fe_{3.8}Ga_{1.2}O_{12}$ has $T_c \simeq 130$°C.

Bobeck et al. (1970) found that mixed (solid solution) rare earth garnets, wherein the rare earths had significantly different ionic radii, had good growth-induced uniaxial anisotropy. Giess et al. (1973a) were able to produce garnet films with enough growth-induced K_u to support submicron diameter bubbles by selecting combinations of the iron garnets—either Sm or Eu paired with the iron garnets of either Yb or Lu. The choice of rare earths very largely determines the bubble velocity which is significantly influenced by magnetic damping (Rossol, 1970). For this reason, highly damping rare earths such as Tb, Dy, and Ho are avoided while Y, Lu, and Gd are favored (Vella-Coleiro et al., 1972).

The garnet films useful for magnetic bubbles are grown epitaxially on nonmagnetic garnets. The lattice parameter match between the film and substrate plays an important role. The film lattice parameter a_f can be calculated from reported values (Geller, 1967) of end member garnets using the additive Vegard relationship. It is further necessary, however, to take account of the increased a_f of LPE films caused by Pb impurity incorporation as discussed by Giess et al. (1972a), Blank et al. (1973), and Robertson et al. (1973). The lattice mismatch between the film and its substrate must be kept small to minimize defects in the film. Besser et al. (1972) found that the unstrained lattice mismatch at room temperature $\Delta a = a_s - a_f$ has the limits $+0.01$ Å for tension and -0.02 Å for compression in the case of vapor-deposited garnet films. Blank and Nielsen (1972) and Tolksdorf et al. (1972) reached the same conclusion for LPE films. Since most films are deposited on $Gd_3Ga_5O_{12}$ substrates grown by the Czochralski method, one usually attempts to make garnet films with $a_f = a_{GGG} = 12.383$ Å.

IV. Growth of Magnetic Bubble Materials

The growth of a magnetic bubble film is dictated by the class of materials being prepared. The amorphous materials are prepared by vapor deposition and in particular by sputtering. There is some indication that these materials may also be prepared by electroplating (Cargill et al., 1974). The garnets are single-crystal films which are grown epitaxially on a single-crystal nonmagnetic garnet substrate. Hence the growth of a garnet film involves two distinct steps: (1) the growth of a suitable substrate and (2) the subsequent growth of a film on it.

1. Preparation of Amorphous Films

The process of film fabrication by sputtering can be divided into three stages: (1) target fabrication (the preparation of a Gd–Co alloy of a selected composition), (2) substrate preparation, and (3) the sputtering or film deposition process.

a. Target Preparation. Targets are prepared by arc melting Gd and Co (or ternary alloys) in an inert atmosphere. This procedure insures that the reaction of Gd with the environment is minimized. The arc-melted alloys are spread on a target plate with the aid of the arc until a relatively uniform target is obtained. A much more uniform target can be fabricated if the alloy is hot pressed in an inert atmosphere. The problem with targets prepared in this manner is the large surface to volume ratio that is invariably present in a powder. Hence reaction of the powder with the environment is likely to be more of a problem than in an arc-melted alloy. Our experience indicates that contaminated targets lead to poor films as evidenced, for example, by high coercivities.

b. Substrate Preparation. Although the nature of the substrate has little effect on the magnetic properties of a film, the surface cleanliness can influence the defect density of the final film. For example, dust particles can lead to pin holes which, in turn, pin magnetic domains. Defects associated with substrates are discussed in Section IV, 1.

c. Sputtering Process. Thin films of amorphous Gd–Co alloys for bubble applications are prepared by rf or dc bias sputtering. Several reviews of the sputtering process are available (Maissel and Glang, 1970; Chopra, 1969). A simple diode configuration is used with a planar target mounted above and parallel to a planar anode plane which also serves as the substrate support. A negative potential of typically 1–2 kV is applied to the target. If an appropriate pressure of an inert gas, e.g., argon, is present, a plasma can be sustained between the cathode (target) and the anode. The region immediately adjacent to the cathode becomes depleted of positive argon ions creating a dark space. The plasma region is essentially electrically neutral and highly conductive so it will not support a large potential difference. Most of the cathode potential is across the dark space. Ions in the plasma drift or diffuse into the vicinity of the dark space and are accelerated across the dark space by the cathode potential. These energetic ions impact the target and dislodge atoms of the target material.

In alloy sputtering it can be shown that in the steady state, the composition of flux leaving the target is the bulk composition of the target. In bias sputtering, however, the growing film becomes a secondary target bombarded by ions accelerated out of the plasma by the bias voltage. Some fraction of the atoms arriving at the substrate are resputtered away and eventually accumulate on unbiased surfaces in the system, typically the walls. One of the constituents usually resputters preferentially so that the composition of the film which accumulates is generally different from the target composition and is a function of the bias voltage. Winters et al. (1969) proposed a model for bias sputtering of alloys. Tarng and Wehner (1971) discussed the effects of bias on film composition qualitatively. Cuomo et al. (1974) derived the functional dependence of accumulation rate on bias voltage for the case of a single element. The accumulation rate equation can be extended to treat composition as a function of bias voltage as shown below.

The accumulation rate, AR, in bias sputtering is given by:

$$AR = F - L - R \tag{10}$$

where F is the atom flux from the target, L is the flux lost via side scatter, geometric losses, or nonsticking to the substrate, R is the flux resputtered from the substrate. It is convenient to restate Eq. (10) as

$$AR = SF - R \tag{11}$$

where S is a generalized sticking coefficient. The flux from the target is:

$$F = E_t j_t^+ \tag{12}$$

where E_t is the sputtering yield defined as the number of atoms dislodged per incident ion. The yield depends on the incident ion, the target atom, and the target potential. j_t^+ is the ion flux incident on the target.

A similar expression can be written for the resputtering flux:

$$R = E_s j_s^+ \tag{13}$$

where E_s is the yield at the substrate bias potential, j_s^+ is the ion flux on the substrate.

The yields and ion fluxes can be expressed in terms of electrode voltages so that the accumulation rate can be written as a function of the bias voltage and target voltage (see Cuomo et al., 1974a).

The alloy composition is obtained by treating the accumulation rate of each constituent independently. The assumptions in this approach are probably only valid for simple alloy systems. The Co/Gd ratio in a film is given by the ratio of the accumulation rate of cobalt to the accumulation rate of Gd.

$$\frac{\text{Co}}{\text{Gd}} = \frac{A_{\text{Co}}}{A_{\text{Gd}}} = \frac{S_{\text{Co}}F_{\text{Co}} - R_{\text{Co}}}{S_{\text{Gd}}F_{\text{Gd}} - R_{\text{Gd}}} \tag{14}$$

We normalize this expression to the Co/Gd ratio of a film deposited without resputtering which should be the bulk Co/Gd ratio of the target to obtain

$$\frac{\text{Co/Gd}}{(\text{Co/Gd})_{\text{target}}} = \frac{1 - R_{\text{Co}}/S_{\text{Co}}F_{\text{Co}}}{1 - R_{\text{Gd}}/S_{\text{Gd}}F_{\text{Gd}}} \tag{15}$$

Restating R and F as functions of bias voltage V_b and target voltage V_t gives

$$\frac{\text{Co/Gd}}{(\text{Co/Gd})_{\text{target}}} = \frac{1 - K_{\text{Co}}[(V_b - V_{\text{Co}})/V_t]}{1 - K_{\text{Gd}}[(V_b - V_{\text{Gd}})/V_t]} \tag{16}$$

where V_{Co} and V_{Gd} are the threshold voltages for resputtering of Co and Gd, respectively. The constants K_{Co} and K_{Gd} depend on system geometry, the dependence of yield on voltage, and the generalized sticking coefficient. The threshold for resputter of cobalt is much higher than that of Gd and cobalt's yield is much lower, so in the low bias regime we neglect the resputtering of cobalt. The threshold for resputtering of Gd is also small enough to be neglected giving a very simple expression

$$\frac{\text{Co/Gd}}{(\text{Co/Gd})_{\text{target}}} = \frac{1}{1 - K(V_b/V_t)} \tag{17}$$

The bias voltage V_b is the effective bias, the potential difference between an argon ion in the plasma and the surface of the growing film. The effective bias is related to the applied bias V_a by the plasma potential V_p.

$$V_b = V_a - V_p \tag{18}$$

Note that the applied bias is negative and the plasma potential is positive with respect to ground so that a larger plasma potential gives rise to a larger effective bias. The Co/Gd ratios in films sputtered under various conditions are shown in Fig. 8. The constant was obtained by fitting Eq. (17) to the

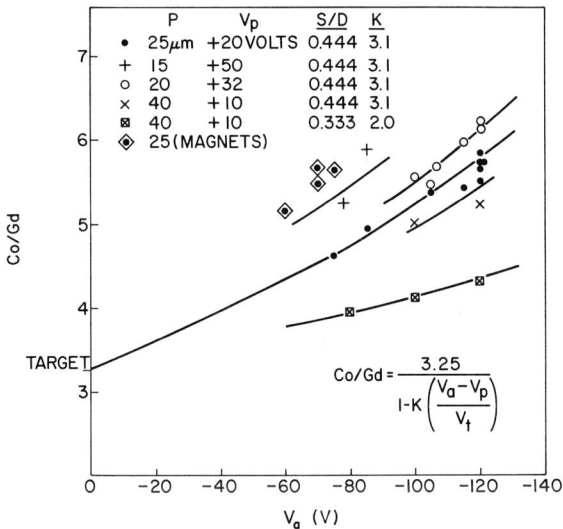

FIG. 8. Co/Gd ratio as a function of applied bias voltage (dc level). Solid lines are a fit to Eq. (17). Data points were obtained by microprobe analysis.

observed Co/Gd ratio obtained at 25-μm pressure and −105 V applied bias. Note that when the system geometry is changed, indicated by the ratio of the target-substrate spacing S_p to the target diameter D, the constant K_{Gd} changes.

Control over the composition of the film provides control over $4\pi M_s$, T_c, compensation temperature, and exchange stiffness. The deposition conditions also determine the magnitude of the uniaxial anisotropy. The anisotropy arises primarily from pair ordering in the growing film. The mechanism of pair ordering is believed to be site selection of the adatom during growth. Site selection is favored by conditions of high resputtering, low deposition rates, and moderate substrate temperatures. It has been shown, for example, that the wall energy is a function of bias voltage (Chaudhari et al., 1973a). These data suggested that the K_u increases with increasing bias voltage, i.e., higher resputtering rates. Also Cronemeyer (1974) showed by ferromagnetic resonance (FMR) measurements that K_u is high in films prepared under high bias conditions. It has also been shown that the anisotropy decreases as the deposition rate increases. The anisotropy also decreases as the substrate temperature is decreased below room temperature. These results are consistent with the pair ordering model. High deposition rates or low substrate temperature inhibit site selection and would be expected to decrease K_u.

2. Preparation of Single-Crystal Garnet Films

Epitaxial garnet films have been prepared by a variety of techniques. Irrespective of the process of film fabrication, the satisfactory growth of an epitaxial film requires a single-crystal and nonmagnetic garnet substrate. The growth of a "good" bubble film therefore involves three basic steps: (1) the growth of a single-crystal substrate material, (2) the preparation of a suitable substrate from the bulk crystal, and (3) finally the growth of an epitaxial film. We shall describe the three processes and consider the liquid phase epitaxy (LPE) process in some detail since this is the preferred mode of film growth.

a. Growth of Single-Crystal Substrate. The most commonly used substrate for garnet film is the gadolinium gallium garnet—$Gd_3Ga_5O_{12}$ (GGG). Other substrates such as neodymium gallium garnets have also been used in the growth of special types of garnet films (Plaskett et al., 1974). The emergence of the magnetic bubble technology has led to the development of high quality garnet substrates. Although the basic crystal pulling technique is similar to that used, for example, in the silicon technology, the garnets have required pushing this to higher temperatures of growth.

The rare earth gallium garnets melt in the range of 1500 to 1800°C with the melting point increasing as the rare earth ion decreases (Brandle and Valentino, 1972). Boules are pulled by the Czochralski technique from melts contined in rf heated iridium crucibles (Linares, 1964). The growth atmosphere is pure dry N_2 sometimes containing a small amount ($\sim 2\%$) of O_2 (Brandle et al., 1972) to suppress gallium suboxide formation. Excessive growth temperature or high oxygen pressure, however, can lead to Ir particle inclusions in crystals. O'Kane et al. (1973) used a computer-controlled system to obtain good crystal diameter uniformity during growth and thereby avoided the formation of dislocations. Rotation rates of 40 to 80 rpm maintain a flat growth interface and prevent (211) facet formation. Facetting results in a "core" defect which is a strained region along the boule axis. Pull rates are from 2 to 8 mm/hr. High rotation rates induce striations and necessitate the use of slower pull rates. We shall return to these considerations again when we correlate defects in films with the growth of the substrate and film deposition. We note, however, that crystals with a diameter of 3 in. have been grown.

b. Substrate Preparation. Crystals are generally pulled in the [111] direction of the cubic garnet. After the growth of a crystal is completed, it is mounted for slicing into wafers. Before this a final crystal orientation determination is done on an X-ray goniometer stage to ensure that the [111] axis is normal to the plane of the film. The wafers are sliced with a diamond saw or a wire saw using a slurry of SiC (O'Kane et al., 1973). The wafers are subsequently polished in the final step with a mechanical and chemical polish. After polishing, the substrates are thoroughly cleaned and kept in a dust-free environment.

c. Epitaxial Film Growth. Epitaxial films have been grown by sputtering, chemical vapor deposition (CVD), and LPE. The latter process is the one currently used. However, the other two processes have yielded epitaxial films and we shall therefore describe them also.

(i) *Sputtering of garnet films.* The sputtering process used to prepare garnet films was first described by Sawatzky and Kay (1968, 1969a,b). They used a substrate maintained at 500°C and found that the composition of the garnet film was different from the target. Decreasing the substrate temperature decreased the deviation from the desired stoichiometric composition. Cuomo et al. (1974) took advantage of this observation, and by using various sputtering techniques they were able to achieve a film with the desired composition. They also found that a garnet substrate even at 450°C did not produce an epitaxial film. Rather, the film was amorphous and paramagnetic. Subsequent heat treatments as low as 650°C and up to 950°C produced an epitaxial film where the crystallization at the film-substrate interface progressed throughout the rest of the film. Films produced by this process showed magnetic bubble domains. However, the quality of the film was relatively poor and a large density of pinning sites was present.

(ii) *Chemical vapor deposition of garnets.* Epitaxial films of garnets have been produced by this process by several investigators, e.g., Mee et al. (1969), Stein (1970), Taylor and Sadagopan (1971), Robinson (1973), and Cowher et al. (1974). The principal advantage of this process was believed to be the ease with which it could be scaled for large-scale manufacture of epitaxial garnet films. However, the garnet films suitable for magnetic bubbles required a large number of components and this led to considerable difficulty in preparing films of a uniform composition and thickness.

The hydrolysis or oxidation of metal halide vapors is the reaction most often used for the CVD growth of single-crystal garnet films. Anydrous metal chlorides are vaporized and transported with a He or Ar carrier gas to a reaction and growth chamber containing substrates at a temperature of over 1100°C, where the chlorides react with O_2, and sometimes H_2O, to form garnet epitaxially on the substrates. Good device quality films have been made by CVD, but high growth temperatures preclude the formation of growth anisotropy in garnets. The halide CVD film technique is limited presently to producing garnets with stress-induced anisotropy. Lower temperature growth is possible with an organometallic approach (Cowher et al. 1974). It is perhaps appropriate to note here that magnetic bubble films of garnets have been grown hydrothermally by Brochier et al. (1972). This technique also allows growth at lower temperatures than CVD.

(iii) *Liquid phase epitaxy (LPE) film growth.* In a comparatively short period of time, the LPE film growth techniques has been developed for garnets to the degree where large numbers of films can be grown reproducibly (Hewitt *et al.,* 1973). Much of this progress was achieved with technology established for bulk crystal growth from fluxed melts. For a detailed review of the garnet LPE film growth literature, the reader is referred to Giess and Ghez (1974).

Most garnet films for bubble devices are grown by isothermal dipping with axial rotation. Levinstein *et al.* (1971) showed that garnet films can be grown isothermally from $PbO-B_2O_3$ fluxed melts which can be supercooled by tens of degrees to achieve supersaturation. By avoiding the necessity of imposing temperature gradients and heat fluxes, it is possible to obtain very good temperature control in this isothermal mode of growth. Giess *et al.* (1972a) extended this technique to include axial rotation of substrates held in a horizontal plane. This growth geometry, depicted in Fig. 9, develops a hydrodynamic flow pattern in the melt such that the diffusion boundary layer at the film–melt interface achieves a steady state.

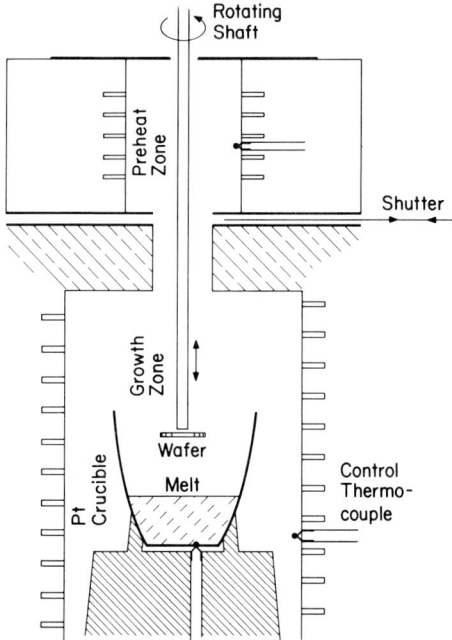

FIG. 9. Growth geometry for isothermal growth of epitaxial garnet films. The substrate is attached to the rotating shaft by three Pt wire fingers. It is preheated to the growth temperature in the preheat zone with the shutter closed. The rotating substrate is then lowered into the melt for the prescribed growth period.

Substrates are generally held in a holder consisting of three Pt wire fingers (Tolksdorf et al., 1972). This wire holder arrangement has a low heat capacity and is easily cleaned. Substrates are preheated to the growth temperature either by being held directly over the melt or by being held in a separate preheat zone directly above the growth zone as shown in Fig. 9. The latter method avoids the possibility of having PbO vapor condense on the substrate surface (Chaudhari, 1972).

A substrate can be heated to the growth temperature (800 to 1000°C) in as little as 2 min without cracking from thermal shock. Often as much as 10 min is allowed for preheating to insure thermal equilibration of the substrate holder assembly. The rotating substrate is then lowered into the melt for the prescribed growth period, typically between 5 sec and 5 min. At the end of the growth period, the substrate is carefully raised out of the melt and is spun rapidly (>600 rpm) while held directly above the melt. This spin-off procedure removes any melt droplets that would continue to grow film during cooling to room temperature if left adhering to the substrate. Finally the film and holder assembly are cleaned in warm dilute acetic acid and detergent solutions followed by a water rinse.

The kinetics of this LPE process have been studied in considerable detail. Giess et al. (1972a) showed that film growth rate goes as the square root of rotation rate and a steady state (constant rate) mode of growth is achieved when the substrate is rotated. Moreover, if there is no substrate rotation, growth rate decays with time t such that film thickness is proportional to $t^{1/2}$. Ghez and Giess (1973) developed a unified expression for the growth rate as a function of time and rotation rate, which is derived from a model involving diffusion through a stagnant boundary layer and a first-order interfacial reaction. The model predicts the observed initial transient state which decays rapidly (exponentially) into a steady state during which the growth rate is constant for the case of a nonzero substrate rotation rate. The diffusion and interfacial reaction constants for this model are of the order of 10^{-5} cm^2/sec and 4000 cm/sec, respectively. Rode (1973) and Blank et al. (1973) treated a model involving just diffusion. Here the diffusion constant is of the order of 10^{-7} cm^2/sec.

V. Anisotropy in Magnetic Bubble Films

1. GENERAL CONSIDERATIONS

Magnetic anisotropy can be described phenomenologically in terms of four principal mechanisms: (1) magnetocrystalline, (2) shape-induced, (3) stress-induced, and (4) atomic short-range order (single ion or pair ordering are special cases of this). Magnetocrystalline anisotropy can be related to the point symmetry of a crystal via its crystal field. It therefore reflects the symmetry of the crystal. For example, hexaferrites with a unique c-axis can have a uniaxial anisotropy. However, neither the garnets nor the amorphous alloys can have uniaxial anisotropies due to magnetocrystalline effects. Garnets are cubic and the amorphous materials have, by definition, no crystal structure. The second general mechanism of anisotropy, i.e., shape-induced anisotropy, can also be ruled out in the case of the amorphous and garnet materials. For a uniform distribution of material the maximum anisotropy that can be produced is determined by the shape of a magnetic material. Magnetic bubble films are thin relative to the other two dimensions of the film. Hence the demagnetizing field favors in-plane magnetization rather than the observed perpendicular easy axis. It could be argued that the bubble films have microscopic rods where the lateral dimension of the rod is small compared to the thickness of a film. Electron microscopic examination of amorphous films and garnets (Herd and Chaudhari, 1973) shows no evidence for such rods in the as-deposited films. This leaves the stress-induced and short-range order effects which we therefore conclude to be the sources of anisotropy in these two materials. Since the amount of contribution of the two is different for the two classes of materials it is convenient to consider each class separately.

2. Amorphous Materials

Stress-induced anisotropy need not play a significant role in amorphous materials. This was demonstrated by using a rock salt substrate and a Gd–Co film (Cuomo *et al.*, 1974b). The domain pattern was examined with the film on the substrate and after removal from the substrate, and no change was detected. Had stress been a significant factor, the anisotropy energy would have been changed. From Eq. (1) we note that a change in the anisotropy energy results in a change in the fundamental length and this in turn leads to a change in the stripe width (see Fig. 4). Since no change in stripe width was detected, it was concluded that the stress-induced component is insignificant compared to the contribution from the last remaining mechanism.

When a material has a known crystal structure, the positions of the atoms can be readily located. For example, in an Fe–C alloy the iron atoms are located at the corners and at the body center position of a cube. For a dilute C alloy the C atoms occupy interstitial positions and distort the lattice. However, the number of C atoms along the three axes of the cube are statistically equal in number, and hence the structure is macroscopically cubic. If now a uniaxial stress is applied along one axis of a cube, the distribution of C atom is no longer statististically equal and more C atoms are present along the axis of tension which leads to a tetragonal structure (Snoek ordering).

In a magnetic alloy, which has a cubic structure, the distribution of pairs of atoms can be similarly influenced by an external magnetic field. For example the Fe–Ni alloys can be magnetically annealed so that the number of Fe–Ni pairs is no longer statistically equal. This results in an easy axis of magnetization along the direction of the applied magnetic field. The number of pairs that align themselves along the magnetic axis is determined by thermal equilibrium at the temperature of heat treatment and follows Boltzmann statistics.

In the case of amorphous films, a similar line of reasoning can be used to explain the origin of magnetic anisotropy. Rather than apply an external field, the deposition process introduces an anisotropy in composition or structure. The change in composition is believed to occur as follows. As Gd and Co atoms arrive they have some lateral mobility (parallel to the surface) and hence can find sites that are energetically (or geometrically) favorable. Since the mobility is confined only to the surface of the growing film, the atoms below the surface do not move over atomic distances and hence the structure characteristic of an amorphous material is obtained. However, the growing film contains a built-in anisotropy from the way the atoms distribute themselves on the surface.

The second structural mechanism relies on a form of self-shadowing. This is more concerned with the deposition process per se and, in principle, predicts that anisotropy can be generated even in a single-component system. Rather than compositional ordering as in the case of Fe–Ni alloy, this model more closely simulates what happens in a material with a uniaxial crystal structure. It is argued that during the deposition process the spacing between the atoms measured parallel to the plane of the films is on an average larger than spacing normal to it. This is a consequence of atomic self-shadowing. The extent to which such a structure is available is a function of many variables. One of the more notable variables is substrate temperature. A low substrate temperature means no relocation; hence the atoms stick where they land, and long columnar growth is anticipated. Too high a temperature leads to rapid relaxation into arrangements that show little anisotropy. Neither the composition nor structural models have been quantitatively verified.

3. Anisotropy in Garnets

Garnets intrinsically have cubic magnetic anisotropy K_1. Hence, they can only support bubbles if a uniaxial anisotropy is induced either by mechanical stress operating through magnetostriction or by a growth mechanism. For a garnet (cubic structure) film oriented (111), the stress-induced anisotropy energy, K_σ, is given by

$$K_\sigma = -\tfrac{3}{2}\lambda\sigma \quad (\text{erg}/\text{cm}^3) \tag{19}$$

where λ is the (111) magnetostrictive coefficient, which is typically -2.4×10^{-6} for YIG (Iida, 1967), and σ is the biaxial (planar) stress. Stress is induced in a film by choosing a composition with a room temperature lattice parameter a_f slightly different from that of the substrate a_s so that there exists a lattice mismatch

$$\Delta a = a_s - a_f \quad (\text{Å}) \tag{20}$$

A range of Δa from -0.02 to $+0.01$ is tolerable in garnet films before their physical perfection degrades. To provide uniaxial anisotropy, Δa must be positive if λ is negative. Eq. (19) can be rewritten

$$K_\sigma \simeq \tfrac{1}{3}\Delta a \lambda 10^{12} \tag{21}$$

These considerations are especially important for films grown at high temperatures ($>1100°C$) where growth-induced K_u anneals out. Mee *et al.* (1969) have given a rather complete analysis of stress in garnet films grown by chemical vapor deposition in (100) and (110) as well as (111) orientation.

Growth-induced anisotropy K_g is not understood on a quantitative basis. Indeed, there are probably several mechanisms contributing to K_g. The situation is further complicated by the coexistence of K_1 and K_σ with K_g in all real systems.

Bobeck *et al.* (1970) demonstrated that the K_g in bulk-grown garnet platelets is sufficient to yield magnetically stable bubble devices. They further showed K_g to be related to the growth direction. Mixed rare earth iron garnets with a net $\lambda \simeq 0$ also had K_g. Thus, when K_g is present in a sufficient amount, stress-free materials can be used to fabricate devices. It was found that K_g was greatest when the two rare earths had the greatest difference in ionic radii and the K_g existed only in those portions of crystals which grew at low ($<1100°C$) temperatures. Kurtzig and Hagedorn (1971) annealed K_g out of both bulk- and LPE-grown garnets at 1300 to 1200°C in 1 to 4 hr, respectively. Giess and Cronemeyer (1973) grew a $Eu_{0.65}Y_{2.35}Fe_{3.8}Ga_{1.2}O_{12}$ film composition on a series of garnet substrates with different lattice parameters. They found $K_g \simeq 8000$ erg/cm^3, for $\Delta a = 0$ and were able to show that K_u ranged from 4000 to 12000 erg/cm^3 as Δa changed from -0.015 to $+0.015$. This change in K_u with Δa is a consequence of K_σ arising from lattice mismatch, which is superimposed on K_g. Here the λ derived from Eq. (21) agrees well with that calculated additively from the end member garnet λ values.

Several theories have been proposed to explain the origin of K_g. For mixed rare earth garnets, K_g has been attributed to directional ordering of or preferential site occupation by the rare earth ions (Rosencwaig and Tabor, 1972; see also Callen, 1973). These theories involving rare earth ions are supported by the intensity of the thermal annealing treatment required to attenuate K_g. If K_g were related to only iron ions rather than rare earth ions, then K_g would anneal more easily. It is possible, however, that the diffusion distance during annealing is greater than nearest neighbor spacing in which case the intensity of annealing argument is invalid. Furthermore, K_g is found in single rare earth garnets and in garnets with only diamagnetic ions other than iron. Akselrad and Callen (1971) proposed that K_g can be explained by distortion of the local symmetry around tetrahedral iron ions caused by preferential site ordering among their tetrahedral neighbors when gallium, for example, is partially substituted for iron. They further suggest a similar distortion of tetrahedral symmetry by ordering of dodecahedral ions should contribute in mixed rare earth garnets. Sturge et al. (1973) have had moderate success in fitting anisotropy data for europium-containing garnets to a theory where site preference of about 20% was found. Gyorgy et al. (1973) considered the interplay of many possible factors, including defects, but favored the site preference picture of induced anisotropy. Indeed, the work of Heilner and Grodkiewicz (1973) on the compositional dependence of K_1 and K_g in mixed garnets showed K_g to reach a maximum for 50:50 mixtures and to go to zero for the end member compositions. Stacy and Rooymans (1971) considered a crystal field mechanism based on oxygen vacancy ordering as the origin of K_g. They estimated that an ordered oxygen vacancy concentration of 10^{-3} could account for observed K_g effects. In LPE films both K_g and Pb content have been shown to increase with decreasing growth temperature by Giess et al. (1972b) and by Blank et al. (1973). This correlation of K_g and Pb has lead Plaskett et al. (1974) to speculate on the possibility of Pb causing a defect ordering similar to that reported by Stacy and Rooymans (1971) as an origin of K_g.

VI. Defects in Films

Any inhomogeneity, structural or chemical, in a film containing a magnetic bubble is undesirable. If the inhomogeneity leads to a change in magnetic properties within a local region, a magnetic bubble may either be attracted or repelled by it. We shall call such a local inhomogeneity a defect. Defects in magnetic bubble films are detected by examining the motion of a magnetic domain spatially. Where the motion is observed to be jerky or a domain is pinned, that area of a film is considered to be defective. Such a definition of a defect is therefore an operational one. Depending upon the drive field, which moves a domain, an inhomogeneity may or may not be detected. However, from a practical standpoint this definition of a defect is satisfactory since the drive fields or conditions of examination are chosen in such a way as to detect those defects that would normally interfere with the operation of a device. We now consider some of the more commonly observed defects. For details of the origin and elimination of these defects and their detection schemes, we refer to the original articles and reviews cited in the following sections.

1. Defects in Amorphous Films

Defects in these films are introduced either through the substrate or during the sputtering process. Surface roughness of and surface scratches on a substrate can lead to inhomogeneous magnetic properties during film growth. As a result magnetic domains are pinned or show a "foggy" appearance indicative of high coercivities. Similarly, dust particles or other chemical and topological perturbations of the surface can lead to defects. Pin holes are generally observed on a dirty substrate. Defects associated with substrate processing and handling can be virtually eliminated by clean room conditions.

Defects can also be introduced during the sputtering process. For example, an inhomogeneous target can give spatially varying chemical compositions. The gradients in magnetic properties pin bubbles or induce nonisotropic response. Spalling of the target during film growth can be another source of defects. This can be eliminated by using a sputtering system where the film is above the target rather than below it, as is usually the case. Incorporation of gas during sputtering can lead to defects by introducing composition fluctuations or alternatively by reacting with the constituents of the film resulting in a microscopically inhomogeneous film. Inert gases do not appear to have this deleterious effect. However, the presence of gases such as O_2, N_2, and H_2 can result in high coercivity films.

2. Defects in Garnets

The three stages of film fabrication during which defects can be introduced are (1) growth and preparation of a substrate, (2) film growth, and (3) handling and device processing. In epitaxial films defects such as dislocations or, in general, strain inhomogeneities are almost entirely associated with the substrates. A prerequisite for the growth of a defect-free epitaxial film is a defect-free substrate. Dislocations in garnet crystals generally arise from two sources. The first is associated with the presence of precipitates, principally iridium precipitates coming from the crucible material used in the growth of the crystal. Due to differential thermal expansion between the precipitates and the garnet, crystal dislocation loops are generated in the crystal. If these loops are cut by the free surface of the substrate, the dislocation line is subsequently continued into the growing epitaxial film. An example is shown in Fig. 10 (Matthews et al., 1973). These dislocations pin magnetic bubbles as, for example, shown in Fig. 11 (Argyle and Chaudhari, 1973). Dislocations are also introduced if the crystal is nonuniformly cooled

FIG. 10. Dislocations in a garnet substrate extending into the epitaxial film.

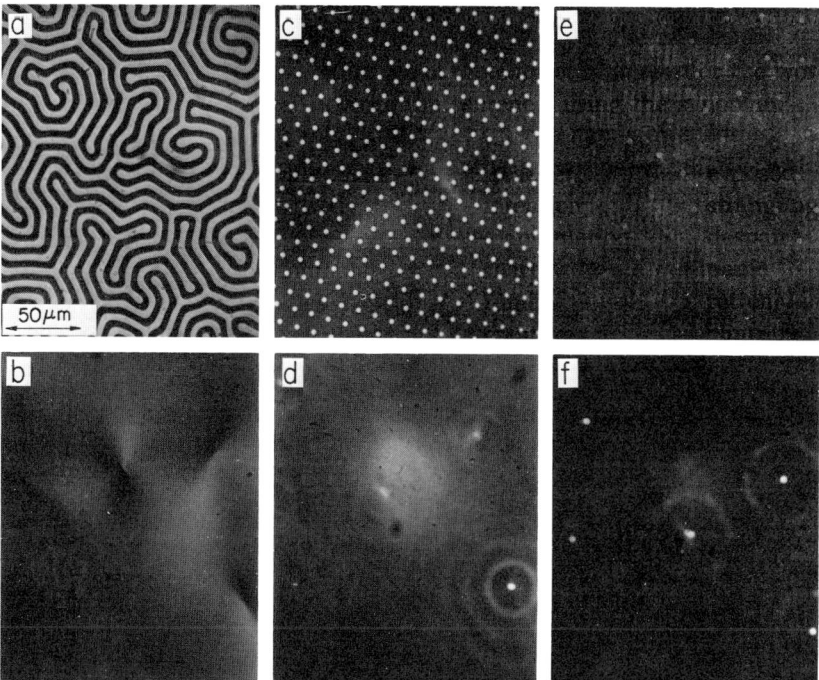

FIG. 11. Magnetic domains interacting with dislocations in an epitaxial garnet film. (a) Stripe pattern of demagnetized state. (b) Birefringence pattern for magnetically saturated garnet. (c) and (d) High density static and dynamic bubble array, respectively. (e) and (f) Low density static and dynamic bubble array, respectively (Chaudhari, 1972).

after its growth or if there are large concentration fluctuations as in coring. The use of controlled atmosphere during the growth of crystals and controlled rotation rates can eliminate virtually all of the dislocations or strain gradients associated with concentration fluctuations. The controlled atmosphere prevents the incorporation of the iridium into the melt.

Once a crystal has been grown it is usually oriented and sliced into wafers which are polished by a combination of mechanical and chemical means. The preparation of a substrate surface can lead to defects via the formation of scratches or microvoids. The scratch is readily revealed by etching in a hot phosphoric–sulphuric acid mixture or by growing an epitaxial film in which case the interaction of the magnetic domain reveals it. An example is shown in Fig. 12.

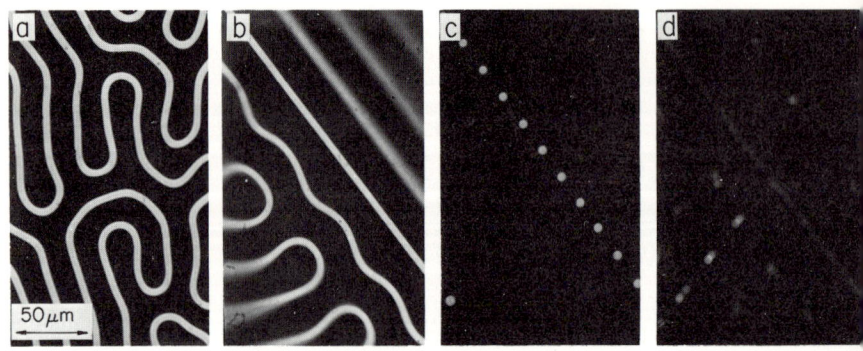

Fig. 12. Interaction between magnetic domains and a scratch in the substrate used to grow an epitaxial garnet film (Argyle and Chaudhari, 1973).

Assuming we have a defect-free substrate perfectly polished, a bubble film may still show defects. These are introduced during the handling of a substrate before a film is grown on it. Dust particles can lead to precipitation of undesirable compositions in an epitaxial film and to pinholes in amorphous films. Similarly residues on the substrate from improper cleaning conditions can lead to defects in a film. Cleanliness and clean room conditions are therefore desirable in magnetic bubble film preparation (Hewitt *et al.*, 1973).

We now consider defects introduced during the growth of a film. Epitaxial garnet films are usually grown from a melt that contains volatile flux agents. A relatively cold substrate may therefore contain vapor deposits of, say, PbO in $PbO-B_2O_3$ flux. An epitaxial film grown on this substrate contains numerous defects at the substrate film interface. These defects can be eliminated by incorporating a preheat zone in which the substrate is heated to the melt temperature so that it is not a particularly favored spot for condensation of the PbO vapor. As the garnet film is grown from a supersaturated solution which is thermodynamically unstable, it is imperative that there be no preferred nucleation site other than the substrate surface for the precipitation of the epitaxial garnet film. Too large a supersaturation leads to copious heterogeneous nucleation on impurity centers which can be present in a typical melt. Too low a supersaturation tends to accentuate any nonuniformity in the growth of a film. For more details on the supersaturation we refer to the section on epitaxial film growth (Section IV, 2). At the end of a film growth sequence, defects known as mesas can be introduced. These are associated with a droplet of the melt which is left on the film as it is removed from the melt. During removal of the film from the melt, the liquid in the droplet precipitates leading to a circular-shaped growth which is thicker than

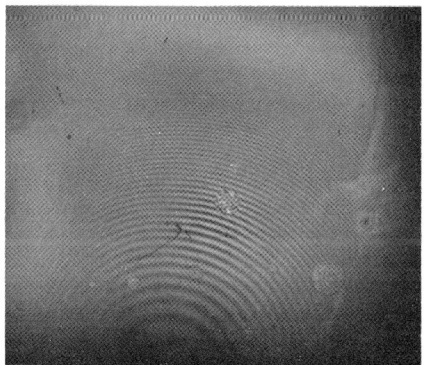

FIG. 13. Interference photograph of a typical mesa on a epitaxial garnet film produced by a droplet of melt left on the film as it was removed from the melt.

the average thickness of the film. During cooling of such a droplet, first the garnet precipitates out followed by the flux along with other oxides which did not deposit epitaxially. Washing the film in an actetic acid solution dissolves the PbO, and the remaining precipitates fall away except for those pinned by the growing mesa. A typical mesa along with a precipitate in its center is shown in Fig. 13.

VII. Summary and Conclusions

Magnetic bubble technology is one of the forerunners among various new technologies for the information storage industry. Its low cost and relatively high performance make it a viable candidate for conventional disc storage. It is one of the solid state alternatives for the conventional mechanical storage areas. The demands placed on materials by this potential technology have generated a considerable amount of basic research. In this article we have sketched out some of the basic magnetic requirements for a bubble material. We have discussed some of the trade-offs that have to be made in order to arrive at optimum conditions dictated by device considerations.

Using this as a background we have considered two classes of materials—the amorphous Gd–Co alloys and the single-crystal cubic garnets—in some detail. These two apparently dissimilar materials are currently the strongest candidates in magnetic bubble technology. The selection, growth, and properties of these materials are described in some detail. We conclude that the selection and growth of these two class of materials is reasonably well understood. However, the static and dynamic magnetic properties of these materials are still incompletely understood. In particular the dynamic properties and origin of magnetic anisotropy are areas in which much remains to be done. A very brief review of this field is presented, and it highlights the lack of clear-cut understanding of the origin of anisotropy in amorphous materials and the many suggestions put forward to explain anisotropy in cubic garnets. We have also reviewed the origin and elimination of defects in these two classes of materials. We are generally optimistic that in this area the few remaining defects can be eliminated. This view may change as the demand for larger substrates increases. However, it is our opinion that there is no intrinsic reason why defects cannot be eliminated in the larger diameter substrates.

References

Akselrad, A., and Callen, H. (1971). *Appl. Phys. Lett.* **19**, 464.
Argyle, B. E., and Chaudhari, P. (1973). *In* "Magnetism and Magnetic Materials" (C. D. Graham, Jr. and J. J. Rhyne, eds.), AIP Conf. Proc. No. 10, p. 403. Am. Inst. Phys., New York.
Besser, P. J., Mee, J. E., Glass, H. L., Heinz, D. M., Austerman, S. B., Elkins, P. E., Hamilton, T. N., and Whitcomb, E. C. (1972). *In* "Magnetism and Magnetic Materials" (C. D. Graham, Jr. and J. J. Rhyne, eds.), AIP Conf. Proc. No. 5, pp. 125–129. Am. Inst. Phys., New York.
Blank, S. L., and Nielsen, J. W. (1972). *J. Cryst. Growth* **17**, 302.
Blank, S. L., Hewitt, B. S., Shick, L. K., and Nielsen, J. W. (1973). *In* "Magnetism and Magnetic Materials" (C. D. Graham, Jr. and J. J. Rhyne, eds.), AIP Conf. Proc. No. 10, pp. 256–270. Am. Inst. Phys., New York.
Bobeck, A. H. (1967). *Bell Syst. Tech. J.* **46**, 1901.
Bobeck, A. H., and Scovil, H. E. D. (1971). *Sci. Amer.* **224**, 78.
Bobeck, A. H., Fischer, R. F., Perneski, A. J., Remeika, J. P., and Van Uitert, L. G. (1969). *IEEE Trans. Magn.* **5**, 544.
Bobeck, A. H., Spencer, E. G., Van Uitert, L. G., Abrahams, S. C., Barns, R. L., Grodkiewicz, W. H., Sherwood, R. C., Schmidt, P. H., Smith, D. H., and Walters, E. M. (1970). *Appl. Phys. Lett.* **17**, 131.
Bobeck, A. H., Blank, S. L., and Levinstein, H. J. (1972). *Bell Syst. Tech. J.* **51**, 1431.
Bonner, W. A., Geusic, J. E., Smith, D. H., Van Uitert, L. G., and Vella-Coleiro, G. P. (1973). *Mater. Res. Bull.* **8**, 1223.
Brandle, C. D., and Valentino, A. J. (1972). *J. Cryst. Growth* **12**, 3.
Brandle, C. D., Miller, D. C., and Nielsen, J. W. (1972). *J. Cryst. Growth* **12**, 195.
Brochier, A., Coeure, P., Ferrand, B., Gay, J. C., Joubert, J. C., Mareschal, J., Vigue, J. C., Martin-Binachon, J. C., and Spitz, J. (1972). *J. Cryst. Growth* **13/14**, 571.
Calhoun, B. A., Giess, E. A., and Rosier, L. L. (1971). *Appl. Phys. Lett.* **18**, 287.
Callen, H. (1972). *In* "Magnetism and Magnetic Materials" (C. D. Graham, Jr. and J. J. Rhyne, eds.), AIP Conf. Proc. No. 5, p. 71. Am. Inst. Phys., New York.
Cargill, G. S., Gambino, R. J., and Cuomo, J. J. (1974). *IEEE Trans. Magn.* **10**, 803.

Chang, H., Chen, T. C., and Tung, C. (1973). *Natl. Comput. Conf., AFIPS Conf. Proc.* **42**, 413.
Chaudhari, P. (1972). *IEEE Trans. Magn.* **8**, 333.
Chaudhari, P., Cuomo, J. J., and Gambino, R. J. (1973a). *IBM J. Res. Dev.* **17**, 66.
Chaudhari, P., Cuomo, J. J., and Gambino, R. J. (1973b). *Appl. Phys. Lett.* **22**, 337.
Chopra, K. L. (1969). "Thin Film Phenomena." McGraw-Hill, New York.
Cowher, M. E., Sedgwick, T. O., and Landermann, J. (1974). *J. Electron. Mater.* **3**, 621.
Cronemeyer, D. C. (1974). *In* "Magnetism and Magnetic Materials" (C. D. Graham, Jr. and J. J. Rhyne, eds.), AIP Conf. Proc. No. 18, pp. 85–89. Am. Inst. Phys., New York.
Cuomo, J. J., Gambino, R. J., and Rosenberg, R. (1974a). *J. Vac. Sci. Technol.* **11**, 34.
Cuomo, J. J., Chaudhari, P., and Gambino, R. J. (1974b). *J. Electron. Mater.* **3**, 517.
de Jonge, F. A., and Druyvesteyn, W. F. (1972). *Festkoerperprobleme* **12**, 531.
Gambino, R. J., Chaudhari, P., and Cuomo, J. J. (1974). *In* "Magnetism and Magnetic Materials" (C. D. Graham, Jr. and J. J. Rhyne, eds.), AIP Conf. Proc. No. 18, pp. 578–592, Am. Inst. Phys., New York.
Geller, S. (1967). *Z. Kristallogr.* **125**, 1.
Ghez, R., and Giess, E. A. (1973). *Mater. Res. Bull.* **8**, 31.
Ghez, R., and Giess, E. A. (1974). *J. Cryst. Growth* **27**, 221.
Gianola, U. F., Smith, D. H., Thiele, A. A., and Van Uitert, L. G. (1969). *IEEE Trans. Magn.* **5**, 558.
Giess, E. A., and Cronemeyer, D. C. (1973). *Appl. Phys. Lett.* **22**, 601.
Giess, E. A., and Ghez, R. (1974). *In* "Epitaxial Growth" (J. W. Matthews, ed.), pp. 183–213. Academic Press, New York.
Giess, E. A., Argyle, B. E., Calhoun, B. A., Cronemeyer, D. C., Klokholm, E., McGuire, T. R., and Plaskett, T. S. (1971). *Mater. Res. Bull.* **6**, 1141.
Giess, E. A., Kuptsis, J. D., and White, E. A. D. (1972a). *J. Cryst. Growth* **16**, 36.
Giess, E. A., Argyle, B. E., Cronemeyer, D. C., Klokholm, E., McGuire, T. R., O'Kane, D. F., Plaskett, T. S., and Sadagopan, V. (1972b). *In* "Magnetism and Magnetic Materials" (C. D. Graham, Jr. and J. J. Rhyne, eds.), AIP. Conf. Proc. No. 5, pp. 110–114. Am. Inst. Phys., New York.
Giess, E. A., Guerci, C. F., Kuptsis, J. D., and Hu, H. L. (1973a). *Mater. Res. Bull.* **8**, 1061.
Giess, E. A., Cronemeyer, D. C., Ghez, R., Klokholm, E., and Kuptsis, J. D. (1973b). *J. Amer. Ceram. Soc.* **56**, 593.
Goodenough, J. B., and Longo, J. M. (1970). *In* "Landolt-Börnstein New Series" (K.-H. Hellwege, ed.), Part a, pp. 126–275. Springer-Verlag, Berlin and New York.
Gyorgy, E. M., Sturge, M. D., Van Uitert, L. G., Heilner, E. J., and Grodkiewicz, W. H. (1973). *J. Appl. Phys.* **44**, 438.
Hagedorn, F. (1974). *In* "Magnetism and Magnetic Materials" (C. D. Graham, Jr. and J. J. Rhyne, eds.), AIP Conf. Proc. No. 18, pp. 222–226. Am. Inst. Phys., New York.
Heilner, E. J., and Grodkiewicz, W. H. (1973). *J. Appl. Phys.* **44**, 4218.
Herd, S. R., and Chaudhari, P. (1973). *Phys. Status Solidi A* **18**, 603.
Hewitt, B. S., Pierce, R. D., Blank, S. L., and Knight, S. (1973). *IEEE Trans. Magn.* **9**, 366.
Hu, H. L., Hatzakis, M., Giess, E. A., and Plaskett, T. S. (1973). *Intermagn. Conf., Washington, D.C.* Pap. No. 26-5.
Huber, D. L. (1970). *In* "Landolt-Börnstein New Series" (K.-H. Hellwege, ed.), Part a, pp. 315–360. Springer-Verlag, Berlin and New York.
Iida, S. (1967). *J. Phys. Soc. Jpn.* **22**, 1201.
Kooy, C., and Enz, W. (1960). *Philips Res. Rep.* **15**, 7.
Kryder, M. H., Ahn, K. Y., Almasi, G. S., Keefe, G. E., and Powers, J. V. (1974). *IEEE Trans. Magn.* **10**, 825.

Kurtzig, A. J., and Hagedorn, F. B. (1971). *IEEE Trans. Magn.* **7**, 473.
Landau, L., and Lifshitz, E. (1935). *Phys. Z. Sowjetunion* **8**, 153.
Levinstein, H. J., Licht, S., Landorf, R. W., and Blank, S. L. (1971). *Appl. Phys. Lett.* **19**, 486.
Lin, Y. S., and Keefe, G. E. (1973). *Appl. Phys. Lett.* **22**, 603.
Linares, R. C. (1964). *Solid State Commun.* **2**, 229.
Maissel, L. I., and Glang, R. (1970). "Handbook of Thin Film Technology." McGraw-Hill, New York.
Malozemoff, A. P. (1972). *Appl. Phys. Lett.* **21**, 149.
Matthews, J. W., Klokholm, E., Sadagopan, V., Plaskett, T. S., and Mendel, E. (1973). *Acta Metall.* **21**, 203.
Mee, J. E., Pulliam, G. R., Archer, J. L., and Besser, P. J. (1969). *IEEE Trans. Magn.* **5**, 717.
Nesbitt, E. A., and Wernick, J. H. (1973). "Rare Earth Permanent Magnets." Academic Press, New York.
Nielsen, J. W., and Dearborn, E. F. (1958). *J. Phys. Chem. Solids* **5**, 202.
O'Kane, D. F., Sadagopan, V., Giess, E. A., and Mendel, E. (1973). *J. Electrochem. Soc.* **120**, 1272.
Orehotsky, J., and Schroeder, K. (1972). *J. Appl. Phys.* **43**, 2413.
Plaskett, T. S., Klokholm, E., and Cronemeyer, D. C. (1974). *In* "Magnetism and Magnetic Materials" (C. D. Graham, Jr. and J. J. Rhyne, eds.), AIP Conf. Proc. No. 18, pp. 75–79. Am. Inst. Phys., New York.
Rhyne, J. J., Pickart, S. J., and Alperin, H. A. (1972). *Phys. Rev. Lett.* **29**, 1562.
Robertson, J. M., Wittekoek, S., Popma, T. J. A., and Bongers, P. F. (1973). *Appl. Phys.* **2**, 219.
Robinson, McD. (1973). *J. Cryst. Growth* **18**, 143.
Rode, D. L. (1973). *J. Cryst. Growth* **20**, 13.
Rosencwaig, A., and Tabor, W. J. (1972). *In* "Magnetism and Magnetic Materials" (C. D. Graham, Jr. and J. H. Rhyne, eds.), AIP Conf. Proc. No. 5, p. 57. Am. Inst. Phys., New York.
Rossol, F. C. (1970). *Phys. Rev. Lett.* **24**, 1021.
Sandfort, R. M., and Burke, E. R. (1971). *IEEE Trans. Magn.* **7**, 358.
Sawatzky, E., and Kay, E. (1968). *J. Appl. Phys.* **39**, 4700.
Sawatzky, E., and Kay, E. (1969a). *J. Appl. Phys.* **40**, 1462.
Sawatzky, E., and Kay, E. (1969b). *IBM J. Res. Dev.* **13**, 696.
Slonczewski, J. C. (1972). *Int. J. Magn.* **2**, 85.
Slonczewski, J. C. (1973). *J. Appl. Phys.* **44**, 1759.
Slonczewski, J. C., Malozemoff, A. P., and Voegeli, O. (1973). *In* "Magnetism and Magnetic Materials" (C. D. Graham, Jr. and J. J. Rhyne, eds.), AIP Conf. Proc. No. 10, p. 458. Am. Inst. Phys., New York.
Smith, A. B., Kestigian, M., and Bekebrede, W. R. (1974). *In* "Magnetism and Magnetic Materials" (C. D. Graham, Jr. and J. J. Rhyne, eds.), AIP Conf. Proc. No. 18, pp. 167–171. Am. Inst. Phys., New York.
Stacy, W. T., and Rooymans, C. J. M. (1971). *Solid State Commun.* **9**, 2005.
Stein, B. F. (1970). *J. Appl. Phys.* **41**, 1262.
Sturge, M. D., LeCraw, R. C., Pierce, R. D., Licht, S. J., and Shick, L. K. (1973). *Phys. Rev. B* **7**, 1070.
Tabor, W. J., Bobeck, A. H., Vella-Coleiro, G. P., and Rosencwaig, A. (1972). *Bell Syst. Tech. J.* **51**, 1427.
Takahashi, M., Nishida, H., Kobayashi, T., and Sugita, Y. (1973). *J. Phys. Soc. Jpn.* **34**, 1416.
Tao, L. J., Gambino, R. J., Kirkpatrick, S., Cuomo, J. J., and Lilienthal, H. (1974). *In* "Magnetism and Magnetic Materials" (C. D. Graham, Jr. and J. J. Rhyne, eds.), AIP Conf. Proc. No. 18, pp. 641–645. Am. Inst. Phys., New York.

Tarng, M. L., and Wehner, G. K. (1971). *J. Appl. Phys.* **42**, 2449.
Taylor, R. C., and Sadagopan, V (1971). *Appl. Phys. Lett.* **19**, 361.
Thiele, A. A. (1970). *J. Appl. Phys.* **41**, 1139.
Tolksdorf, W., Bartels, G., Espinosa, G. P., Holst, P., Mateika, D., and Welz, F. (1972). *J. Cryst. Growth* **17**, 322.
Van Uitert, L. G., Smith, D. H., Bonner, W. A., Grodkiewicz, W. H., and Zydzik, G. J. (1970). *Mater. Res. Bull.* **5**, 455.
Vella Coleiro, G. P., and Tabor, W. J. (1972). *Appl. Phys. Lett.* **21**, 7.
Vella Coleiro, G. P., Smith, D. H., and Van Uitert, L. G. (1972). *Appl. Phys. Lett.* **21**, 36.
Wijn, H. P. J. (1970). *In* "Landolt-Börnstein New Series" (K.-H. Hellwege, ed.), Part b, pp. 547–647. Springer-Verlag, Berlin and New York.
Williams, C. M., and Schindler, A. I. (1973). *J. Appl. Phys.* **44**, 5575.
Winters, H. F., Raimondi, D. L., and Horne, D. E. (1969). *J. Appl Phys.* **40**, 2996.
Wolfe, R., and North, J. C. (1972). *Bell Syst. Tech. J.* **51**, 1436.
Wolfe, R., North, J. C., and Lai, Y. P. (1973). *Appl. Phys. Lett.* **22**, 683.

Author Index

Numbers in parentheses are reference numbers and indicate that an author's work is referred to although his name is not cited in the text. Numbers in italics show the page on which the complete reference is listed.

A

Abelès, F., 13, 66, 75, 76(12), 78(12), *143*, 240(232a), *261*
Aboaf, J. A., 31(114), 35(114), *67*
Abrahams, S. C., 264, 267, 276, 287, *294*
Ackerman, P. W., 33(118), *67*
Aharoni, S. M., 8(38), *65*
Ahilea, E., 5(19), *65*
Ahn, K. Y., 266, *295*
Airy, G. B., 16, *66*
Aitchison, R. E., 9, 31(43), 33(43), 35(43), 36(43), *65*
Akselrad, A., 288, *294*
Alexander, F. B., 50(228), *70*
Alexander, R. W., 237(223), *261*
Alexopoulos, K., 219(176), *259*
Alibert, C., 22(93), *67*
Allen, E. A., 5(22), *65*
Allen, F. G., 47(223), *70*
Almasi, G. S., 266, *295*
Alperin, H. A., 273, 274, *296*
Alsford, R. W., 5(17), *65*
Amans, R. L., 59(266), *71*
Andelin, J. P., Jr., 136(50), *144*
Anderegg, M., 193(93), *257*
Anderson, G. S., 50(226), *70*
Antypas, G. A., 46(206), *70*
Apfel, J. H., 105(20), 119(29), 132(39), 143(54), *144*
Arai, T., 34(157, 161), 35(157), *68*
Arakawa, E. T., 176(40, 41, 42, 43, 44, 45), 178(41), 189(79), 220(180), 221(183), 225(191), 226(191), 227(205), 228(205), 229(205), 237(222), 243(240), *256, 257, 260, 261*
Archer, J. L., 282, 287, *296*
Argyle, B. E., 275, 288, 290, 292, *294, 295*
Arvin, M. J., 28(106), *67*
Ashley, J. C., 167(27), 174(34), 179(54), *256, 257*

Austerman, S. B., 276, *294*
Axelrod, N. N., 209(132), *258*

B

Bachner, F. J., 116(26), *144*
Bädeker, K., 1, *64*
Baer, Y., 217(166), *259*
Bagchi, A., 203(121), *258*
Barker, A. S., Jr., 154(12), 236(218), 238 (226, 227), *256, 261*
Barns, R. L., 264, 267, 276, 287, *294*
Bartels, G., 276, 284, *297*
Bartholomew, R. F., 33(120), *68*
Bassett, G. A., 4(9, 10), *65*
Bates, B., 132(40), *144*
Bauer, E., 215(153), 217(153), *259*
Baumeister, P., 74(9, 10), 79(10), 82(10), 86(10), 102(10), 123(9), 124(9), 127(9), 141(53), 142(53), *143*(53), *144*
Beaglehole, D., 226(193), *260*
Bekebrede, W. R., 270, *296*
Bell, R. J., 237(223), *261*
Bennet, A. J., 159(15), *256*
Bennett, H. E., 21, *67*, 227(201), *260*
Bennett, J. M., 21, *67*
Bernert, B., 188, *257*
Berning, P. H., 16(75), *66*, 74, 75, 76(11), 78(11), 79(8), 82, 91(8), 95, 116, *143*
Besser, P. J., 276, 282, 287, *294, 296*
Binias, G., 178(51), *257*
Bird, V. M., 227(197), *260*
Birkhoff, R. D., 176(40, 43, 44, 45), *256, 257*
Bjork, R. H., 244, *261*
Blandenet, G., 39, *69*
Blank, S. L., 271, 276, 283, 284, 288, 292, *294, 295, 296*
Bloch, A. N., 240(233), *261*
Blocher, J. M., 31(111), *67*
Boardman, A. D., 159(17), *256*

Bobeck, A. H., 264, 265, 267, 271, 275, 276, 287, *294*, *296*
Bodesheim, J., 254(251), *261*
Boersch, H., 174, 219(177), 220(179), *256*, *259*
Boicy, J. H., 32(116), 33(116), *67*
Boles, J., 133(45), 143(45), *144*
Bolland, T. K., 194(97), *258*
Bongors, P. F., 276, *296*
Bonner, W. A., 275, 276, *294*, *297*
Bonnet, M., 51(252), 52, 62(252), *71*
Borburgh, J. C. M., 244(243), *261*
Bordure, G., 22(93), *67*
Borrelli, N. F., 22(97, 98, 99), 23(97, 98, 99), 26, *67*
Bos, J. G., 33(134), *68*
Bösenberg, J., 186(67), 187(67), 190(67), 197(104, 105), 236(220, 221), 237(221), 240(221), 241(220, 221), *257*, *258*, *261*
Bosnell, J. R., 56(261), *71*
Boyd, D. R., 51(248, 249), *71*
Bradley, D. J., 132(40), *144*
Bradshaw, A. M., 217(167, 168), *259*
Brambring, J., 186(65, 66), 195(98), *257*, *258*
Brandle, C. D., 281, *294*
Bratton, R. J., 28(105a), *67*
Braundmeier, A. J., Jr., 176(41, 42), 178(41), 221(183), 243(240), 254(253), *256*, *260*, *261*
Brochier, A., 282, *294*
Broida, H. P., 198(111), *258*
Brown, W. R., 176, *256*
Bruns, R., 153(9), 242(9), *255*
Buck, T. M., 47(223), *70*
Bueche, F., 8(39), *65*
Burke, E. R., 265, *296*
Bürker, U., 179(52), 220, *257*, *260*
Burstein, E., 154(11), 240(232), *256*, *261*
Busch, G., 217(166), *259*
Bylander, E. G., 47(224), *70*

C

Calhoun, B. A., 275, *294*, *295*
Callen, H., 288, *294*
Callott, T. A., 217(162), *259*
Cambel, A. B., 46(204), *70*
Campbell, D. S., 7(34), 8, 31, *65*, *67*
Canfield, L. R., 227(195), *260*

Cargill, G. S., 277, *294*
Carlson, D. E., 12(61), 33(60, 151, 152), 60(273), *66*, *68*, *71*
Cawthorne, C., 167(24), *256*
Celli, V., 245(245), *261*
Chang, H., 265, *295*
Chase, W. R., 45(199), 46(199), *69*
Chaudhari, P., 271, 273, 275, 278, 280, 284, 285, 290, 291, 292, *294*, *295*
Chaurasia, H. K., 43(195), 44, *69*
Chen, T. C., 265, *295*
Chen, W. P., 240(232), *261*
Chen, Y. J., 240(232), *261*
Chernyaev, V. N., 39(169), 40(169), *69*
Chiba, K., 47(222a), *70*
Chopra, K. L., 4(12, 13), 5(15, 16), 6(12, 23), 42(181), *65*, *69*, 277, *295*
Chou, N. J., 31(114), 35(114), *67*
Chow, S. L., 6(24), *65*
Christy, R. W., 236(219), *261*
Coeure, P., 282, *294*
Colbert, W. H., 43(189, 192), *69*
Conwell, E. M., 240(234), *261*
Cook, H. D., 56(260), 57, 58, 62(260), *71*
Cornely, R., 6(25), 43(25), *65*
Costich, V. R., 98(14), 110(24), 111(24), 123 (14), 124(14), 135(14), 136(46), *143*, *144*
Cowan, J. J., 225(191), 226(191), *260*
Cowher, M. E., 282, *295*
Cox, J. T., 19(78), *66*
Cox, R. E. L., 45, *69*
Cram, L. S., 220(180), *260*
Creuzburg, M., 210, 212, 214(146), 218(174), *259*
Cromer, D. T., 12(56), *66*
Cronemeyer, D. C., 275, 280, 281, 287, 288, *295*, *296*
Crowell, J., 198, 219(175), 229(209), *258*, *259*, *260*
Cunningham, R. D., 22(90), 23(90), 27, *67*
Cuomo, J. J., 51(242), 54(242), *70*, 271, 273, 274, 277, 278, 280, 282, 285, *294*, *295*, *296*

D

Daniels, J., 202(116), 205(129), 206, 214(147), *258*, *259*
Daude, A., 178(60), 227(204), *257*, *260*

AUTHOR INDEX 301

Davis, N. O., 176(43), *256*
Dearborn, E. F., 275, *296*
Deitch, R. H., 54, *71*
de Jonge, F. A., 266, *295*
Dennison, B. J., 12(58), 33(58, 130), *66*, *68*
Dennison, D., 74, 132(12), *143*
Dent Glasser, L. S., 11(52), *66*
DeWit, J. H. W., 28, *67*
DiNardo, R. P., 179(55), *257*
Dobberstein, P., 174(32), 220(179, 181), 221 (185, 187), 227(200), 229, *256*, *259*, *260*
Dockerty, R. C., 22(86), 23(86), 24, 25, *67*
Doi, K., 47(218), 48(218). *70*
Donaldson, E. E., 46(206), *70*
Doremus, R. H., 15(66, 67), *66*, 198(109), *258*
Dorn, R., 217(160), *259*
Doyle, W. T., 198, *258*
Druyvesteyn, W. F., 266, *295*
DuBois, A., 56(264), 58(264), 62(264), *71*
Dufour, C., 74, 132(5), *143*
Duke, C. B., 203(121, 123, 123a), *258*
Dunstädter, H., 50(232), *70*
Duthler, C. J., 198(111), *258*

E

Eastman, D. E., 217(169), *259*
Economu, E. N., 146(4a), *255*
Eguiluz, A., 159(16), *256*
Ejiri, A., 186(68), 187(70), *257*
Eldrige, H. B., 174(33), *256*
Elkins, P. E., 276, *294*
Elliott, D., 35(165), 36(165), 37(165), *69*
Elson, J. M., 154(11), 245(245), *256*, *261*
Emeric, A., 190(85), *257*
Emeric, N., 190(85), *257*
Emerson, L. C., 167(27), 176(44), 179(54), *256*, *257*
Endriz, J. G., 227(203), 228(203), *260*
Ennos, A. E., 43(185), *69*
Enz, W., 264, *295*
Espinosa, G. P., 276, 284, *297*
Evans, J. H., 167(25), *256*

F

Faith, W. N., 203(122), 215(152), *258*, *259*
Fan, J. C. C., 116(26), *144*

Fano, U., 227, *260*
Feibelman, P. J., 159(17, 18), 203(121), *256*, *258*
Feist, W. M., 31, 36(113), 37(113), 39(113), *67*
Fellenzer, H., 202(141), *259*
Felt, R. M., 32(116), 33(116), *67*
Ferrand, B., 282, *294*
Ferrell, R. A., 146(2), 150(2), 173, 210(133), 255, *256*, *258*
Feshbach, H., 84(16), *143*
Fetisow, E. P., 192(87), *257*
Feuerbacher, B., 178(80), 186(69), 187(69), 188(77), 189(78), 190(80), 193(93), 227(202), *257*, *260*
Fine, J., 46(208), *70*
Fischer, A., 33(131), 35(131), 36(131), 37, *68*
Fischer, B., 238(228, 229), *261*
Fischer, R. F., 275, *294*
Fistul, V. I., 51(243, 244), *70*, *71*
Fitten, B., 178(80), 190(80), 193(93), *257*
Foley, G. H., 116(26), *144*
Fonstad, C. G., 27, *67*
Forstmann, F., 192(89), *257*
Frank, I., 174, *256*
Fraser, D. B., 56(260), 57, 58, 62(260), *71*
Fridrich, E. G., 59(267), *71*
Fritzsche, D., 174(32), 219(177), 220(179), *256*, *259*
Fuchs, R., 159, 162(19), 163(19), 167(26), 192(91), 198, 255, *256*, *257*, *261*
Fujimoto, F., 166(21), 218(21, 173), *256*, *259*
Fujiwara, K., 60(270), *71*
Fulton, E. J., 167(24), *256*
Furman, S. A., 136(49), *144*
Furukawa, Y., 35(163a), *68*
Furuuchi, D., 47(221), 48(221), 62(221), *70*
Fuschillo, N., 6(25), 45(25), *65*

G

Gaiser, R. A., 33(122), *68*
Gallaway, W. S., 227(197), *260*
Gallus, G., 20(80a), *67*
Gambino, R. J., 51(242), 54(242), *70*, 271, 273, 274, 277, 278, 280, 282, 285, *294*, *295*, *296*
Garfinkel, H. M., 33(120), *68*
Gay, J. C., 282, *294*
Geffcken, W., 74, 132(1), 141, *143*, *144*

Geiger, J., 204(127), 210, 212(134), *258*
Gelber, R. M., 116(27), 132(39), *144*
Geller, S., 275, 276, *295*
Georgescu, V., 33(156), 34(156). *68*
Gerasimova, L., 51(247), *71*
Gesell, T. F., 189(79), 227(205), 228(205), 229, 257, *260*
Geusic, J. E., 276, *294*
Ghez, R., 283, 284, *295*
Giallorenzi, T. G., 54(256), *71*
Gianola, U. F., 275, *295*
Giess, E. A., 264, 275, 276, 281, 283, 284, 287, 288, *294*, *295*, *296*
Gilham, E. J., 9, 43(188), 62(40, 41), *65*, *69*
Gillery, F. H., 50(229), *70*
Gilles, J. F., 42(179), *69*
Ginzburg, W., 174, *256*
Glang, R., 42, *69*, 133, *144*, 277, *296*
Glass, H. L., 276, *294*
Godwin, R. P., 188(77), *257*
Goland, A. W., 179(55), *257*
Golay, M. J. E., 129(35), *144*
Golovcenco, I., 33(137), *68*
Gomer, R., 34(158), *68*
Goodenough, J. B., 275, *295*
Goodman, J., 60(272), *71*
Goodman, L. A., 3(4), *64*
Gress, R. W., 31(115), *67*
Griffin, A., 159(17), *256*
Grodkiewicz, W. H., 264, 267, 275, 276, 287, 288, *294*, *295*, *297*
Groth, R., 33(141), 35, 37, 62(141), *68*
Guerci, C. F., 276, *295*
Gundlach, K. H., 218(171), *259*
Gurney, R. W., 9, *65*
Gyorgy, E. M., 288, *295*

H

Haacke, G., 58, 59(265a), 62(265a), *71*
Haaijman, P. W., 9(44), *65*
Haayman, P. W., 33(138), 35, *68*
Hadley, L., 13, *66*, 74, 100, 102(18), 113, 115, 132(2), *143*
Hagedorn, F. B., 270, 287, *295*, *296*
Hagström, S. B. M., 188(75), *257*
Hahn, R., 141(52, 53), 142(53), 143(52, 53), *144*

Halherbe, A., 132(42), *144*
Hall, D. G., 254(253), *261*
Hamakawa, Y., 35(163b), *68*
Hamilton, T. N., 276, *294*
Hamm, R. N., 225(191), 226(191), 227(205), 228(205), 229(205), 237(222), *260*, *261*
Hammer, D. C., 176(44), *256*
Hampe, A., 227(200), *260*
Hang, K. W., 12(60), 33(60, 152), *66*, *68*
Hannay, N. B., 30(110), *67*
Hanson, W. F., 176(40), *256*
Harris, J., 159(17), *256*
Harrison, D. H., 132(41), 141(53), 142(53), 143(41, 53), *144*
Harrison, M. J., 192(90), 193, *257*
Hartman, T. E., 7(31), *65*
Hartstein, A., 240(232), *261*
Harvey, G. A., 14, *66*
Hass, G., 13, 19(78), *66*, 74, 100, 102(18), 113, 115, 129(6), 130, 131, *143*, *144*, 227(195), *260*
Hattendorff, H. D., 176(46), 177(46), 178(46), 179(46), *257*
Hattori, M., 182(63), 186(63), *257*
Hatzakis, M., 264, *295*
Hayes, R. E., 5(17), *65*
Heavens, O. S., 16(74, 76), 17(74), *66*
Hecq, M., 50, 56(264), 58, 62(264), *70*, *71*
Hedgecock, N. E., 6(24), *65*
Heilner, E. J., 288, *295*
Heinz, D. M., 276, *294*
Heitmann, D., 179(56), 180(58), 221(189), 223(189a), *257*, *260*
Helwig, G., 50(231), *70*
Herd, S. R., 4(14), 5(14), *65*, 275, 285, *295*
Herickhoff, R. J., 176(40, 45), *256*, *257*
Herman, E. S., 6, 7(30), *65*
Hermansen, A., 74, *143*
Hewitt, B. S., 276, 283, 284, 288, 292, *294*, *295*
Hill, R. M., 6(27, 28), *65*
Hirsch, A. A., 5(19), *65*
Hoenig, S. A., 45(203), *69*
Hoffman, D. M., 43(193), 44, 45(201), 46(201), 47(197), 48(197), 49(197), *69*
Hoffmann, I, 190(81), 195(101), 197(101), *257*, *258*
Hoffmann, K. W., 218(172), *259*
Holland, L., 2, 43(186), 47(212), 50, 51(186), 62(186), *64*, *69*, *70*, *114*, *144*

Holst, K., 239(231), 241(231), 242(231), *261*
Holst, P., 276, 284, *297*
Holzl, J., 218(172), *259*
Honcia, G., 119(30), *144*
Honig, R. E., 47(222), *70*
Hori, Y., 47(218), 48(218), *70*
Horn, K., 217(160), *259*
Hornauer, D., 241(236), 242, 243(239), 244, 254(252), *261*
Horne, D. E., 278, *297*
Horsley, B., 33(132), 35, 37, *68*
Horstmann, C., 252(247a), *261*
Houston, J. E., 22(87), 23(87), *67*
Hu, H. L., 264, 276, *295*
Huang, F. H., 50(225), *70*
Huber, D. L., 275, *295*
Hunderi, O., 190(86), *257*
Huntington, H. B., 50(225), *70*
Hutley, M. C., 227(197), *260*

I

Ibach, H., 215, 216, 217(156, 157, 160), *259*
Iida, S., 287, *295*
Ikuzo, I., 47(218), 48(218), *70*
Imai, I., 34(157), 35(157, 164), 36(164), 37(164), *68*, *69*
Isaki, L., 217(158), *259*
Ishida, K., 218(173), *259*
Ishiguro, K., 34(157), 35(157), *68*
Ito, H., 49(224a), *70*

J

Jacquemin, J. L., 22(93), *67*
Jaffe, M. S., 59(267), 60(268), *71*
Jandus, J., 47(216), *70*
Janssen, G. H., 33(138), 35(138), *68*
Jarzebski, Z. M., 30, *67*
Jasperson, S. N., 227(199), 228(199), *260*
Jedlicka, E., 47(216), *70*
Jezequel, G., 178(60), *257*
Johnson, P. B., 236(219), *261*
Johnson, S. E., 198(111), *258*
Jones, H., 7(36), *65*
Jordan, L. K., 215(150), *259*

Jost, R., 37(165a), *69*
Joubert, J. C., 282, *294*
Junge, A. E., 33(145, 147, 150), *68*
Juranek, H. J., 245(245), *261*

K

Kajiyama, K., 35(163a), *68*
Kanazawa, H, 158(14a), *256*
Kane, J., 39(171, 172), 40, 41(170, 171, 172, 174), 62(171, 172), *69*
Kapitza, H., 237(225), 238(225), 243(225, 239), 244(239), *261*
Karakashian, A. S., 244(242), *261*
Kasprzak, L., 4(14), 5(14), *65*
Kaw, P., 15(69), *66*
Kay, E., 282, *296*
Keefe, G. E., 266, 271, *295*, *296*
Kennedy, D. I., 5(17), *65*
Kern, W., 39(170, 171, 172), 40(170, 171, 172), 41(170, 171, 172), 62(170, 171, 172), *69*
Kestigian, M., 270, *296*
Kienel, G., 20(80a), *67*
Kim, H., 31(114a), *67*
Kimbara, A., 190(84), *257*
Kirkpatrick, S., 273, 274, *296*
Klemperer, O., 190(82), *257*
Kliewer, K. L., 159, 162(19), 163(19), 192(91), 255, *256*, *257*, *261*
Klokholm, E., 275, 281, 288, 290, *295*, *296*
Kloos, T., 150(7), 201(115), 202(120), 203(120), 206(7), 207(120), 208, 209(7), 255, 258
Knight, S., 283, 292, *295*
Kobayashi, T., 271, *296*
Koch, H., 34(159), 35(159, 163), *68*
Koedam, M., 5(21), *65*
Koffyberg, F. P., 22(94), 30, *67*
Kofstad, P., 9, 10(45), *66*
Kohn, W., 146(6), *255*
Kohnke, E. E., 22(87), 23(87), *67*
Kokkinakis, T., 219(176), *259*
Komaki, K., 166(21), 218(21, 173), *256*, *259*
Komnik, Y. F., 2, *64*
Kondo, R., 60(271), *71*
Koonce, S. E., 51(239), *70*
Kooy, C., 264, *295*
Korzo, V. F., 39, 40, *69*

Köstlin, H., 37, *69*
Kovener, G. S., 237(223), *261*
Kowalczyk, S. P., 217(165, 169), *259*
Kramer, D. A., 47(222), *70*
Krane, K. J., 203(120a), *258*
Krebs, K., 119(30), *144*
Kreibig, U., 190(83), 198(110), 218(110), *257, 258*
Kretschmann, E., 153(10), 155(13), 177(47), 181, 187(71, 73), 196, 197(103), 223(189d), 229(210), 231(47, 215), 233, 235(10), 242 (102), 245(103, 245), 246(212, 246, 247), 247(103, 249), 250(103), 251, 252, 253, 254, *256, 257, 258, 260, 261*
Krishnamurthy, B. S., 15(68), *66*
Kröger, E., 174(35), 176(35), 179, 197(103), 201, 210, 223(189d), 229(210), 245(103), 247(103), 250(103), *256, 258, 259, 260*
Kruglova, A. V., 33(142, 143), *68*
Kryder, M. H., 266, *295*
Kryzhanovskii, B. P., 22(82), 27(82), 33(142, 143, 144), 47(222b), *67, 68, 70*
Kuhnert, R., 178(51), *257*
Kunioka, A., 50(236a), *70*
Kunz, C, 202(119), 204(125, 126), 205(126), 209(126), 211(138, 139), *258, 259*
Kuo, C. Y., 60(269), *71*
Küppers, J., 215, *259*
Kuptsis, J. D., 276, 283, 284, *295*
Kurtzig, A. J., 287, *296*
Kushihashi, A., 60(270), *71*
Kuznetsov, A. Y., 33(142, 143, 153), *68*

L

Ladwig, H., 33(136), *68*
Lagarde, Y., 39(172a), *69*
Lai, Y. P., *297*
Laibowitz, R., 4(14), 5(14), *65*
Laitinen, H. A., 31(114a), 35(165), 36(165), 37(165), *67, 69*
Lakshmanan, T. K., 50(235), *70*
Landau, L., 270, *296*
Landermann, J., 282, *295*
Landmann, U., 203(123), *258*
Landorf, R. W., 283, *296*
Lang, N. D., 146(6), *255*
Langkowski, J., 203(120b), *258*

Lappe, F., 50(233), *70*
Larson, D. C., 7(32), 8(32), *65*
Law, J. T., 47(223), *70*
Le Bas, J., 5(18), *65*
LeCraw, R. C., 288, *296*
Lehmann, H. W., 51, 52, *70*
Leibowitz, D., 45(201), 46(201), *69*
Lely, J. A., 33(134), *68*
Lemberg, H. L., 240(233), *261*
Lems, W., 37(165a), *69*
Levin, E. M., 11(49, 50, 51), 20(51), *66*
Levine, J. D., 46, *70*
Levinstein, H. J., 271, 283, *294, 296*
Ley, L., 217(165, 169), *259*
Ley, R. P., 28(104), 29, 30, *67*
Licht, S. J., 283, 288, *296*
Lichtman, D., 46(207), *70*
Lieberman, M. L., 50, 51(237), *70*
Lifshitz, E., 270, *296*
Lilienfeld, J. E., 174(32), *256*
Lilienthal, H., 273, 274, *296*
Lin, Y. S., 271, *296*
Linares, R. C., 281, *296*
Lindau, I., 188(75), 193(92), 194, *257, 258*
Livesay, R. G., 33(117), *67*
Loch, L. D., 22(84), 27(84), 28, *67*
Longo, J. M., 275, *295*
Lopez-Rios, T., 240(232a), *261*
Lucas, A. A., 167(23), 213(142, 143), *256, 259*
Lucas, M. S. P., 42(180, 182), 43(182), *69*
Ludeke, R., 217(158), *259*
Lüth, H., 217(160), *259*
Lyashenko, S. P., 34(160), 35(160), 36(160), *68*
Lyford, E., 33(117), *67*
Lytle, W. O., 12(59), 33(59, 145, 147, 150), *66, 68*

M

McAlister, A. J., 182(62), 186(62), *257*
McAvley, J. W., 33(140), *68*
McCarthy, S. L., 236(219a), *261*
McCloud, H. A., 20(80), *66*
Macdonald, J., 116, *144*
McFeely, F. R., 217(165, 169), *259*
McGuire, T. R., 275, 288, *295*
McKetta, J. J., Jr., 12(55), *66*

Macleod, H. A., 136, *144*
McMaster, H. A., 33(121), *68*
McMurdie, H. F., 11(49, 50, 51), 20(51), *66*
McPhedran, R. C., 227(198), *260*
Maczek, C., 241(235), *261*
Mader, S., 7(35), *65*
Mahan, G. D., 217(169), *259*
Maier, R. L., 127(31, 32), 143(31, 32), *144*
Maissel, L. I., 4, 7, 42, *64*, *69*, 277, *296*
Makishima, S., 22(85), 23(85), *67*
Malliaris, A., 8(37), *65*
Malozemoff, A. P., 271, *296*
Mar, R. W., 22(95), 27, *67*
Marchal, M., 51(252), 52, 62(252), *71*
Marcotte, V. C., 31(114), 35(114), *67*
Marcuse, D., 254, *261*
Mareschal, J., 282, *294*
Mark, H. F., 12(55), *66*
Marley, J. A., 22(86, 99), 23(86, 99), 24, 25, *67*
Marriott, J. G., 32(116), 33(116), *67*
Marshall, N., 238(228, 229), *261*
Martin, R. J., 169(28), *256*
Martin-Binachon, J. C., 282, *294*
Marton, J. P., 22(90), 23(90), 27(90), *67*
Marvin, A., 245(245), *261*
Mateika, D., 276, 284, *297*
Mathews, J. W., 290, *296*
Matsunami, H., 49(224a), *70*
Mattox, D. M., 53(253b), *71*
Mayer, H., 218(172), *259*
Maystre, D., 227(198), *260*
Meaburn, J., 136(48), *144*
Medrud, R. C., 50, 51(237), *70*
Mee, J. E., 276, 282, 287, *294*, *296*
Mehta, R. R., 50(236), 51(236), 52, *70*
Melnyk, A. R., 192(90), 193, *257*
Mendel, E., 281, 290, *296*
Mendenhall, H. E., 35(162), *68*
Menzel, D., 217(167), *259*
Meyer, D. T., 42, 43(183), *69*
Mie, G., 198(106), *258*
Miller, D. C., 281, *294*
Miller, W. R., Jr., 209(132), *258*
Miloslavskii, V. K., 34(160), 35(160), 36(160), *68*
Minn, S. S., 6(26), 43(184, 187, 191), *65*, *69*
Mochel, J. M., 33(133), *68*
Molzen, W. W., 51(253), 52, 62(253), *71*

Moore, H., 33(117), *67*
Morgan, D. F., 22(88), 27, *67*
Morgan, W. L., 43(189, 190, 192), *69*
Morse, P. M., 84(16), *143*
Moser, F., 15(72), *66*
Mott, N. F., 7(36), 9, *65*
Muller, H. K., 51(245, 246), *71*
Murayama, Y., 53, *71*
Murphy, J. A., 31(115), *67*
Murr, L. E., 5(20), *65*
Myers, H. P., 190(86), *257*

N

Naegele, J., 47(220), *70*
Nagasawa, M., 22(85), 23(85), *67*
Natta, M., 167(22), *256*
Naylor, P. H., 240(233), *261*
Nesbitt, E. A., 273, *296*
Neugebauer, C. A., 4(6, 7), 6, 7(29), 43(194), *65*, *69*
Ngai, K. L., 146(4a), *255*
Nielsen, J. W., 275, 276, 281, 284, 288, *294*, *296*
Nikolaeva, I. N., 51(247), *71*
Nilsson, P. O., 188(75), 193(92), 194, *257*, *258*
Nishida, H., 271, *296*
Nishino, T., 35(163b), *68*
North, J. C., 271, *297*
Novikov, V. M., 33(119), *68*
Nozik, A. J., 58, *71*

O

Oelhafen, P., 217(163), *259*
Offret, S., 43(184, 187), *69*
Ohring, M., 4(14), 5(14), *65*
O'Kane, D. F., 281, 288, *291*, *296*
Okaniwa, H., 47(222a), *70*
Okatov, M. A., 33(144), *68*
O'Neill, J. J., Jr., 46(210, 211), 50(227), 54 (227), *70*
Oo, K., 49(224a), *70*
Oohashi, T., 54, *71*
Orehotsky, J., 273, *296*
Orel, E. N., 47(222b), *70*
Orlowski, R., 237(224), 245(224), *261*
Oswald, R. G., 217(162), *259*

Othmer, D. F., 12(55), 66
Otto, A., 202(141), 206(131), 210, 212(135, 140), 230(213, 214), 234(217), 241(235, 237), 254(251), 258, 259, 260, 261
Oxley, J. H., 31(111), 67

P

Pafomov, V. E., 159, 174(20), 256
Palatnik, L. S., 2, 64
Pankratz, J. M., 51(251), 52, 62(251), 71
Paranjape, B. V., 159(17), 256
Paranjape, V. V., 15(68), 66
Pardee, W. J., 217(169), 259
Pashley, D. W., 4(8), 65
Pazdzerskii, V. A., 15(70), 66
Peaker, A. R., 33(132), 35, 37, 68
Perneski, A. J., 275, 294
Peters, F. G., 51(239), 70
Peterson, A., 22(89), 27(89), 67
Peterson, R. E., 130(38), 144
Petit, R., 223(189c), 260
Pettit, R. B., 202(117), 212(117), 258
Philpott, M. R., 240(234a), 261
Pickart, S. J., 273, 274, 296
Piedmont, J. R., 47(224), 70
Pierce, R. D., 283, 288, 292, 295, 296
Plaskett, T. S., 264, 275, 281, 288, 290, 295, 296
Pockrand, I., 223(189b), 224(190), 226(194), 244(241), 260, 261
Pollak, R. A., 217(165, 169), 259
Poloniak, E. S., 54(258), 56(258), 62(258), 71
Pope, R. A., 45(203), 69
Pope, T. P., 136(50), 144
Popma, T. J. A., 276, 296
Poppa, H., 4(11), 65
Porteus, J. O., 203(121, 122, 123a), 215(152), 258, 259
Portier, E., 50, 70
Powell, C. F., 31, 67
Powell, C. J., 205(128), 214(144), 258, 259
Powers, J. V., 266, 295
Pressau, J. P., 50(229), 70
Preston, J. S., 9, 43(188), 47(214), 50(230), 62, 65, 69, 70
Pulliam, G. R., 282, 287, 296

Q

Queisser, H. J., 238(228), 261
Quinn, J. J., 159(16), 256

R

Radeloff, C., 174(32), 256
Raether, H., 146(4, 5), 153(9), 186(65), 195 (98), 196, 197(104), 202(116, 120), 203(120, 120a, 120b), 204(5), 207(120), 211(137), 214(4, 145, 146), 215(149), 223(189a, 189b), 227(206), 231(215), 233, 237(224), 239(231), 241(231, 236), 242(9, 102, 231), 243(239), 244(239), 245(224), 247(249), 255, 256, 257, 258, 259, 260, 261
Raimondi, D. L., 278, 297
Rairden, J. R., 43(194), 69
Ramsey, J. R., 130(38), 144
Randlett, M. R., 4(13), 42(181), 65, 69
Randlett, R., 6(23), 65
Rasigni, G., 15(65), 66
Raymond, R. F., 12(58), 33(58, 130), 66, 68
Ready, D. W., 31(113), 36(113), 37(113), 39(113), 67
Reddaway, S. F., 22(96), 23(96), 25, 26, 67
Reed, T. B., 10(47), 66
Reiss, K. H., 33(124), 68
Remeika, J. P., 28(103), 67, 275, 294
Reynolds, F. L., 46(205), 70
Rezek, J., 6(24), 65
Rhodin, T. N., 6, 7(30), 65
Rhyne, J. J., 273, 274, 296
Rice, S. A., 240(233), 261
Rilee, E. W., 33(116a), 67
Ripken, K., 188(76), 257
Ritchie, R. H., 146(1, 3), 149, 150(1), 158, 174(33), 179(54), 198, 201, 219(175), 221 (183, 184, 186), 225(191), 226(191), 229 (209), 237(222), 245(245), 250(250), 255, 256, 257, 258, 259, 260, 261
Robertson, J. M., 276, 296
Robbins, C. R., 11(49, 50, 51), 20(51), 66
Robin, S., 178(60), 227(204), 257, 260
Robinson, McD., 282, 296
Rode, D. L., 284, 296
Roe, D. W., 33(128), 68
Rohatgi, A., 33(135), 35, 37(135), 68

Romanov, Y. U., 217(164), 218(164), *259*
Romeijn, F. C., 9(44), *65*
Rooymans, C. J. M., 288, *296*
Rosenberg, R., 278, 282, *295*
Rosencwaig, A., 271, 288, *296*,
Rosier, L. L., *294*
Rossol, F. C., 275, 276, *296*
Rouard, P., 15(65), *66*
Rowe, J. E., 215(155), 216, 217(156, 157, 159), *259*
Roy, R., 11(52, 54), *66*
Ruppin, R., 194(96), *258*
Rupprecht, G., 47(213), *70*
Rusu, G. I., 33(137), *68*
Rusu, M., 33(137), *68*
Ryabova, L. A., 39(167, 168), 40(167, 168), *69*
Ryan, J. D., 32(116), 33(116), *67*

S

Sadagopan, V., 281, 282, 288, 290, *295, 296, 297*
Sakai, Y., 50(236a), 60(271), *70, 71*
Samson, S., 27, *67*
Sandfort, R. M., 265, *296*
Sasaki, T., 34(157), 35(157), *68*, 186(68), *257*
Sauerbrey, G., 174(32), 219(177), 220(179, 181), 221(187), 227(200), *256, 259, 260*
Saunders, A. E., 33(146), *68*
Sauter, F., 192(88), *257*
Savary, A., 178(60), 227(204), *257, 260*
Savitskaya, Y. S., 39(168), 40(168), *69*
Sawatzky, E., 282, *296*
Scheer, M. D., 46(208), *70*
Scheibner, E. J., 215(150), *259*
Schilling, J., 214(148), 215(148, 149), *259*
Schindler, A. I., *297*
Schlesinger, M., 6(24), 22(90), 23(90), 27(90), *65, 67*
Schlüter, M., 177, *257*
Schmalfeld, B., 176(50), 178(50), 186(50), 187(50), *257*
Schmidt, P. H., 264, 267, 276, 287, *294*
Schmidt, R. N., 130(37), 132, *144*
Schmüser, P., 204(124), 206(124), 207(124), 211, *258*
Schnatterly, S. E., 227(199), 228(199), *260*
Schönwald, J., 154(11), *256*

Schreiber, P., 195, 196, 197, *258*
Schroeder, D. J., 74, 82(6), 132(6), *143*
Schröder, E., 195(100), 196, 197, 228(207), *258, 260*
Schroeder, H., 74(7), 129(7), 130(7), 131(7), *143*
Schroeder, K., 273, *296*
Schulz, D., 186(72), 187, *257*
Schweizer, H. P., 39(170, 171, 172), 40(170, 171, 172), 41(170, 171, 172), 62(171, 172), *69*
Scott, G. D., 5(22), *65*
Scovil, H. E. D., 264, 265, *294*
Seah, M. P., 215(151), *259*
Sedgwick, T. O., 282, *295*
Seifert, H. G., 219(177), *259*
Sevier, K. D., 200(112), *258*
Shafer, M. W., 11(54), *66*
Shalabutov, Y. K., 45(202), *69*
Sherwood, R. C., 264, 267, 276, 287, *294*
Shick, L. K., 276, 284, 288, *294, 296*
Shionoya, S., 22(85), 23(85), *67*
Shirley, D. A., 217(165, 169), *259*
Shubin, L. D., 47(224), *70*
Shuermeyer, F. L., 45(200), 46(200), *69*
Siddall, G., 43(186), 50, 51(186), 62(186), *69*
Sigsbee, R. A., 43(194), *69*
Sihvonen, Y. T., 51, *71*
Sinclair, W. R., 51, *70*
Silcox, J., 180(59), 202(117), 212(117), *257, 258*
Silin, V. P., 192(87), *257*
Singh, H. P., 5(20), *65*
Skibowski, M., 186(69), 187(69), 188(77), *257*
Slack, L. H., 33(116a, 135), 35(135), 37(135), *67, 68*
Slonczewski, J. C., 269, 271, *296*
Smith, A. B., 270, *296*
Smith, D. H., 264, 267, 275, 276, 287, *294, 295, 297*
Smith, N. V., 217(161), *259*
Smith, P., 77(13), 84(13), *143*
Smith, R. C., 47(224), *70*
Sobajima, S., 47(222a), *70*
Sohler, W., 241(237), *261*
Sondheimer, E. H., 7(33), 8(33), *65*
Sosniak, J., 50(228), *70*
Spence, W., 47(219), 48(219), *70*

Spencer, E. G., 28(103), 67, 264, 267, 276, 287, 294
Spicer, W. E., 217(161), 227(203), 228(203), 259, 260
Spitz, J., 39(172a), 69, 282, 294
Stacy, W. T., 288, 296
Stanford, J. L., 227(201, 206), 228(208), 260
Steele, S. R., 31(113), 36(113), 37(113), 39 (113), 67
Stefan, V., 33(137), 68
Stein, B. F., 282, 296
Steinmann, W., 176, 178(37), 179(37, 52, 53), 186(69), 187(69), 190(81), 195(101), 197, 220, 227(202), 241(235), 256, 257, 258, 260, 261
Stern, E. A., 146(2), 150(2), 153(8), 174(36), 182(62), 186(62), 210(133), 220(178), 221, 225, 245(248), 247, 255, 256, 257, 258, 259, 260, 261
Stettmaier, K., 195(101), 197(101), 258
Stewart, J. E., 227(197), 260
Stillinger, D. W., 51(239), 70
Stockdale, G. F., 12(60), 33(60, 152), 66, 68
Stone, J., 88, 90(17), 143
Stuke, J., 50(234), 70
Sturge, M. D., 288, 295, 296
Sturm, K., 158(14c), 256
Sugita, Y., 271, 296
Sugiyama, I., 47(222a), 70
Summitt, R., 22(97, 98, 99), 23(97, 98, 99), 26, 67
Sunjic, M., 213(142, 143), 259
Suzukawa, T., 33(127), 68
Suzuki, H., 182(63), 186(63), 257
Swan, J. B., 205(128), 212(140), 258, 259

T

Tabata, O., 39, 40, 69
Tabor, H., 129(134), 144
Tabor, W. J., 271, 288, 296, 297
Takagi, N., 47(222a), 70
Takahashi, M., 271, 296
Talwalker, A. T., 31(115), 67
Tanaka, T., 49(224a), 70
Tao, L. J., 273, 274, 296
Tarng, M. L., 278, 297
Tarnopol, M. S., 33(149), 68

Taylor, R. C., 282, 297
Teng, Y., 221, 225, 244(242), 260, 261
Teshima, R., 159(17), 256
Thelen, A., 105(19), 110, 143, 144
Thiele, A. A., 267, 269, 271, 275, 295, 297
Thomas, L. D., 33(123), 68
Thompson, K. T., 5(22), 65
Thomson, S. J., 14, 66
Tien, P. K., 169(28), 256
Tinder, R. F., 46(206), 70
Title, A. M., 136(50), 144
Tittel, H. O., 176(49), 178(49), 186(49), 257
Toigo, A., 245(245), 261
Tolansky, S., 16(73), 66
Tolksdorf, W., 276, 284, 297
Tousey, R., 227(195), 260
Tripp, K. F., 130(36), 131(36), 144
Trounson, E. P., 176(39), 256
Tsui, D. C., 218(170), 259
Tung, C., 265, 295
Turbadar, T., 234, 260
Turner, A. F., 74(7), 79(8), 82, 91(8), 95, 106(21), 116, 129(7), 130(7), 131(7), 143, 144
Turner, D. T., 8(37), 65
Twietmeyer, H., 238(230), 239(230), 261

U

Ufford, C., 110, 144
Ulrich, R., 169(28), 256
Urbach, F., 15, 66

V

Vainshtein, V. M., 51(243, 244, 247), 70, 71
Valentino, A. J., 281, 294
Van Cakenberghe, J. L., 42(179), 56(264), 58(264), 62(264), 69, 71
van den Berg, P. M., 244(243), 261
van der Linden, P. C., 33(138), 35(138), 68
Van Oosterhout, G. W., 9(44), 65
Van Uitert, L. G., 264, 267, 275, 276, 287, 288, 294, 295, 297
Vaynshetyn, V. M., 51, 70
Veas, F., 5(22), 65
Veeneman, D., 33(138), 35(138), 68

Vella-Coleiro, G. P., 271, 276, *296*, *297*
Verwey, E. J., 9, *65*
Vigue, J. C., 282, *294*
Vijh, A. K., 10(48), *66*
Vincent, C. A., 22(91, 92), 27(91, 92), *67*
Vincent, R., 180(59), 202(117), 212(117), *257*, *258*
Viserian, I., 33(156), 34(156), *68*
Viverito, T. R., 33(116a, 135), 35, 37(135), *67*, *68*
Voegeli, O., 271, *296*
Vogel, S. F., 50(236), 51(236), 52, *70*
Vojnovich, T., 28(105a), *67*
von Blanckenhagen, P., 219, *259*
von Festenberg, C., 180(57), 202(116), *257*, *258*
von Fragstein, C., 190(83), *257*
Vonogradova, V. V., 33(154), *68*
Vorob'eva, O. V., 33(154), *68*
Voss, W. A. G., 43(195), 44, *69*
Vossen, J. L., 42, 46(177, 210, 211), 50(177, 227), 51(177), 53(254), 54(177, 227, *257*, *258*), 55, 56, 57, 62(257, 258, 259, 263), 64(274), *69*, *70*, *71*

W

Waber, J. T., 12(56), *66*
Waghorne, R., 56(261), *71*
Wagner, C., 9, *66*
Wagner, D., 158(14b), *256*
Wagner, W. E., 33(146), *68*
Walters, E. M., 264, 267, 276, 287, *294*
Wartenbury, E. W., 33(118), *67*
Watanabe, H., 47(215), *70*
Waylonis, J. E., 227(195), *260*
Weast, R. C., 11(53), 12(53), *66*
Webb, M. B., 6, 7(29), *65*
Weber, W. H., 236(219a), *261*
Wehner, G. K., 278, *297*
Weiher, R. L., 28(102, 104), 29, 30, *67*
Weinrich, A. R., 43(189), *69*
Weller, J. F., 54(256), *71*
Welz, F., 276, 284, *297*
Wendel, G., 33(125), *68*
Wendelken, I. F., 203(123a), *258*
Wendt, J. F., 46(204), *70*
Wernick, J. H., 273, *296*

Wessel, P., 176(39), *256*
West, E. J., 54(256), *71*
Weston, D. G. C., 22(92), 27(92), *67*
Whitcomb, E. C., 276, *294*
White, E. A. D., 276, 283, 284, *295*
Widmer, R., 51, 52, *70*
Wijn, A. P. J., 275, *297*
Wilems, R. E., 250(250), *261*
Wille, H., 179(53), *257*
Williams, B. E., 9(41), 43(9), 62(9), *65*
Williams, C. M., *297*
Williams, M. W., 221(183), 227(205), 228(205), 229(205), 237(222), *260*, *261*
Williams, V. A., 51, *71*
Winters, H. F., 278, *297*
Wittekoek, S., 276, *296*
Woeckel, E., 221(187), *260*
Woelke, C.D., 51(249), *71*
Wolfe, R., 271, *297*
Wolter, H., 74, *143*
Wright, D. A., 22(88, 96), 23(96), 25, 26, 27, *67*
Wyekoff, R. W. G., 22(100), 28(100), 30(100), *67*
Wyrobisch, W., 217(168), *259*

Y

Yamada, K., 182(63), 186(63), *257*
Yamaguchi, S., 186(64), *257*
Yamaguchi, T., 190(84), *257*
Yamanaka, S., 54, *71*
Yamane, Y., 33(127), *68*
Ying, S. C., 159(16), *256*
Yoshida, S., 190(84), *257*
Yoshikawa, A., 60(271), *71*

Z

Zacharias, P., 188, 194(95), 198(110), 202(118), 218(110), *257*, *258*
Zaromb, S., 19, 20, *66*
Zavracky, P. M., 116(26), *144*
Zellmer, D. L., 35(165), 36(165), 37(165), *69*
Zeppenfeld, K., 202(116), *258*
Zunick, M. J., 33(129), *68*
Zurheide, M., 186(72), 187, *257*
Zydzik, G. J., 275, *297*

Subject Index

A

Admittance, in interference filter theory, 75–77
Admittance matching, 74
Aluminum film, isotransmittance curves for, 118
Amorphous films
 defects in, 289
 for magnetic bubbles, 277–279
Amplitude reflection, in interference filter theory, 75–77
Anisotropy, in magnetic bubble films, 284–288
 see also Magnetic anisotropy
Antireflection coatings, for semiconductor oxide films, 19–20
ATR method
 in light excitation of nonradiative surface plasmons, 233–234
 for low energy surface plasmons in semiconductors, 238
Augmented double-cavity filter, 127–128
Augmented spaces, design of filters with, 106–110

B

Bandpass filter design
 bandpass filter attributes and, 93–94
 in metal–dielectric interference filters, 93–102
Bandpass filters
 attributes of, 93–94
 choice of metals for, 95–102
 thickness of layers in, 102
 transmittance of, 122–123
Bandwidth broadening, in multiple casting filters, 126
Binary oxides, oxygen vacancy and, 10
Bloch lines, in magnetic domain dynamics, 269–271
Blocking filters, in metal–dielectric filter production, 140–142

C

Cadmium oxide, electrical and optical properties of, 30
Cadmium oxide films, sputtering of, 1
Characteristic matrix method, for metal–dielectric interference filters, 77–78
Charge oscillations, plasmas as, 145–147
Chemical etching, in transparent conducting films, 61
Chemical vapor deposition, 31–41
 chloride hydrolysis in, 31–38
 pyrolysis in, 38–41
Chlorides, hydrolysis of in film deposition, 31–38
Coupling, by grating, 221
Curie point, for magnetic material, 268

D

Dark mirror, in three-cavity interference filter, 74
Dielectric function, determination of for metals, 234–238
Doped oxides, in semiconductor oxide films, 10–12
Doping level, optimum, 11

E

Electrical conductivity, light transmission and, 2
Electroless plating, in gold film processing, 60
Electron beam evaporation, 45–46
Electron energy losses, in nonradiative surface plasma excitation, 199–219
Electrons
 excitation by, 219–223
 excitation of nonradiative surface plasmons by, 199–219
 excitation of radiative surface plasmons by, 173

SUBJECT INDEX

reflection of at surface, 213-214
transmission and reflection of, 214
Equivalent layer spacers, MDM filter design and, 110-112
Evaporation of filaments, electron emission and, 45

F

Fabry-Perot interferometer, 74, 88-90
  radiant transmittance of, 88
  transmittance of symmetrical multilayer in, 90-92
Ferrell mode, for radiative surface plasmons, 159-163
Filament evaporation, electron emission and, 45
Film boundaries, scattering at, 8
Film coalescence, with nucleation-modifying layers, 45
Film deposition process, 30-60
  amorphous substrates and, 42
  chemical vapor deposition, 31-41
  film thickness control in, 133
  in filter production, 132-133
  miscellaneous types of, 59-60
  nucleation-modifying layers in, 42-43
  oxide target sputtering, 53-59
  properties related to, 63
  reactive evaporation, 47-49
  reactive sputtering, 49-53
  sheet resistivity and, 62
Films
  semiconductor oxide, see Semiconductor oxide films
  transparent conducting, see Transparent conducting films
Filter, one-M, see One-M filter
Filter production techniques, 132-142
  cementing and addition of blocking filters in, 140-142
  and deposition techniques for metals, 132-134
  optical monitoring in, 136-140
  reflection or transmission monitoring in, 134-136

G

Garnet films
  defects in, 289-293
  liquid phase epitaxy for, 283-284, 287-288
  single-crystal, 281-284
  sputtering of, 282
Garnets
  chemical vapor deposition of, 282
  for magnetic bubbles, 275-276
Gold films
  isotransmittance of, 114
  nucleation of, 43
Grating, coupling by, 221
Guided light modes, 168-170
  excitation of, 241-242
Guided light waves, roughness factor in, 254

H

Hexaferrites, for magnetic bubbles, 275
Horizontal Bloch line, in magnetic domain dynamics, 269-270

I

ILEED, 215
Indium oxide, electrical and optical properties of, 28-30
Indium oxide films
  from pyrolysis of indium acetylacetonate, 39
  reactively sputtered, 51
  typical properties of, 41
Induced transmission filter, 74
Interference, in thin films, 16-18
Interference filters
  metal-dielectric, see Metal-dielectric interference filters
  multistage, 74
Interisland separations, trapping centers for, 6
Ion plating process, reactively sputtered films and, 53
Island films, as transport conductors, 6
Island structure, model of, 6

L

LEED, 215–216
Light, excitation of surface plasmons by, 180–199
Light absorption, in transparent conducting films, 13–16
Light emission
 and coupling by statistical surface roughness, 219–221
 via roughness, 219–223
Light waves, excitation of surface plasmons by, 223–229
Liquid phase epitaxy, film growth by, 287–288
Lithium film, reflectivity of, 236

M

Magnesium fluoride spacer, in filter production, 138
Magnetic anisotropy, 284–285
 amorphous materials and, 285–286
 in garnets, 286–288
Magnetic bubble films, 263–294
 see also Magnetic bubbles
 anisotropy in, 284–288
 defects in, 288–293
 in information storage industry, 293–294
 inhomogeneity in, 288–290
 materials selection for, 265–266
 static properties in, 266–269
Magnetic bubble materials, growth of, 276–284
Magnetic bubbles
 amorphous film preparation for, 277–280
 amorphous materials for, 273–275
 anisotropy energy required for, 268
 crystalline materials for, 275–276
 defined, 263
 information stored in, 264
 magnetic theory and, 266–272
 materials for, 272–276
 performance and cost factors for, 265–266
 single-crystal garnet films for, 281–284
 size of, 263–264
 velocity of, 269
Magnetic material, Curie or Néel point for, 268

Magnetic stripes, vs. magnetic bubbles, 267
Magnetic theory, 266–272
 domain dynamics in, 269–272
Matching stacks, 113–114, 119
MDM filters, 88–90, 113
 augmented, 125–127
 cementing of, 140–142
 design of, 102–104
 equivalent layer spacers and, 110–112
 maximum transmittance of, 109
 monitoring of, 138
 optical constants vs. wavelength for, 104
 phase shift in, 103
 reflectance and transmittance of, 137
 spectral transmittance of, 107–108
 wedged, 105–106
MDM structure, in reflection filters, 130
MDMDM filters, 113
MDMDMDM filters, 124, 143
Metal–dielectric interference filters, 73–143
 admittance and amplitude reflection in, 75–77
 bandpass filter design and, 93–102
 basic theory of, 75–93
 characteristic matrix for, 77–78
 Fabry–Perot type, 74–75, 88–90
 future developments in, 142–143
 MDM type, see MDM filters
 multiple cavity and other designs of, 124–128
 net flux ratio for, 79–82
 net flux ratio for assembly of layers, 82–86
 net flux ratio maximum value in, 86–87
 net flux ratio properties in, 87–88
 one-M filter design for, 112–114
 production of, 132–142
 radiant absorptance in, 81
 radiant reflectance and transmittance in, 78–79
 reflection filters as, 128–132
 single-cavity type, 102–112
Metal–dielectric–metal interference filters, see MDM filters
Metal films
 see also Metal–dielectric interference filters; Thin films
 bulk resistivity ratio for, 8
 carrier scattering sites in, 7
 continuous, 7–9

defects in, 7
impurities in, 8
optical properties of, 13-14
resistivities of, 8
size effect in, 7
transparent conductivity, *see* Transparent conductivity films
Metal oxides, *see* Semiconductor oxide films
Multilayer interference filters, applications of, 74
Multiple cavity filters, 124-128
augmented double-cavity, 127-128
transmittance and absorbance curves for, 128

N

Néel point, for magnetic material, 268
Net flux ratio
for assembly of layers, 82-86
maximum value of, 86-87
for metal–dielectric interference filters, 79-80
properties of, 87-88
for single layer in metal–dielectric interference filters, 81-82
Nonhomogeneous films, plasma resonance absorption in, 190
Nonradiative surface plasmons
see also Surface plasmons
angular dependence of differential excitation in, 204
asymmetric layer system for, 207-208
asymmetry of surface loss intensity and, 210-213
coating of with plasma film, 209
and coupling dispersion of symmetric layer system, 206-207
coupling of by grating, 221-223
dispersion curve of, 226
dispersion due to finite thickness of coating, 208-209
electron energy losses and, 199-219
excitation of, 199-255
excitation of by electrons, 215-223
excitation of by evanescent light waves, 230-255
and excitation of guided light modes, 241-242

excitation probability of surface losses for, 200-201, 215
light scattering and, 245-248
nonnormal incidence in, 209-210
and optical constants or dielectric functions of metals, 234-238
photoelectron emission by, 240-241
probability for exciting surface losses in, 204-206
and reflection of electrons at surface, 213-215
retardation in, 202
roughness effects in, 242-244
small spheres and, 218-219
surface coatings and, 205
Nucleation modifying layers
film coalescence with, 45
metal films deposited over, 42

O

One-M filter
admittance vs. wavelength for, 120-121
configuration for, 116
defined, 112
examples of, 117-124
matching stacks and, 113-114, 119
simplest type of, 113
spectral absorbance for, 122
spectral transmittance of, 123
One-M filter design, 112-124
procedure for, 116-117
three-layer symmetrical systems and, 112-116
Optical constants, determination of for metals, 234-238
Optical monitoring, in filter production, 136-140
Orthoferrites, for magnetic bubbles, 275
Oxide films, semiconductor, *see* Semiconductor oxide films
Oxide materials, in semiconductor oxides, 9-12
Oxides
binary, 9-10
doped, 10-12
Oxide target sputtering, 53-59
hollow-cathode source in, 50
Oxygen vacancy, in crystal, 10

P

Photoelectron emission, by nonradiative surface plasmons, 240–241
Photoelectrons, registration of plasma resonance by, 188–190
Plasma boundary, very thin films in, 238–240
Plasma films
 radiative plasma oscillations and, 168
 resonant light emission of, 176
Plasma radiation, theory of, 173–176
Plasma resonance
 in reflection at thin films, 188
 registration of by photoelectrons, 188–190
 in small spheres, 198–199
Plasma resonance absorption, 180–186
 fine structure of, 191
 in nonhomogeneous films, 190
 by optical excitation of volume plasmons, 191–192
Plasma resonance emission, 195–198
Plasma slab
 asymmetric layer system and, 154–157
 symmetric layer system and, 149
Plasmons, 164–166
 radiative and nonradiative, see Radiative plasmons
 surface, See Surface plasmons
Potassium films, plasma resonance emission from, 197
Potassium foil, absorption of p-polarized light by, 193
Poynting vector, 79–80
Prism method, in light excitation of nonradiative surface plasmons, 233–234
Pyrolysis, in chemical vapor deposition, 38–41
Radiant absorptance, for metal–dielectric interference filters, 81
Radiant reflectance and transmittance, for metal–dielectric interference filters, 78–79
Radiant transmittance, defined, 78–79
Radiative surface plasmons
 electron excitation of, 173–180
 excitation of, 171–199
Ferrell mode for, 159–163
 interaction of with nonradiative plasmons, 164–166
 light and, 180–199

photoelectron current produced by, 189
plasma films and, 168
plasma radiation and, 193
roughness and, 164–166, 248–250
Reactive evaporation, in film deposition, 47–49
Reactive sputtering, 49–53
 compound synthesis in, 50
 ion plating process and, 53
Reflectance filter, spectral transmittance of, 131
Reflection, in thin films, 16–18
Reflection filters, 128–132
 antireflection coating for, 130
Reflection or transmission monitoring, in filter production, 134–135
Refractive index, for semiconductor oxide films, 20
Resonant light emission, of thin plasma films, 176
Rhenium films, rf sputtering in preparation of, 45
Richardson equation, 6
Roughness
 in leakage of guided light waves, 254
 light emission via, 219–223
 light scattering by, 245–248, 250–255
 in nonradiative surface plasmons, 242–244

S

Scattering
 on rough silver surfaces, 250
 in semiconductor oxide films, 21–22
 by silver film, 252–253
Semiconductor oxide film, 2
 see also Transparent conducting films
 absorption phenomena in, 14–15
 antireflection coatings for, 19–20
 binary oxides in, 9–10
 doped oxides in, 10–12
 electrical and optical properties of, 22–30
 electrical conduction in, 9–12
 film side absorption in, 17
 impurities in, 12
 influence of substrate in, 12
 refractive index variation in, 20
 scattering and surface morphology of, 20–22

substrate side absorption in, 17
Semiconductors, low energy surface plasmons in, 238
Silver films
  dispersion relations of, 156
  light intensity scattering by, 253
  measured reflectivity of, 235
  number of photons on, 177
  plasma resonance emission from, 195–196
  radiation pattern for, 252
  reflection minima for, 237
  transmittance spectra of, 194
Single-cavity filter design
  augmented spacers and, 106–110
  equivalent layer spacers and, 110–112
  MDM filter and, 102–104
  wedged filters and, 105–106
Small spheres
  fast-electron experiments on, 218
  plasma resonance effects in, 198–199
Spray hydrolysis, in chemical vapor deposition, 32
Sputtering
  electron source in, 45
  of garnet films, 282
  for magnetic bubbles, 266
  of oxide targets, 53–59
  positive ions in, 46
  reactive, 49–53
Stannic oxide crystals, resistivity and carrier concentration of, 27
Stannic oxide films
  conductivity vs. temperature for, 28
  donor activation energy for, 25
  doped, 36
  doping levels in, 37
  electrical and optical properties of, 22–26, 34
  Hall mobility for, 24
  hydrolysis in preparation of, 34, 36
  indium oxide and, 39–40
  preference over indium oxide films, 40
  pyrolysis in preparation of, 41
  reactively sputtered, 51
  refractive index of, 25
  undoped, 34
Stannous oxide, electrical and optical properties of, 22–26
Surface charge oscillations, 145

Surface losses
  asymmetry in intensity of, 210–213
  probability and position of, 200–204
  total probability for exciting, 204–206
Surface plasma oscillations, 145–255
  see also Surface plasmons
  plasma resonance emission and, 195–198
Surface plasmons, 145–171
  as charge oscillations, 146–147
  compared with guided light modes, 168–171
  coupling of by grating, 224–227
  coupling by statistical surface roughness, 227–229
  dispersion relations of, 157–158
  electron energy losses and, 180
  excitation of by light, 180–199, 223–229, 233–255
  existence of, 145
  low energy, 238
  nonradiative, 149–159, 164–166
  nonradiative surface plasmon excitation and, 199–255
  radiative, see Radiative surface plasmons
  reflectance deficit of, 228–229
  in spheres and on voids, 166–167
  surface fields of, 169
Surface roughness, coupling by, 219–221, 227–229
  see also Roughness
Symmetrical multilayer, transmittance of, 90–92

T

Target heating, in oxide target sputtering, 54
Thermionic emission theory, 6
Thin films
  see also Metal films
  evaporation and sputtering of, 42–47
  light absorption in, 15
  nonradiative surface plasmons on, 149–159
  nucleation-modifying layers in, 9
  on plasma boundary, 238–240
  plasma resonance in reflection at, 188
  quantum effects in, 15
  radiative surface plasmons on, 159–163
  reflection, refraction, and interference in, 16–19

Thin metal films
  aircraft industry need for, 1
  deposition process for, 3
  discontinuous, 4–7
  electrical conduction in, 4–9
  first reported observation of, 1
  island films as, 6
  island size and interisland spacing in, 4
  oxidation state of, 2
  postoxidation of, 47
  properties of, 3
  "thinness" in, 5–6
Thin plasma films, resonant light emission of, 176
Three-layer symmetrical systems, 112–116
Transition radiation, 174
Transparence, electrical conductivity and, 2
Transparent conducting films, 1–64
  application of materials and processes in, 60–63
  chemical etching in, 61
  film deposition processes for, 30–60
  high pressure regime and, 50
  light absorption in, 13–16
  light loss in, 13
  oxide target sputtering and, 53–59
  parameters obtained through various deposition processes, 63
  reactive evaporation of, 47–49
  thin metal films as, 45
  transparence of, 13–22
  transparence vs. conductivity in, 60
Trapping centers, for interisland separators, 6
Tunneling, between allowed states, 7

U

Ultraviolet filters
  reflection monitoring system for, 135
  spectral transmittance of, 111–112

V

Vertical Bloch lines, in magnetic domain dynamics, 270–271
Volume plasma oscillations, 146

W

Wedged filters, design of, 105–106
Windshield deicing, transparent electrical heaters for, 1

QC
176
A1
H3
v.9
1977

JUN 8 1977